Introducing Property Valuation

Michael Blackledge

Routledge
Taylor & Francis Group

LONDON AND NEW YORK

First published 2009
by Routledge
2 Park Square, Milton Park, Abingdon, Oxon OX14 4RN

Simultaneously published in the USA and Canada
by Routledge
270 Madison Ave, New York, NY 10016

Routledge is an imprint of the Taylor & Francis Group, an informa business

Typeset in Sabon by
HWA Text and Data Management, London
Printed and bound in Great Britain by
MPG Books Group, UK

British Library Cataloguing in Publication Data
A catalogue record for this book is available from the British Library

Library of Congress Cataloging-in-Publication Data
Blackledge, Michael.
  Introducing property valuation / Michael Blackledge.
     p. cm.
  1. Valuation. I. Title.
  HF5681.V3B53 2009
  333.33´2–dc22   2008050957d

ISBN10: 0–415–43476–9 (hbk)
ISBN10: 0–415–43477–7 (pbk)
ISBN10: 0–203–87617–2 (ebk)

ISBN13: 978–0–415–43476–8 (hbk)
ISBN13: 978–0–415–43477–5 (pbk)
ISBN13: 978–0–203–87617–6 (ebk)

Contents

Illustrations

Figures

Tables

Cases

Acknowledgements

All material quoted from RICS publications is reproduced by permission of the Royal Institution of Chartered Surveyors, which owns the copyright.

Material from J. Law (ed.), *Oxford Dictionary of Business and Management* (1990) reproduced by permission of Oxford University Press.

Extracts from G. Parsons, *The Glossary of Property Terms* (2004) reproduced by permission of Estates Gazette, Copyright Elsevier Ltd.

Crown Copyright material reproduced under the terms of the Click-use Licence by permission of Office of Public Sector Information (OPSI).

Disclaimers

All the examples provided in this book are intended to present solutions to realistic problems, but are nonetheless hypothetical or fictional and any similarity to the exact facts of any specific actual property, investor or investment is coincidental.

The publisher and the author make no representations or warranties with respect to the accuracy or completeness of the contents of this work and specifically disclaim all warranties, including, without limitation, warranties of fitness for a particular purpose. No warranty may be created or extended by sales or promotional materials. Whilst this text provides the underlying theory and principles of property valuation, no responsibility is accepted for any financial loss incurred from following the guidance provided. The advice and strategies contained herein may not be suitable for every situation and constitute general guidelines only and do not represent to be advice on any particular matter. This work is sold with the understanding that neither the author nor publisher is herein rendering valuation, legal, accounting or other professional services. Neither the publisher nor the author shall be liable for damages arising herefrom or from any errors or omissions. No reader or purchaser should act on the basis of material contained in this publication without first taking professional advice appropriate to their particular circumstances.

The fact that an organisation or website is referred to in this work as a citation and/or potential source of further information does not mean that the author or publisher endorses the information or services the organisation or website may provide or recommendations it may make. Further, readers should be aware that internet websites listed in this work may have changed or disappeared between when this work was written and when it is read, that only Acts of Parliament and Statutory Instruments have the force of law and that only courts can authoritatively interpret the law.

PART 1

Background

1

Economic context

In this chapter ...

- What is property valuation and why might property need to be valued?
- The different types of value that may apply to a property and distinguishing between the terms value, price, worth, cost and market value.
- How the economic forces of supply and demand and the special characteristics of the property market determine the price and value of property.
- Why land use develops in recognisable patterns and how this may affect types of property within a region and its value.
- What constitutes the investment and property markets?
- The range of UK taxes that affect property and their basis of assessment.
- How town and country planning legislation and decisions affect land allocation and use; how this influences values and why this can be justified.

1.1 Why is a valuation needed?

What does 'property' mean when referring to *property valuation*? In English law, goods and belongings owned by a person or legal body are termed *personal property* whereas land and buildings are *real property*. Sometimes, to make this distinction clear, land and buildings are termed *real estate*, a phrase long used in the US and increasingly adopted in the UK. Thus this book is concerned with the valuation of real property, real estate or land and buildings.

There are many possible reasons for valuing property, such as:

- to buy or sell;
- to let or take a lease or agree a rent review;
- to assess tax or business rates payable;

- for insurance;
- to obtain a compensation payment;
- to borrow money using the property as 'security';
- to show its value as a fixed asset on a company balance sheet;
- to develop or redevelop.

Some of these values need assessment on a frequent or recurring basis, others only very occasionally. All create opportunities for a property valuer to employ his or her professional skills and expertise to provide the required figure and advice to a client.

1.2 What types of property value are there?

There are many types, including:

- freehold value
- leasehold value
- asset value
- alternative use value
- annual value
- before and after value
- break-up value
- book value
- compulsory purchase value
- depreciated value
- deprival value
- development value
- divorce value
- exchange value
- existing use value
- fair value
- forced sale value
- going concern value
- gross development value
- hope value
- market value
- marriage value
- mortgage value
- permitted development value
- ransom value
- rateable value
- rental value
- residual value
- site value
- speculative value

- surrender value
- tax value
- value in use
- value to the owner
- zone A value.

Many of these could apply to a specific property at the same time – and all are likely to be different figures. Therefore, to ask 'what is the value of this building?' is a meaningless question. A valuer must know which specific value or values he or she is required to find; and before proceeding must clearly define and firmly agree this in writing with a client. Subsequent chapters explain the most important and frequently requested of these value types.

1.3   Important terms and concepts: value, price, worth, cost and market value

What is value? A dictionary definition of the noun is: 'amount of commodity, money, etc. considered equivalent for something else; material or monetary worth of thing; worth, desirability, utility, quantities on which these depend' (Coulson *et al.* 1975: 932). How relevant is this to a property value and as the definition twice refers to 'worth' is this the same thing as value? The short answers are that the definition only partly applies and no, worth may not always be the same as value when applied to property.

Real estate certainly meets the criteria that any good or service must possess to have value in *economic* terms, which are:

- *utility* – usefulness to potential buyers; the greater its potential for use for different purposes, the greater its utility;
- *scarcity* – this does not mean that it literally has to be very scarce, merely that the supply is limited and insufficient to meet total demand;
- *demand* – this has to be effective, so that there are potential buyers who wish and are *able* to purchase;
- *transferability* – ownership has to be able to be transferred otherwise it cannot be sold.

A property valuer's definition of *value* could be *the present price for the rights to receive income and/or capital in the future*. What does this mean? There are three aspects to the definition: *present price*, *capital* and *income*.

Present price is what it is worth *today*. This is a vital aspect of property valuation. All values are calculated at a specific date (the 'valuation date') and are only valid for a limited period after that day. How long this validity lasts will depend on the state of the market. In a strongly inflationary market where prices are rising by large percentages over short time periods, the value calculated today could have changed in as short a period as one or two months and no longer be valid. It is essential therefore to establish *when* a valuation was or is to be carried out, as it is a statement of value at that date only.

Present price should be assessed objectively, that is without bias or favour to a particular person's viewpoint. This type of price is usually expressed as *present value*, which is a term used frequently by valuers and forms an essential ingredient of all valuation theory and formulae. Most values calculated by property valuers are present values as at a stated valuation date.

Capital is a one-off lump sum receipt, obtainable from say the sale or mortgage of the property. *Income* indicates a sum of money receivable at regular intervals over time. Property income comes from *rent* or *interest* payments. Rent is the payment made by a tenant to a landlord for use and occupation and is usually expressed as the amount of money involved each year or 'per annum'. Interest payments are made on mortgage loans advanced against the security offered by a real property.

Value, by itself, can be a subjective concept in that a property will have different values at any point in time according to the purpose for which it is being valued and the circumstances of the party for whom it is being valued. However, normally when value is assessed subjectively, from a specific person or organisation's viewpoint, it will be referred to as a calculation of *worth*.

In referring to the earlier writings of William N. Kinnard Jr, Nick French concluded that:

> In the language of economics used by Kinnard, worth can be considered as value in use, whereas price or market value can be considered as value in exchange. As Kinnard stated: 'Market value can be regarded as the price that a willing buyer would pay, and a willing seller would accept, with each acting rationally on the basis of available market information, under no undue pressure or constraint, with no fraud or collusion present'. (French 2004: 83)

There is a long history of distinguishing between forms of value. '

> Aristotle was the first to distinguish between 'value in use' and 'value in exchange', …[but] the defining economic text relating to value was Adam Smith's *The Wealth of Nations* published in 1776. However, much of the discussion in his text brings together the theories and economic writings of economists from the preceding 200 years (French 2004: 83).

The RICS provide standard definitions of the most important words and phrases used in property valuation in their *Red Book* (Royal Institution of Chartered Surveyors 2007d). Three fundamental definitions are *valuation*, *market rent* and *market value*. The *Red Book*, together with these and other definitions, is considered in detail in Chapter 9 (Section 9.1).

Distinctions need to be made between the words *value*, *price* and *worth*, which have similar meanings in everyday use, but have different ones within the context of property valuation (RICS and IPF 1997: 7):

- *price* – the actual observable exchange price in the open market;
- *value* – an estimate of the price that would be achieved if the property were to be sold in the (open) market;
- *worth* – a specific investor's perception of the capital sum that he/she would be prepared to pay (or accept) for the stream of benefits which he/she expects to be produced by the investment (in other words a subjective rather than objective assessment of value).

There are three basic motives why people and organisations spend money on property. These are:

- *investment* – a return on capital funds. The basic aim is to obtain growth on the invested sum so that this amount becomes larger with time.

- *occupation* – for the occupier's own use and benefit for residential or business purposes.
- *speculation* – in the hope of making a profit on expenditure by taking a calculated risk on the spent money based on the premise that in the future a considerably larger sum will be recouped. However, speculation involves risk and the size and likelihood of financial gain is far more uncertain than on an investment.

Expenditure may be for any one, two or all three of these reasons. The motives behind the spending will largely determine what the worth of a property will be to each individual investor.

Additionally, there is a definite difference in meaning between *value* and *cost* in relation to property valuation. A basic definition of *cost* from a valuer's viewpoint could be: *a measure of (past) expenditure*.

Cost is usually an expression of what has been paid for a commodity. For example, purchasers of a piece of land can say that it cost them £1 million to purchase it. They are referring to a past event. It is possible to refer to costs that have not yet been incurred, but in that case, an estimate is really being made of what the expenditure will be – it is not until after the money has been spent that it can be said with certainty what the cost of the item was. The definitions of value and cost thus involve the three verbal tenses, in that value refers to a *present* worth of rights to *future* capital and/or income, whereas cost relates to a *present* expression of an expenditure generally incurred in the *past*.

Cost is often confused with value and many laymen assume they are the same thing. For instance, people may erroneously believe that if they have just purchased an item then the price they paid, or cost, would represent the market value of that item at the time. This may not be so; they may have obtained a bargain or, conversely, through their lack of knowledge of the market, they may have paid more than its true market value. Whilst these general principles hold true for most goods and services, they are particularly valid in the case of land and buildings. To equate cost with value is at best an unreliable equation, and in many cases, it can be very wrong.

A ridiculous example, to help prove this point, would be constructing a high specification office block at a cost of millions in a completely inaccessible location, such as the middle of the Sahara Desert. Its cost is colossal, and yet its value would be negligible. Indeed, it may have no value at all simply because there would be no demand for such a property due to its total impracticability – and this is the whole basis of the difference between value and cost. The forces of supply and demand determine value. When there is little or no demand, then the property will have little value, however much it costs to construct or acquire. Conversely, if there is extremely good demand, and particularly if this is coupled with low or restricted supply, the property will have a very high value, which can far exceed its cost.

All items, whether they are goods and services or land and buildings, will have a market value or price at which they will be expected to sell. In economics, this is usually referred to as the *equilibrium price*, in that it is the point where demand is equal to supply and thus the system is in equilibrium. This will produce the open market value, which cannot be found from costs alone since not all the factors that make up the effective demand and supply are then being taken into account.

There are numerous types of costs relating to land and buildings. In many cases, the adjective describing the cost is the same as that for value, but the meanings are different, as are the monetary sums involved. Examples would be development value and development cost. Development value is the estimated market value that a completed project will command. Development cost is the total expenditure required to complete the project, such as construction costs, professional fees and interest charges on borrowed finance.

It is especially unreliable to use a cost approach to valuing as the size or age of a building increase. An increase in size will not usually lead to value rising in proportion to the added costs. With an old building, there is the difficulty of deciding how much to deduct from the costs of new construction to reflect the age and obsolescence of the structure to be valued.

The conclusion is that cost and value are rarely equal to each other but sometimes value is based on the loose assumption that they are related for properties that have no readily accessible open market value. Due to its unreliability, this method of valuing, considered in Chapter 15 and known as the 'contractors method' or 'Depreciated Replacement Cost' (DRC) approach, is only used when no other method can be or in support of a value arrived at by another method.

1.4   Supply and demand as determinants of price and value

There is an old cliché that there are only three factors that determine the value of a property, which was adopted as the title of the very popular UK Channel 4 television programme presented by Phil Spencer and Kirsty Allsop, namely *Location, Location, Location*. When determining value, *where* a property is situated is usually of considerably more importance than *what* the property is. Locations in short supply and with high demand will command high market values.

Tim Harford provides a brilliantly simple explanation of the effects of supply and demand and the importance of location (Harford 2007: 5–24). As he makes clear by using the example of upmarket apartments in London and New York, scarcity and bargaining strength are the driving forces within any market that determine prices, and land and buildings are no different in this respect to other goods and services. Where there is a shortage of supply and high demand, prices are driven upwards.

Harford also explains how legal restrictions, such as the 'green belt' in the UK, can make a resource even scarcer. Property development is closely limited by use of planning restrictions within this area to retain more open space and help protect the environment. He questions whether the existence of such a 'green belt' around London is the reason why London property prices are high, but agrees it is a contributing factor as it further diminishes the supply of available land for development (Harford 2007: 18).

In economic terms, the property market is an imperfect one. Each property is different. An increase in supply cannot be made to match a rise in demand very easily, if at all. Both supply and demand are inelastic. Seldom do both parties to a transaction have full knowledge of the market and they may not be acting at 'arms length'. They may be under undue pressure to buy or sell.

The total supply of land in any given location or country is fixed, or perfectly inelastic. Surface area available cannot physically be increased to meet demand. It is true that in some areas low-lying land has been or could be reclaimed from the sea, but conversely in other coastal areas, land is continually being eroded by the tides and lost. The small gain on the one hand is offset by the loss on the other, and therefore for all practical purposes the surface area of the world and any particular country or region within it remains fixed.

Given that there are only so many square kilometres or miles of land available in a given location, the only changes that can take place in supply are in the allocation of land between the various uses or in the intensity of use. It would be possible, for example, to use a larger proportion of the land surface for industry and this would provide some elasticity to the industrial land supply curve. However, if this were done, then it would leave a decreased area available for the other land uses. An increase in land usage for industry would therefore require a corresponding reduction in some other land use or uses to compensate.

Example

A town has a surface land area of 100,000 hectares. Of this total, 10,000 hectares is currently used for industry and 30,000 for agriculture. It has been decided to double the land provision for industry at the direct expense of the agriculture, thus leaving 20,000 hectares available for each use. The supply and demand changes are diagrammatically represented in Figures 1.1 and 1.2.

Using the economist's proviso of 'other things being equal', although in reality they seldom are, these changes will lead to a decrease in the value of industrial land per hectare and an increase in the value of agricultural. Even if demand has not altered, agricultural land will therefore now command higher prices on the open market merely because it has become in short supply. The interaction of the forces of supply and demand on the open market have created this rise in value, which has been caused by the totally fixed size of land available.

In Figure 1.1, the decrease in supply of agricultural land from 0Q to 0Q1 results in price rising from 0P to 0P1 given that demand DD remains unchanged. In Figure 1.2, the increase in supply of industrial land causes the price to drop from 0P to 0P1.

In practice, when news of the release of agricultural land for industrial development is announced, the demand for it is likely to rise, which could match or exceed the shift in supply, so that the price may not decrease and may even increase from level 0P. Thus 'other things' are unlikely to remain 'equal' and when they do not a different conclusion may occur.

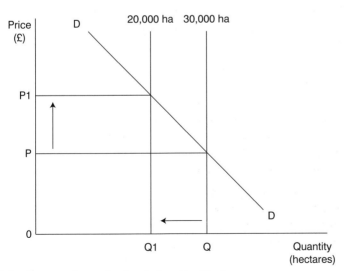

Figure 1.1   Decrease in supply of agricultural land in the town

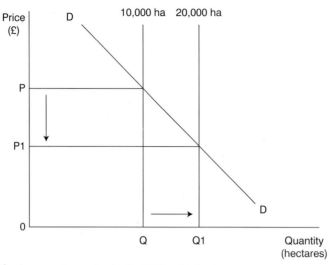

Figure 1.2   Increase in supply of industrial land in the town

Given that there cannot be an increase in the size of the land itself, the other possible way of increasing the space available for any specified land use is to increase the intensity of usage. This may be achieved in a vertical direction by constructing taller buildings with more floor levels, so that for every hectare of land devoted to that land use, the amount of floor space available for that use can be increased. Of course, this is not a remedy available to all land uses; but the multi-storey approach is a possible solution for commercial land uses such as offices or for residential by building flats rather than houses. The other alternative is to fit more buildings onto a site and sacrifice the quantity of open space retained.

Through utilising either or both of these policies, additional surface land area does not need to be sacrificed to increase the supply for the use. As the effective supply for all land uses is increased by this method, the town enjoys greater wealth from its fixed land area. There are limits though as to how far either solution can be taken. Development which is so dense that buildings cover every square metre of available land is not environmentally acceptable. In a city or town, there is a maximum height of building that is either physically feasible or aesthetically desired. However, these factors can be stretched to extremes, as witnessed by the skylines of such cities as New York, Hong Kong and London, where large numbers of high-rise buildings have been constructed close together to maximise the usage of the very limited land area available in the city centre.

Within any category of land use, the levels of supply for competitive properties will affect the value of any particular premises. For instance, if there is an inadequate supply of modern offices in an area, then substantial rents may be obtainable for old offices or converted houses. Conversely, if the supply of modern offices were adequate to meet the office demand, then it would be difficult to let older buildings, even at very much reduced rent levels.

Whatever the level of supply, its interaction with demand will determine market prices. As Harford (2007: 62) points out, these prices 'reveal information'. The parties to a transaction usually have a choice whether to buy or sell at a particular price. Should they decide to conclude a deal it will be because at that point of time each considers the agreed figure an acceptable compromise between both their perceptions of what the property is worth from a 'selling' or 'buying' viewpoint. It is this principle that underpins the comparison method of valuation, which is considered in Chapter 10. Observation and analysis of market transactions is a fundamental concept used by property valuers to provide evidence in support of the validity of their valuations.

1.5   Land use principles

The pattern of land use in any urban area is a reflection of competition for sites between various uses operating through the forces of supply and demand. In the long term, an activity will tend to locate at the place that gives it the greatest relative advantage. For businesses this will be the profit-maximisation location and for consumers the utility or facility-maximisation location.

The person or organisation willing to pay the highest price for a site is the one most likely to occupy and use it. Any competitors who are unable or unwilling to match this price will be competed away. Unless the market is modified by government policy or legislation, sites in urban areas will tend to be devoted to the use that produces the highest value, which will largely be determined by accessibility, complementarity and intensity of use:

- *Accessibility* – to transportation systems, markets, other similar users, labour supply, etc.
- *Complementarity* – which leads like and some unlike users to group together. Although different uses, offices and shops are usually found together in a city centre as they complement each other. For example, the workers travelling into the offices will use the shops at lunch times and after work. Different shops will group together to gain the benefits of passing trade and from the increased range and choice presented to the consumer by that shopping centre. The greater the number of shops and the more diverse their sizes, types and range of products offered, the greater the number of people who will travel to and use the centre. This increases the potential pool of customers for all the retailers.
- *Intensity of use* – the more intense the permitted use for the site, then generally the higher will be its value. Those sites which enjoy the greatest accessibility and complementarity will have the highest demand and will therefore need to be used intensively to try and satisfy as much of this demand as possible. However, intense use is not always possible or desirable due to site conditions, planning restrictions, type of use and social and environmental considerations.

How land uses are distributed within an urban area and the effect of location on value have been the subjects of much research and discussion by economists, philosophers and sociologists over more than two centuries. Some of the leading writers include:

- Adam Smith (1723–1790)
- David Ricardo (1772–1823)
- J.H. Von Thünen (1783–1850)
- John Stuart Mill (1806–1873)
- Karl Marx (1818–1883)
- Henry George (1839–1897)
- Robert Ezra Park (1864–1944) and Ernest Watson Burgess (1886–1966)
- Walter Christaller (1893–1969)

Adam Smith is generally regarded as the founder of political economics and, as previously indicated, his publication in 1776 began the serious study of value and its causes. In his *On the Principles of Political Economy and Taxation*, published in 1817, Ricardo established that land was not price determining but was price determined by whatever factor use was based upon it. This was due to the inability of the supply of land to change to meet increased demand, unlike supplies of capital and labour. He reasoned that 'The price of wheat is not high because the price of land is high, the price of land is high because the price of wheat is high.' This led him to develop the theory of 'economic rent' being a payment made in excess of the 'supply price', which is the minimum reward necessary to retain a factor of production in its present use. In modern terms for land, this would be the difference between value in its existing use and value in its next-best remunerative use, which will be determined by market forces and planning permission.

Von Thünen's *The Isolated State* in 1826 put forward the theory of concentric circles of value developing in urban areas, such that the value of land and rents is determined by distance from and transport costs to the city centre. Mill's *Principles of Political Economy*, published in 1848, had a similar impact to Smith's and was read as a standard work on economics for the following sixty years. Marx's *Das Kapital* (1867–95) considered that its economic base determined the political, legal and social structure of a society and land ownership was the key factor in this. Henry George expounded the argument for the taxation of the economic rent element of land. His philosophy was that land was not created by man but is a gift of nature and as such is owned by the whole community. Thus, any value created by a particular use belongs to that community.

In 1925, Park and Burgess developed the concentric ring theory. They concluded that accessibility, values and density declined with increasing distance from the city centre and five circular zones become established. From the centre, these are: the central business area, zone of transition, factory and low income housing zone, middle and high income housing zone and the outer commuter zone. Christaller's 1933 publication, *Central Places in South Germany*, expounded the central place theory, which was that *travel time* rather than distance was the main location-determining factor.

As identified by some of these writers, most modern urban areas share a broadly similar pattern of land uses, with variations and differences in each individual town or city. These five zones or regions are: the central business district or zone, the zone of transition, suburban area, rural–urban fringe and rural area (Lean and Goodall 1966: 146–52).

Central Business District (CBD) or Zone (CBZ)

This is the core of the city and is the area that has the highest levels of accessibility and complementarity. It is the central point from which the transportation routes radiate. In most cases, the CBD is in the geographical centre of the town, but this does not necessarily have to be so.

The central area is relatively small sized and, coupled with intense demand from users due to the advantages of its location, it will enjoy peak land values. The scarcity of land together with these high values will produce the greatest intensity of use of land in the urban area, which results in high-rise buildings.

Commercial uses that most benefit from high accessibility and complementarity, like offices, retail and certain leisure uses such as theatres, will congregate in this area. The other uses that benefit from and need to be in this area are major public buildings such as museums, main libraries, town halls and central administrative offices. This intense competition for space and the high land values will severely restrict the amount of residential property in this area, and those that do exist will command high values, particularly if in good condition.

The zone of transition

This zone immediately surrounds the CBD and can be termed the inner-city area. It possesses relatively high land values and was created from the expansion of the CBD. It usually consists of a mix of completely new buildings or old buildings being converted, rehabilitated or redeveloped. It is the oldest residential district and often consists of luxury residences or low-income multi-family dwellings at relatively high densities due to the high land values. Radial transport routes out of the centre offer reasonable accessibility for a number of other users.

Suburban area

Land values and intensity of land use are much lower than in the previous two areas. The majority user is residential at moderate densities and associated complementary uses including open space and recreational areas. Development tends to be low-rise with the possible exception of regional centres within the area. With lower land values, there is less pressure for high-rise development to maximise usage of the available sites.

Rural–urban fringe

This is the surrounding countryside where single-family homes mix with agriculture. Other than those employed in local agriculture, persons living here are in higher -income groups and choose to live here as space is relatively unlimited, which allows larger dwellings to be built. Lower income groups are discouraged by the high costs of commuter travel back to town.

To compete with residential use, usage of land for agriculture needs to be intense, with such uses as market gardening being common. In Great Britain, these areas of the country have usually formed part of the Green Belts established around the major cities, and in particular London, since the late 1940s as a means of preserving open space and countryside within close proximity to the city and to prevent the further spread of the urban conurbation beyond these established limits. These were set up in response to the continual outward growth of cities in the 1920s and 1930s that resulted in 'ribbon development', where buildings were constructed along all the major roads radiating from the city to the surrounding towns and areas. This created the effect when travelling along such a route of never seeing any countryside, as the buildings from one area of the city merged with those in the next, and with those in successive towns.

Rural area

This is the open countryside beyond the outer boundary of the city. It is largely devoted to agriculture, forestry, heath land and other open space with few buildings. In the UK, many such areas are designated as National Parks, Areas of Outstanding Natural Beauty or Sites of Special Scientific Interest, with strict controls on any form of development being permitted within them. Land values in rural areas are the lowest for the locality in that they have minimum accessibility and can offer no complementarity of uses for commercial purposes.

The size of the five zones, and their exact shape will vary from city to city, town to town and region to region. In many cases, an area or areas radiating from one centre will overlap with those from a nearby conurbation, so that all five zones do not exist in that particular district. In addition, small regional centres can be situated within the boundaries of a major city, and these too establish their own zones of land use that overlap those of the city itself. In this way an area of land may be within the suburban area of the city, but form part of the central business district of the regional town.

Establishing the economic area of land use within which a property is located is an essential factor in understanding the economic, social, political and geographical factors that exist and help determine the levels of supply and demand for a particular property type and thus influence its value.

1.6   The investment and property markets

A 'market' can be considered as any effective arrangement for bringing buyers and sellers into contact with one another. It does not need to be a single place. Indeed, markets can be on a local, national or international basis.

The investment market brings the supply of existing and new investments coming onto the market into equilibrium with the demand for investments backed by money. It is the effective supply and demand interacting that produces an equilibrium price within the market. This equilibrium price can be expressed in terms of capital value and/or yields.

At any one time, some individuals or institutions will have capital and income in excess of their current expenditure whilst others have expenditure plans in excess of their current capital and income. The investment market exists for the mutual benefit of both, collecting the surplus funds of some to make them available to others who require them in exchange for forms of investment.

Surplus funds can be accrued due to a deliberate act of saving to provide for future expenditure or uncertainties or can simply be 'saved' as they are just not needed to satisfy all present expenditure wants. In either case, the level of saving has a direct bearing on the level of investment, and the factors that will influence the level of savings in an economy, especially interest rates, will affect investment.

The property market exists wherever property transactions can take place. This contrasts with the market for financial securities where stock exchanges around the world have fixed times for trading. Prices determined on the exchanges are freely available at any time and are published nationally each day. There is no such arrangement with property prices. Deals to buy, sell and rent property can be made anywhere at any time and are not always reported. Property transactions can remain confidential as successful buyers and sellers may be reluctant to disclose how much they paid or received for a property. This results in future buyers and sellers not always knowing what the level of demand and supply and thus the expected market value will be.

Unlike financial securities, every property is different. Correct valuation of each can be complex. This is why investors should take advice from professionally qualified valuers, whose knowledge and expertise will enable them to provide a reasoned valuation on which to make purchase and sale decisions.

1.7   Taxation and its effects

The UK taxes that affect real property and property valuation are:

- Income Tax
  - Payable by individuals on earned (remuneration from employment and self-employment) and 'unearned' (from investments) income, including rents received from property.
  - 'Income' is deemed as any non-capital sum received during year of assessment (irrespective of whether received irregularly or not more than once during that year).
- Corporation Tax
  - Operates in similar tax way to Income Tax, but charged on the annual taxable profits of companies which are resident in the UK.
- Capital Gains Tax (CGT) and Corporation Tax on capital gains
  - Potentially payable when a capital gain (as opposed to an income sum) is made on the sale or deemed disposal of an asset such as land and buildings.
  - Does not apply to the sale of a person's main or sole residence or a limited number of other investments, including all government stocks ('gilts').
- Value Added Tax (VAT)
  - An indirect tax added to the value of goods and services supplied, including building work and other property-related services.
  - Payable on rents received from property or on the sale of property only if the landlord/ vendor has formally elected to opt to tax the property.
  - Certain supplies are exempt from the tax, others are zero-rated (that is tax is 'charged' at nil per cent) and the remainder are standard rated at the prevailing normal rate.
- Inheritance Tax (IHT)
  - Payable on receipt of money or assets through gift by will on death of donor or given during donor's lifetime but donor dies within seven years of gift being made.
  - The capital value of all property at the date of death is added to the value of the deceased's estate and the tax deducted from the estate or paid by the recipient where a lifetime transfer has previously been made.
- Stamp Duty Land Tax (SDLT)
  - Payable by a purchaser of land and buildings and applies to both freehold and leasehold purchases, as well as the rent and premium on the grant of a new lease.
  - It is an addition to the costs of acquisition of property.
- Business Rates and Council Tax
  - Paid to local authorities and assessed on an annual value of business property (rates) and a capital value of residential property (Council Tax).

Tax capital allowances are the means by which tax relief is given for certain capital expenditure (mainly on fixtures or plant and machinery, but also for industrial and agricultural buildings). These can cut the real cost of commercial buildings, which may be reflected in their yields (Chidell 2007).

The above gives only a very brief idea of the effects each tax has on property. More information on the scope and content of each tax and the current percentage rates payable can be obtained from HM Revenue and Customs website (http://www.hmrc.gov.uk/). The appropriate allowances and adjustments that need to be made in property valuations for the effects of some of the taxes are considered below (see Sections 5.3, 6.8, 11.13 and 11.14).

1.8   Influence of town and country planning legislation and policies on property values

Ratcliffe (1974: 4) defined town planning as

> a reconciliation of social and economic aims, of private and public objectives. It is the allocation of resources, particularly land, in such a manner as to obtain maximum efficiency, whilst paying heed to the nature of the built environment and the welfare of the community. In this way planning is therefore the art of anticipating change, and arbitrating between the economic, social, political, and physical forces that determine the location, form, and effect of urban development. In a democracy it should be the practical and technical implementation of the people's wishes operating within a legal framework, permitting the manipulation of the various urban components such as transport, power, housing, and employment, in such a way as to ensure the greatest benefit to all.

Left to a free market price mechanism, solely determined by the interaction of supply and demand, land could be allocated between competing uses in an efficient way to the detriment of the community.

> The private sector developer seeking to maximise his personal profit frequently neglects the provision of both social services and public utilities. The very need for planning arose out of the inequality, deprivation and squalor caused by the interplay of free-market forces and lack of social concern prevalent during the nineteenth century. Furthermore, unplanned, these forces combine to produce the fluctuating booms and slumps that epitomise private sector instability. (Ratcliffe 1974: 5)

Town planners, acting on behalf of the community, can thus wield a very significant influence on allocation of land usage and the intensity of that use. In the UK, this power is primarily given to the local planning authorities, who are the county, unitary and district councils. They are subject to statutory regulations and guidelines enacted by Parliament and administered by Communities and Local Government, a government department headed by the Secretary of State for Communities and Local Government. Current planning legislation is consolidated in the Town and Country Planning Act 1990.

The planning authorities have the power to:

1. allocate land for uses
2. permit or restrict changes of use
3. regulate intensity of use and
4. permit or restrict new development.

These powers are vested in these public bodies so that development can be regulated for the overall benefit of the community and the country having taken account of financial, social, political, economic and environmental factors. In this way, the allocation of land can be balanced between competing uses. It should also ensure any non-profitable uses, such as parks and open spaces can be sufficiently provided, which may not be the case if allocation was left entirely to free-market forces.

When authorising a category of use or a change of use for a property a local planning authority is directed by The Town and Country Planning (Use Classes) Order 1987. Planning permission is generally required for a change from one 'use class' to another.

The main 'use classes' are:

- A1: shops
- A2: financial and professional services
- A3: restaurants and cafés
- A4: drinking establishments
- A5: hot food takeaways
- B1: businesses (offices, light industry)
- B2: general industrial
- B8: storage and distribution
- C1: hotels
- C2: residential institutions
- C3: dwelling houses
- D1: non-residential institutions (schools, libraries, surgeries)
- D2: assembly and leisure (cinemas, swimming baths, gymnasiums)

The decisions and policies of the planning authorities will affect the level of supply for any particular land use and this will have a direct effect on values. To ensure this power is not abused, tight political, ethical, legal and financial controls are imposed by central government and within local government itself.

Progress check questions

- What 'property' is being valued in 'property valuation'?
- What is 'value'?
- Why is this different to 'cost'?
- Why should every valuation be dated?
- How can a single property have different values at the same time?
- What is the difference between capital and income?
- Why may worth, price and value be different sums of money?
- How does the nature of property affect its supply?
- Why is it more likely that high-rise buildings will be situated in or near the city centre?
- How do prices 'reveal information' to valuers?
- What are the main three factors that determine the highest value for a property?
- How does the pattern of land use within an urban area determine the types of property to be found in an area and their values?
- What is the function of an investment market?
- In what ways is property ownership and investment affected by tax legislation?
- How can decisions made by town planners affect the value of property?

Chapter summary

Property valuations are needed for many different purposes and numerous types of value can apply to any property at one time. It is critical that both valuer and client are clear about which value is to be assessed and why.

Value, price, worth, cost and market value may all sound similar descriptions of the same monetary sum, but in the context of property valuation can have quite different definitions. Valuers often find the present value of the right to receive income and/or capital amounts. Seldom is cost used as a measure of value.

Interaction of the market forces of supply and demand will determine prices and market values. By observing and analysing the prices in the market, the valuer can obtain information that will inform future valuations. The inherent nature of land and buildings makes it difficult to change the level of supply in the short term and there may be absolute limits on the quantity of space that can be made available within a city area for any particular use. Accordingly, any change in demand can have a disproportionate effect on price. Land uses within an urban area often form distinct patterns, strongly influenced by land availability and accessibility. The most significant factor in determining the value of a property is where it is located. Taxation and planning policies and legislation will also affect values.

Further reading

Balchin, P., Kieve, J.L. and Bull, G.H. (1995) *Urban Land Economics and Public Policy*, 5th edn, Basingstoke: Palgrave Macmillan.

Ball, M., Lizieri, C. and MacGregor, B.D. (1998) *The Economics of Commercial Property Markets*, London: Routledge.

Egan, D. (1995) 'Mainly for students: property cycles explained', *Estates Gazette*, 9547 (25 Nov.): 147–8; repr. in P. Askham and L. Blake (eds), *The Best of Mainly for Students* (London: Estates Gazette, 1999), vol. 2, pp. 117–23.

Evans, A. (2004) *Economics and Land Use Planning*, Oxford: Blackwells.

—— (2004) *Economics, Real Estate and the Supply of Land*, Oxford: Blackwells.

French N. (1998) 'Word play', *Estates Gazette*, 9803 (17 Jan.): 130–2.

Gaskell, C. (1998) 'Words on worth', *Estates Gazette*, 9820 (16 May): 125-8; repr. in P. Askham and L. Blake (eds), *The Best of Mainly for Students* (London: Estates Gazette, 1999), vol. 2, pp. 369–75.

Gilbertson, B., Preston, D. and Howarth, A. (2006) *A Vision for Valuation* (RICS Leading Edge Series), London: RICS. Online: <http://www.rics.org/NR/rdonlyres/BBEBD43B-11CA-4A2E-BDAC-B37BB2505CA1/0/vision_for_valuation.pdf>

Harvey, J. (2000) *Urban Land Economics: The Economics of Real Property*, 5th edn, Basingstoke: Palgrave Macmillan.

Wyatt, P. (2007) *Property Valuation in an Economic Context*, Oxford: Blackwells.

The property valuation profession

In this chapter ...

- What a property valuer does and why.
- Who may require the services of a valuer?
- Why a valuation is different to a building survey.
- How accurate are valuations?
- Why valuers should be independent and objective in their work.
- The role of the Royal Institution of Chartered Surveyors and services it provides to the profession.
- The globalisation of property valuation and likely future developments.

2.1   Skills required by and role of the property valuer

What does a property valuer do? The main task, by definition, is to find the value of a property. To do this, property valuers have to possess and be competent in a diverse range of skills, such as:

- research methods
- calculation
- measurement
- report writing
- negotiation
- law
- management and business finance
- a working knowledge of economics and politics

- a knowledge of building construction
- an awareness of environmental issues.

These are acquired and refined through a lengthy process of academic study, practical experience and lifelong learning (LLL), also known as continuing professional development (CPD).

Before valuers can value, they must know exactly what type of value they are seeking to find, for whom they are finding it and for what purpose this valuation is being sought. Without this knowledge, the resultant figure will have no relevance and has the potential to be taken out of context and interpreted in an incorrect manner.

Failure to establish with a client at the outset the exact nature of the instruction, including the full facts of what is required and why and how it will be obtained, will lead to later difficulties. This could range from the provided information being misunderstood, to non-payment of fees, to a possible claim for negligence against the valuer for failing to fulfil a duty of care and undertake the task required in a careful, reasonable and professional manner.

When communicating with clients, valuers should endeavour to use clear, concise and plain English. However, within the property world and between property valuers, a large number of unusual or unique words, phrases or abbreviations are used. Such specialist terminology is not uncommon in most trades or professions. It takes time for newcomers to valuation to become familiar with this terminology and readers are recommended to gain an understanding of it to assist in their study of the subject. The standard reference work that provides the relevant definitions is *The Glossary of Property Terms* (Parsons 2004). Subscribers to the *Estates Gazette's* electronic service can also access this source online (http://www.egi.co.uk/Articles/Glossary. aspx?NavigationID=468).

Property valuers have a range of valuation methods they can use to estimate the value of any type of property. These are covered in detail in Chapters 10 to 15. A valuer will usually use more than one method, or more than one variation of the same method to value a property, to provide a 'checking and balancing' system and ensure greater reliability and accuracy of results.

The valuer's role in general is to advise as to what would be the best figure obtainable for a given property, in the open market, at a specific date. To do this, the valuer must know how the many and varied characteristics of real property can affect value and how changes in social, economic and political factors, in the local, national and international contexts, are likely to influence it. Legislation will have a major impact on assessment of value and the valuer must have a good working knowledge of the relevant law to be able to undertake the required valuations correctly.

Purposes for which a valuation may be required include sale or purchase, rent to be paid or demanded, the amount of mortgage which could be advanced on a security, calculation of compensation payable or receivable, assessment of taxation or rating and the advisability of investment. The intention of the valuation, together with the circumstances and requirements of the client requesting the assessment, can greatly influence the value. As a result the valuer may provide each of a number of clients concerned with the one property a different valuation, or indeed, different valuations to the same client on the same property depending on the definition of values being sought.

In addition to valuation appraisals, valuers may also advise on the policy to be adopted in the management of properties and investments to at least maintain, and where possible increase, their annual income and capital value. Amongst other things this will involve being proactive in anticipating future trends in the market and suggesting solutions to problems that may arise from them so as to ensure income is maximised and running costs minimised, without detriment to the overall condition of the investments. This management and advisory function can extend to negotiating lease renewals and rent reviews and appraising development and improvement proposals, as well as the management of the physical structure of the buildings. Valuation research specialists can also advise on trends in the market generally and on particular property types in specific locations where required by clients.

There are three main reasons why valuers are employed because of their specialist knowledge:

- the property market is an imperfect one – supply and demand are always changing and are different in each location and for each type of property and information on transactions is often restricted;
- each individual property and the interests therein tend to be unique, or at least never exactly the same as other properties;
- legislation – the complex and inter-related laws relating to property are forever changing, and only a specialist with full knowledge of them, which needs to be constantly updated, can successfully interpret them correctly.

It is because qualified professional valuers make an in-depth study of these matters, and are fully informed of all factors affecting property values, that they can formulate a reasonable and logical opinion on a value for a given property and situation.

The main tasks undertaken by valuers are:

- receiving and confirming instructions;
- inspecting the property and its location;
- liaising with the client's other professional advisers where necessary, such as accountants, lawyers and management consultants;
- researching and analysing all relevant information;
- carrying out all calculations to arrive at a valuation;
- reporting the results of the research and providing the valuation;
- negotiating with the other party's representatives to reach agreement;
- instructing solicitors on behalf of a client or employer;
- providing property advisory and management services.

With a vast range of property types, locations and reasons for valuations to select from, once qualified and working in professional practice most valuers choose to specialise. This may be a geographical and/or property category specialism. So, some valuers will deal with one region of a country or a specific form of property such as retail or industrial only. In addition, the range of professional work undertaken provides other forms of specialism. For instance:

- general practice – encompassing the full range of valuation surveying skills (although it is becoming increasingly unusual for an individual to cover all and every type of work required);

- valuation – of capital values for taxation, insurance, asset value, accounts, loan security, unit pricing, purchase or disposal;
- landlord and tenant – rent reviews, lease renewals and associated dispute resolution matters;
- investment and fund management – sale and purchase, portfolio selection and management, creating and implementing an investment strategy to maintain and enhance the value of property assets;
- property management – includes maintenance and repair, insuring, assessing service charges and landlord and tenant matters;
- research – of the market to advise and more fully inform clients;
- rating – assessment of and/or appeals against rateable values;
- agency – marketing and sale or letting of commercial, industrial or residential property;
- planning, development and regeneration – appraisal of opportunities and impacts, maximising a property's potential, organising and undertaking financing and project management and liaising between the many and varied stakeholders;
- facilities management and management consulting – acting as analyst and adviser to corporate clients on business strategy applied to property or ensuring the delivery of services to properties.

This wide choice of specialist areas of professional work available is reflected in the current organisation of the Royal Institution of Chartered Surveyors' *professional groups* (see Section 2.7).

2.2   Who may require the services of a valuer?

Any person or any organisation that occupies, owns, seeks to own or finances property is likely to require a valuation.

Who owns or seeks to own property? This can broadly be categorised into two sectors:

- Private sector
    - private individuals
    - insurance companies
    - pension funds   } = 'financial institutions'
    - banks
    - industrial and commercial companies
    - property companies
    - property developers and construction companies
    - overseas buyers
- Public sector
    - local authorities
    - public authorities
    - central government departments and agencies
    - charities

Who finances property?

- UK clearing banks and their subsidiaries
- major overseas banks
- merchant banks
- building societies
- pension funds and insurance companies
- venture capitalists
- specialist lending institutions, such as finance corporations.

The services of a valuer can thus be used by anyone with a legal interest in landed property or considering acquiring such an interest. Potential clients can range from global major organisations down to individual citizens. Large numbers of valuers work in private practice for large or small property consultancy or surveying firms, in partnerships or as sole practitioners, and act for clients. Alternatively, many large organisations, including some sections of central government, have their own 'in-house' valuers who act for them on their property matters. There is thus a very wide scope of potential employers. Geographically there are opportunities on a worldwide scale with RICS qualified professionals being recognised internationally.

2.3   Difference between a valuation and a building survey

A valuation is defined by the RICS as 'A member's opinion of the value of a specified interest or interests in a property, at the date of valuation, given in writing. Unless limitations are agreed in the terms of engagement this will be provided after an inspection, and any further investigations and enquiries that are appropriate, having regard to the nature of the property and the purpose of the valuation' (Royal Institution of Chartered Surveyors 2007d: 9).

It is important to appreciate that a valuation is *not* a building survey. Citing *Definitions of Inspections and Surveys of Buildings* by the Construction Industry Council (CIC), the RICS defines a *building survey* as:

> an inspection and assessment of the construction and condition of a building and will not normally include advice on value ... The survey will generally include the structure, fabric, finishes and grounds. The exposure and testing of services are not usually covered. The extent of the survey will be subject to the specific agreement between the surveyor and client and advice on costs of repair will be subject to such agreement. The report may include reference to visible defects and guidance as appropriate on maintenance and remedial measures. The report may recommend that elemental or specialist investigations are undertaken or other specialist advice obtained relating to specific issues. The survey will not normally include intrusive investigation of materials or structure, or inaccessible or hidden areas, unless agreed with the building owner.
>
> (Royal Institution of Chartered Surveyors 2005a: 5)

In the residential property market, three levels of inspection may be undertaken. Despite this choice, around 90 per cent of individuals buying residential property in the UK rely entirely upon the 'level 1' inspection, referred to as a residential mortgage valuation. This is usually undertaken on behalf of an institutional lender, but private buyers and sellers may separately commission such reports.

The case of *Lloyd v. Butler and another* [1990] 47 EG 56 concerned a mortgage valuation where the valuer failed to reach the duty of care expected in such cases. It was concluded that the typical residential valuation inspection should last 20–30 minutes and requires 'someone with a knowledgeable eye, experienced in practice who knows where to look ... to detect trouble or the potential form of trouble'. Although not necessarily obliged to follow up every trail of evidence, the case inferred that the valuer must alert the lender and borrower to the risk where serious defects are suspected and that further investigations should be made before there is a legal commitment made to purchase. One of the conclusions in *Smith v. Bush* [1987] 3 All ER 179 was that if there are no clearly identifiable signs of suspected problems, or a surveyor misses defects because the signs are hidden or concealed, then it will be difficult to demonstrate negligence. On the other hand, *Platform Funding Ltd v. Bank of Scotland plc* [2008] EWCA Civ 930 established that in addition to carrying out his/her inspection and valuation with reasonable skill and care, a surveyor's duty may include other unqualified obligations imposed by the conditions of engagement or required report format.

Effectively falling between a mortgage valuation and a building survey, the Homebuyer Survey and Valuation (HSV) 2005 (commonly known as level 2) is a form of limited contract that tightly defines the extent of inspection within its terms of engagement (Royal Institution of Chartered Surveyors 2005c). The HSV provides a standard pro-forma report using headings covering structural elements, 'urgent' and/or 'significant' defects, a market valuation and an insurance reinstatement cost assessment. This format replaced the HBSV (Home Buyers Survey and Valuation) in 1997, which in turn had superseded the former HBR (House or Flat Buyers Report) in 1993. A new version, known simply as the RICS Homebuyer Report (HBR) will be used from July 2009. The amount of time spent on site is typically 90 minutes (SAVA HSV Benchmarking).

A high proportion of the problems that arise on this type of survey are simply because the parties are not sufficiently clear about the nature and scope of the inspection, with some clients mistakenly expecting the HSV to be effectively the same as a building survey, which it is not. A Residential Building Survey (commonly known as level 3) is intended to provide the buyer with the most extensive report option and is therefore the most expensive. Formerly called a structural survey, this survey is

a bespoke service suitable for all residential properties and provides a full, detailed picture of their construction and condition. It is likely to be needed if the property is, for example, of unusual construction, is dilapidated or has been extensively altered – or where a major conversion or renovation is planned. Building Surveys are usually tailored to the client's individual requirements. The report includes extensive technical information on material and construction as well as details of the whole range of defects, major to minor.

(Royal Institution of Chartered Surveyors 2005c: 14)

The inspection and report is limited to readily accessible parts of the property, and so unless the client makes specific arrangements with the owner to lift or move furniture, lift floor coverings and take up floorboards, the surveyor will not generally be expected to do so. The surveyor will not normally test the services, but may be instructed to coordinate electricians, heating engineers, etc. The exact extent of liability will depend upon the precise terms and conditions of engagement that the surveyor and client enter into.

The case of *Hacker* v. *Thomas Deal and Company* [1991] 44 EG 173 involved a building survey (known then as a structural survey). The judge stated: 'Bearing in mind that there is a difference between the various types of reports which are produced upon the sale and purchase of houses, and in particular, that this was not a valuation report but a structural survey ... which involves a detailed and thorough inspection of the property to be inspected ... it obviously involves seeing what there is to be seen through the eyes not of the layman but of the expert.'

The surveyor in this case was likened to a detective trying to anticipate problems, and to visualise what could be happening within the areas of the building which are inaccessible to the naked eye, but without actually disrupting the fabric, decorations or the fittings of the house. The judge went on to say that in this situation 'one does not start going into all the little crevices in the hope of finding something unless there is some tell-tale sign which indicates that it would be advisable to do so'. The decision made it clear that the courts expect a more detailed inspection of the fabric of a building, and a deeper knowledge and expertise of construction and defects be employed in the case of building surveys compared to valuations.

2.4   The level of accuracy expected of the valuer

A major concept that should be appreciated is that in many respects property valuation is more of an art than an exact, scientific subject. For all the use of mathematical formulae and calculations, valuers also exercise subjective opinion based on their knowledge of the market and their interpretation of facts. Two valuers, given the same property to value and the same facts to work from, will often arrive at slightly different final values as they have each formed somewhat diverse opinions on the current state of the market and how the information concerning the property should be interpreted.

It is vital that value is calculated after carrying out careful and meticulous research and is based on sound principles, having taken all relevant factors into account and disregarding all irrelevant factors. Doing less than this is unacceptable. Nevertheless, having diligently carried out the valuation functions there are usually some factors that are open to subjective opinion. This may cause the final value suggested by the individual valuer to vary a little from that suggested by another practitioner. Providing they have taken all reasonable care and have based their calculations on proven principles and logical argument, neither valuer in a negotiation, where their final values differ, can necessarily be said to have produced either the 'right' or 'wrong' answer. Moreover, both figures could be reasonable from each side's viewpoint and the final sum agreed would need to be a compromise that would depend on the result of negotiation between the parties.

The valuer would then be expected to negotiate with the other party's valuer or representative, to argue how the given facts should be interpreted and to agree a price or figure acceptable to both sides. In the end, value comes down to what a prospective purchaser or tenant is ready and able to pay and what the prospective vendor or landlord is financially able and willing to accept. Unless a figure satisfactory to both can be achieved, all valuation theory is pointless. There is no use in deciding that a property is worth a certain value if nobody is prepared and able to purchase or rent it at that price.

Crosby and Matysiak (2002) stated that 'The accuracy of any valuation is defined as how close the valuation is to the exchange price in the market place'. A number of court cases have examined the expected degree of accuracy expected in a valuation.

In *Singer and Friedlander Ltd* v. *John D Wood & Co.* (1977) 243 EG 212 Watkins J stated: 'Any valuation falling outside what I shall call the bracket brings into question the competence of the valuer and the sort of care he gave to the task of valuation.' In the light of advice provided by an expert witness, this bracket was put at 10 per cent either side of the figure provided. In addition, Watkins J said that the bracket could be extended to about 15 per cent or a little more, either way in exceptional circumstances. In *Muldoon* v. *Mays of Lilliput Ltd* [1993] 14 EG 100 Judge Zucker QC, again as advised by an expert witness, used a range of 15–20 per cent. In *Axa Equity and Law Home Loans Ltd* v. *Hirani Watson* [1999] EGCS 90 it was held that 10 per cent need not be the normal starting point, and in *Birmingham Midshires Building Society* v. *Richard Pamplin & Co.* [1996] EGCS 3 a variation of 11 per cent was used (Grenfall 2003). Conversely, 'for straightforward valuations, where numerous comparables are readily available, the courts will ordinarily narrow the bracket to 5% either side of the true valuation' (Johnson 2008: 22).

> In deciding whether the bracket should be widened, the court will consider the following factors:
> * Unusual nature of the property
> * Lack of comparables
> * High value
> * Extreme market conditions
> * Restrictive instructions
>
> (Johnson 2008: 22)

The *Valuation Sale Price Variance Report 2006* (RICS and IPD 2006: 1) addressed this issue of accuracy in covering four key European markets of the UK, France, Germany and the Netherlands. It examined 'the correlation between valuations of commercial properties and subsequent actual sale prices'. It was stated that 'the accuracy of valuation is central to pricing and performance measurement of property investment' and addressed the 'two key questions' of 'how much do sale prices differ from valuations' and 'are there any market or structural explanations for the differences observed?'

This report concluded that 'in 2005 there was a common trend for larger properties to be valued more accurately than smaller assets across all four markets'. Within the UK, the 'un-weighted average absolute valuation price difference' for all commercial property was 10.3 per cent. This measured how far each valuation was from the sale price, regardless of whether it was above or below that price and treating each sale as equally important. A further finding was that there was a positive 'direction difference' which indicated that 'a premium over valuation is typically achieved when selling properties', or 'in other words, valuers appear to have taken consistently conservative views in marking property to market'(RICS and IPD 2006: 1).

Similar conclusions were drawn from the 2007 report, which concentrated solely on the UK market. This found that 'There was an average difference of 10.9% between sale prices and this "adjusted" valuation in 2006 for all sales in the IPD UK Annual Databank. In other words, on average, once adjusted for capital growth in the underlying market, sale prices differed by just under 11% from their preceding valuation.' In addition, 'three quarters of all transactions were sold at a price above their preceding adjusted valuation' (RICS and IPD 2007: 6).

The conclusion that may be drawn from this research is that valuers have a tendency to err on the side of caution and possibly slightly undervalue. This is perhaps understandable. Should clients achieve a higher sale price than their valuation they will probably be surprised and pleased, but if they can only sell at a figure less than valuation they will be more likely to both be disappointed and to question the accuracy of the provided valuation. In any event, market data confirms that it is very difficult for a valuation figure to be clear-cut and unequivocal. There is always an element of subjectivity in the assessment of value. Inevitably, this results in an inherent small margin of variation between valuations made by different valuers or between a valuation and the final market exchange price. The key is keeping this margin as small as possible.

2.5 Independence of the valuer

Valuers are expected to act objectively, impartially and without any vested or conflict of interest. 'Members undertaking valuations must act with independence, integrity and objectivity' (RICS 2007d: PS1.6, 16) and 'a commitment to ethical standards' (RICS 2007d: PS1.4, 15).

On occasion there may be the need for 'additional criteria for independence' to be fulfilled. 'For some purposes, statutes, regulations, the rules of regulatory bodies, or client's special requirements, may set out specific criteria that the valuer must meet in order to achieve a defined state of independence.' Where this applies, valuers 'must establish the criteria required and confirm that they meet them in the terms of engagement and the report' (RICS 2007d: PS 1.7, 17).

2.6 The role of the Royal Institution of Chartered Surveyors (RICS) as a regulatory and unifying body for the profession

The Institution of Surveyors was founded in 1868. This later became the Royal Institution of Chartered Surveyors (RICS), which 'has evolved into a renowned international organisation with approximately 120,000 members in 120 countries'. It 'is the leading source of land, property, construction and related environmental knowledge' and exists to 'promote best practice, represent consumers' interests and provide impartial advice to society, businesses, governments and global organisations' (RICS 2007c).

It 'is one of the most respected and high profile global "standards and membership" organisations for professionals involved in land, property, construction and environmental issues. Accountable to both members and the public, RICS has three main roles:

- to maintain the highest standards of education and training;
- to protect consumers through strict regulation of professional standards;

- to be the leading source of information and independent advice on land, property, construction and associated environmental issues' (RICS 2007a).

The major publication of the institution that regulates property valuation is *RICS Valuation Standards*, commonly referred to as the *Red Book* (RICS 2007d). This is considered in more detail in Chapter 9.

2.7   The RICS professional groups

'Launched on 1 January 2001, the seventeen faculties embrace the wide range of specialisms covered by chartered surveyors'. They are 'international in their outlook and are coordinated by individual faculty boards' (RICS 2009a). Faculties changed to professional groups in 2009. Their responsibilities include training and development, 'specialist and research information, guidance and standards, contributing to policy formation and quality control over use of specialist chartered designations' (RICS 2009a). A member of the RICS can join up to four of the groups.
   The group titles are:

- Arts and Antiques
- Building Control
- Building Surveying
- Commercial Property
- Quantity Surveying and Construction
- Dispute Resolution
- Environment
- Facilities Management
- Geomatics
- Machinery and Business Assets
- Management Consultancy
- Minerals and Waste Management
- Planning and Development
- Project Management
- Residential Property
- Rural
- Valuation.

2.8   RICS information resources

The RICS online database www.isurv.co.uk provides 'best practice guidance for surveyors' and includes:

- 'contributions from over 150 experts
- a case law database with expert commentary

- model letters, forms, agreements and precedents
- all relevant RICS Guidance Notes, Information Papers and Standards
- indexed links to surveying information on other websites' (RICS 2007h).

RICS publications on valuation matters are contained within *isurv Valuation*. There are additional separate sections, called 'channels', on Building Surveying, Construction, Commercial Property, Disputes, Environment and Planning. It is regularly updated and its content expanded.

There is a distinction drawn between official RICS information, such as the *Valuation Standards*, and information written specifically for *isurv Valuation* by individual authors. This latter material 'represents a personal viewpoint or expert opinion, and is not the official view' but is aimed at helping 'valuation professionals keep up to date with practice and developments in their area of expertise' (RICS 2007h).

2.9   International dimensions of the profession

Property valuation has become an international profession. The RICS *Red Book* reflects this, incorporating International Standards. The growth of the European Union and the opening up of property markets in eastern Europe and Far East to foreign investment have helped this expansion of the profession and demand for its services. Increasingly global clients undertaking cross-border property investment seek this advice.

In addition to the RICS in the UK and equivalent bodies in other countries, the International Valuation Standards Committee (IVSC) exists to develop and promote conformity of professional practice around the world. Its website summarises its concept and role:

> The International Valuation Standards Committee (IVSC) is a not-for-profit corporation registered in the State of Illinois, with its headquarters in London. It was originally founded in 1981 by the major real estate valuation institutes from the major economies. It has now broadened its membership to include professional associations for valuers of many types of assets, including plant and equipment, minerals, intangible assets and businesses. Its membership represents over fifty different countries. IVSC is committed to the development of a single set of global standards and requirements for the valuation of all assets and liabilities.
>
> (IVSC 2007a: 6)

The eighth edition of the International Valuation Standards (IVS) was published in July 2007 and their contents are incorporated into the wording of the RICS *Red Book* so that compliance with one provides compliance with the other.

The means by which property performance is measured and evaluated is becoming increasingly standardised. Louargand (2007: 24–5) confirms that 'we are seeing a global convergence of measurement standards, with IPD (formerly Investment Property Databank) expanding to cover 21 countries so far'. The same writer also points out that 'the US capital markets are shifting from what was known as the AIMR (Association for Investment Management Research, formerly the Institute of Chartered Financial Analysts) standard to the GIPS (Global Investment Performance Standard) so that assets held in any country by investors from any country will be evaluated on a consistent basis. And we are seeing the adoption of global accounting standards, with RICS leading the way.'

To remain competitive and to seize opportunities that will arise, valuers need to be increasingly flexible in their approach. They must be prepared to acquire an ability in foreign languages to enable them to converse with their overseas clients and business contacts. They must seek to obtain knowledge and experience in the legal, financial and valuation characteristics specific to each foreign property market. As Gilbertson et al. (2006: 2) state, 'Globalisation, increased levels of cross-border economic activity and the emergence of global clients are driving international standards in accounting, banking and valuation. There is a bright future for those valuers who understand the dynamics in their market and anticipate or always respond to change.'

2.10 Likely future developments including automated valuation models (AVMs)

Information technology and the internet have already brought significant changes to the valuation profession, enabling research to be undertaken more speedily and in more depth and reports and calculations prepared more easily. Electronic distance measuring devices, PDAs (personal digital assistants) and laptop computers have sped up the process of measuring, referencing, valuing and reporting on property and recording inspections.

Another area of activity that has grown significantly in recent years is that of property research. Most major private practices and many large companies and financial institutions have specialised departments for carrying out this function, both to provide a general overview of the market and for specific job or client needs. Far more property-specific data and statistics are now available than was the case before the expansion of information technology. This is likely to continue with the increasing demand that all property markets are as 'transparent' as possible, meaning that full and detailed information on all aspects of each one is freely available, in line with other financial markets.

The potential application of computerised neural networks to property valuation was considered in the mid-1990s. Work by Lam (1996) and Connellan and James (1996) suggested use of this technology to eliminate subjectivity and improve accuracy within property valuation. Although not developed further in the UK at that time, the underlying principle has been implemented in Automated Valuation Models (AVMs) that have begun to be used for certain types of residential property valuations in the early years of the twenty-first century.

Johnstone et al. (2004) explain the role of AVMs as follows:

> Traditionally, desktop valuations are produced on a bespoke basis by individual valuers, rather than relying on an automated centralized process. More recently, automated valuation models (AVMs) have been developed to allow for a rapid assessment of a property value. Here, a user inputs the target address and property characteristics (for example,, number of bedrooms, age, etc.) into an automated model. The model will then return a value for the target property. AVMs have been an established feature of the U.S. market for about five years, and are used in approximately 10% of all mortgage originations in the U.S.

The models use regression analysis to 'fit' the subject property details to the comparables and provide a confidence rating of the final recommended valuation. Their use is still quite limited at present, being more likely to be considered suitable for a mortgage advance where the loan-to-value (LTV) is quite low and thus the perceived risk to the lender is correspondingly low. However, there is a potential for AVMs to become more sophisticated and widespread in their use in the future. Al-Khatib (2006) reported that research undertaken by the UK Council of Mortgage Lenders (CML) 'predicted that, within five years, AVMs will account for some 40 per cent of residential mortgage valuations with 50 per cent still based on an inspection, and the remainder being desktop and drive-by valuations'. Downie and Dobson (2008: 6) also stated that 'a quarter of all home sales and more than half of all remortgages in the UK will be based on AVMs by 2011'.

A concern is that human valuers will be replaced by computerised AVMs. However, drawing on the US experience from use of this technology over some twenty years, Downie and Dobson (2008: 8) conclude that 'although a subset of valuations will eventually be carried out electronically, human valuers will retain their current roles and will have opportunities to undertake new roles using AVM outputs. As well as providing valuations for the many properties not amenable to AVMs, their expertise is required to interpret, check and evaluate AVM outputs.'

The other major likely development for the coming years is the effects of climate change on property valuations. The wide-ranging and influential Stern Review (Great Britain 2006) has helped focus attention on sustainability, energy efficiency and risk of flooding and the implications of these on the value of a property are going to be a significant consideration in the future. This point was confirmed by the chairman of the RICS Valuation Faculty who stated that 'valuers need to understand and reflect how markets are responding (today and in the future) to legislation, changes in the built environment and public perceptions of sustainability' (Peto 2007b).

Progress check questions

- What does a property valuer do?
- Why do property valuers need to possess a diverse and extensive range of skills?
- Why is it important for a valuer to have a good understanding and working knowledge of the law?
- Who may need a property valuation?
- Why must a valuer establish clearly at the outset of an instruction what type of value is to be assessed and why the client needs this information?
- Why is a valuation not a building survey?
- Why can there be a difference of opinion over the value of a property between different valuers?
- Why do sale prices sometimes vary from valuations?
- What role does the RICS play in property valuation?
- What types of specialist areas of valuation work are available and how is this reflected in the organisation of the RICS professional groups?
- What particular challenges face property valuers in the next few years?

Chapter summary

Property valuation involves many different tasks and skills. From establishing and confirming what specific value is being sought by the client, through to inspection of the property, researching and analysing all relevant data, calculating and reporting value and liaising with other professional advisors, the range of abilities required is extensive. Most valuers specialise in a specific type of valuation work, a geographical area and/or a category of property. Apart from advising on value, valuers can also be involved in the management of property. Potential clients for valuations range from individual citizens to global organisations.

It is important to distinguish a valuation and a valuation inspection from a building survey and to understand the requirements and limitations of each. Whilst all valuations must be undertaken thoroughly, competently and with all reasonable care, there is often a small variation between values and final sale price or between figures produced by two or more valuers on the same property. This is due to the process requiring a certain amount of subjective assessment and interpretation together with the mathematical appraisal.

Valuers must remain independent and maintain the highest standards of integrity and ethics to make certain the figures they provide are impartial. The RICS acts as a regulatory body to ensure high standards are set and maintained and provides a wealth of material, both printed and on online, to assist in this. The profession has become international in scope and challenges faced in the future come from increased use of information technology, global clients and markets and the effects of climate change on valuation.

Further reading

Brett, M. (1998) *Property and Money,* 2nd edn, London: Estates Gazette.
—— (2002) *Valuation Standards for the Global Market,* London: RICS.
—— (2004) *Property under IFRS: A Guide to the Effects of the New International Financial Reporting Standards,* London: RICS.
Lynch, T. and Clark, K. (eds) (2006) *Real Estate Transparency Index,* Chicago: Jones Lang LaSalle.
Estates Gazette (1996) 'Mainly for students: economic role of the valuer', 9633 (17 Aug.): 82–3.
—— (1998) 'Mainly for students: surveying the field', 9840 (3 Oct.): 176–8.
Royal Institution of Chartered Surveyors (2009) *Consultation Draft Valuation Information Paper: Reflecting Sustainability in Commercial Property Valuations,* London: RICS. Online at: <http://www.rics.org/NR/rdonlyres/07B0710A-7BB4-4F3A-85C5-029571035E64/0/ConsultationDraftVIPReflectingSustainabilityinCommercialPropertyValuations.pdf> (accessed 4 January 2009).
Royal Institution of Chartered Surveyors and Investment Property Databank (2008) *Valuation And Sale Price Report 2008,* London: RICS. Online: <http://www.rics.org/NR/rdonlyres/E2B714B6-C1EB-4299-B3FB-112A9AB461D9/0/430_Valuation_And_Sale_Pricefinal.pdf> (accessed 30 December 2008).
Santo, P. and Hall, B. (2008) 'Question time', *RICS Residential Property Journal* (Jun./Jul.): 9.

CHAPTER

3

Investment

In this chapter ...

- Types of investment and investors.
- Nominal and real returns on an investment, how they are calculated and the importance of distinguishing between them.
- What is the 'ideal' investment and does it exist?
- How risk and reward determine yields.
- What is the reverse yield gap and when has it occurred?
- The benefits and drawbacks of real estate as a form of investment.
- Deciding which investment best suits an investor and how the performance of investments is measured.
- Methods of purchase or sale of land and buildings and their advantages and disadvantages.

3.1   Types of investment

One of the principal reasons for the purchase of landed property (land and buildings) is for investment. Many other types of investment exist, although the majority of funds are placed in three main markets:

1. property or real estate
2. gilts
3. equities.

Individual investors will decide which form of investment best suits their objectives and requirements at a particular moment in time. The forms of investment available include:

1. goods and chattels
2. National Savings Certificates and Bonds
3. bank or building society deposit or savings accounts and saving schemes
4. life assurance policies including property bonds
5. unit trusts and investment trusts
6. precious metals and stones
7. works of art and collections of rare objects
8. currency and commodities
9. stocks and shares:
    a) ordinary or preference shares
    b) debentures
    c) gilt-edged securities and loan stock
10. real estate.

1. Goods and chattels

Clearly, not all such purchases could be considered investments, especially where they are bought for use. By necessity, appropriate goods and chattels must be of a durable nature with a long-term life expectancy. Examples of items that could be so categorised are:

*   jewellery
*   vintage or 'limited-run' motor cars
*   yachts and boats
*   aircraft
*   antique furniture.

By their nature, most durable goods and chattels are 'wasting assets' in that their value decreases with age and the more they are used. They will slowly deteriorate and become less efficient with age and use. This process may be slowed by regular expenditure on maintenance and repair, but will not be halted entirely.

Although this is true of the above examples, there is an exception when the item has acquired a rarity value that outweighs its decrease in value through age and use. A good example of this would be a Supermarine Spitfire fighter aircraft, especially if still in flying condition. Immediately before and throughout the Second World War, these aircraft were mass-produced in thousands. Indeed, at the end of hostilities, most such aircraft were scrapped as they were replaced by modern jet fighters. However, today there are very few of these machines in good condition still existing in the world. Despite, or more likely because of, its considerable age and having been flown for thousands of hours, such an aircraft would today command an extremely high value, far in excess of its original price, due to its rarity.

Most other mass-produced items, nearly seventy years old and having seen considerable use, would have negligible or nil value today. The current rarity and historical importance of the item gives it a high value. The same would be true of a limited-run vintage motorcar, such as a Rolls Royce, which again could well appreciate in value with age, whereas most 'normal' cars would quickly depreciate in value with time.

2. National Savings Certificates and Bonds

These provide a very safe type of investment as the precise purchase price and interest rate are known and usually the redemption value is stated at the time of purchase. They can generally be redeemed at any time. Despite these advantages, they can offer modest returns and little capital appreciation.

Premium Bonds could be considered a form of investment in this category. They are issued by the UK central government but provide no interest payment. Instead, the Bond numbers are entered into a regular lottery, and prizes awarded to the winning numbers, which are selected by a computer popularly known as 'ERNIE'. Obviously, those Bond owners whose number is selected will receive a 'dividend' on their investment, which could be very substantial indeed. Conversely, Bonds could be owned for many years and never be selected to receive a prize. However, the Bonds can be redeemed at any time for their face value. It may therefore be better not to consider Premium Bonds as an investment at all, but rather in the nature of a lottery ticket, where the return of the initial stake money is guaranteed.

3. Bank or building society deposit or savings accounts and saving schemes

These also generally offer a very safe form of investment. Some current accounts can also pay interest, but usually at a lower rate than deposit or savings accounts.

The interest rate is variable, depending on the behaviour of interest rates generally in the economy, but the capital is always secure in nominal terms, providing the institution remains solvent. In the UK in September 2007 a loss of confidence in the stability of Northern Rock, due to difficulties in the world credit market emanating from heavy defaults in sub-prime US mortgage lending, led to huge demand from savers to withdraw their deposits amid fears that their money may be lost if the company become illiquid. These fears were allayed when the UK government promised to guarantee all such savings in the event of this or similar institutions falling into financial difficulties. Subsequent financial turmoil throughout world credit markets and banking in 2008 led to further assurances being issued by US, UK, European and other national governments to underwrite savings placed within major banks and financial institutions in an attempt to maintain stability of the system and prevent panic withdrawals by savers.

When the investor withdraws interest payments from the bank or building society account to provide an annual interest return, there will be no capital appreciation. It will remain at the same nominal sum as that initially deposited. However, liquidity of capital is generally good in that money can be withdrawn from many types of account at nil or short notice, although larger sums and accounts paying higher rates of interest may be require more or even considerable notice.

The interest rate payable at any one time is dependent on the nature of the account held. The principle is that the larger the sum of money invested, and the longer the notice required for its withdrawal, the higher the interest rate that will be paid. There are also accounts where an additional rate may be earned in return for investing a minimum specified sum on a monthly or other regular contractual basis.

4. Life assurance policies

These pay a specified sum of money to a beneficiary on the death of the assured life. Endowment assurance policies pay the sum on death or at the end of an agreed period, whichever occurs soonest.

'With profits' policies provide that the insurance company will increase the size of the guaranteed payment on maturity, depending on the profitability of the company each year over the term of the policy. The amount of this additional bonus payment will vary from year to year, according to the state of the company, and could even be zero. Conversely, they can show growth at or above the level of inflation, thus ensuring that the sum payable on maturity maintains its value in real terms.

A single or more usually annual or monthly premiums are paid for a policy. The level of premium is decided having regard to the age, sex, occupation and state of health of the insured person.

Endowment policies are most frequently taken out as part of a mortgage agreement. An endowment mortgage requires the mortgagor to pay interest only on the sum borrowed during the period of the loan. In addition, the endowment insurance policy should guarantee that if the mortgagor dies during this period, a sum sufficient to repay the loan would be available. Alternatively, this same sum 'with profits' will be paid on maturity of the policy. Theoretically, this will be sufficient to redeem the loan and leave the mortgagor with some additional cash. In practice, there have been cases where the policy has underperformed and provided an insufficient sum to repay the loan.

A property bond is a life assurance policy 'in the form of units in a property investment fund owned and managed by a life assurance company' (Parsons 2004: 202).

5. Unit trusts and investment trusts

Both of these are vehicles that offer an opportunity for small investors, with limited or no knowledge, to invest in the stock market at reduced personal risk. Both types of trust aggregate funds invested by many investors, and invest these on their behalf in equities. The advantage is that the trust can spread the risk by investing in many different companies, whereas the small investor on his or her own may only have sufficient capital to acquire shares in one or two companies. In addition, experts will decide when the trust should buy and sell and select which particular securities should be purchased, relieving the investor of this worry and responsibility.

A unit trust consists of a trustee (such as a bank) who will handle the finances and keep the securities and cash reserves; and a management company who will select the individual securities to be bought and sold. Investors purchase units in the trust at prices initially determined by the trust managers themselves, but which subsequently vary depending on the increase or decrease in value of the trust's investments. There are always two prices quoted, a buying and a selling, for the units. In the US unit trusts are known as mutual funds.

An investment trust is a company in its own right that exists for the purposes of investing its shareholders' capital on the equities market. Shares (not units) in these trusts trade on the Stock Exchange in the same way as other quoted companies, and investors can purchase their shares at the current market price as determined on the Exchange.

Both types of trust offer reasonably secure methods of investing in stocks and shares, as there is an expectation that any losses in one part of the portfolio will be offset by gains in another part. Providing the overall trend of the equities market is upward, the value of the investment in real terms is maintained or improved. However, by its very nature, such a form of investment cannot offer any guarantees and the value of an individual investor's holding can decrease as well as increase in value depending on the performance of the trust managers and the equities market.

6. Precious metals and stones

The main categories are gold, silver, platinum and diamonds, although many other precious stones could form worthwhile investments. In the case of the precious metals, ingots or bars are the appropriate form for investment, not jewellery, although solid coins could have a value above the pure intrinsic worth of the metal, depending on the rarity of the coin type, its condition and year of issue.

For retention of maximum value, precious stones ideally are retained in an uncut state, since poorly cut stones can suffer a serious depreciation of value. Again, usually the value is reduced substantially if the stones are mounted in jewellery, as this restricts the market for their sale, unless the piece has a particular historical or celebrity value.

The market for these types of investments fluctuates considerably, depending on world supply and demand. For instance, there have been both substantial growth and losses in the past in the gold market and therefore there can be a degree of risk attached to such a holding.

7. Works of art and collections of rare objects

Paintings, sculptures, china, miniatures, crystal glass, postage stamps, coins, medals, etc. can all command high values depending on their quality, pedigree and rarity. Some goods and chattels bought for use or precious metals and stones could also fall into this category.

Rarity, by itself, is usually enough to ensure such items retain a high value. However, the history of the item can enhance this value considerably. The first or last painting ever produced by a famous master artist, for example, could command a higher value than that artist's other pieces. The war or sports medal awarded to a nationally famous hero would sell at a higher price than the same type of medal awarded to a lesser-known person.

The market for many of these items can be extremely unpredictable, with bidding far exceeding reserves at auction. Millions of pounds are paid for paintings produced by the 'old masters', and yet they can often still be resold at a profit later. Conversely, artists that were popular fifty years ago may now no longer be so highly regarded, and their work can decrease in value.

This unpredictability of the market, and the high level of technical and expert knowledge needed to spot the fakes and make wise purchases, can deter many investors from this form of investment. There is also the added problem of security from the ownership of such items, although conversely prestige and enjoyment can be derived from the display of the work of art or collection.

8. Currency and commodities

More often traded for short-term speculative gains rather than longer term investment, they form a substantial sector of daily financial dealing on the foreign and stock exchanges.

9. Stocks and shares

Shares represent part-ownership of a company, whereas stock and debentures are forms of loan to companies and to central and local government.

a) Ordinary or preference shares

A share is an equal part of the capital of a company. Each share is equivalent to another similar share in the same company. The shares in public limited companies trade on the Stock Exchange, but shares in private companies are sold privately.

Preference shares are issued at a fixed rate of interest. As the name suggests, holders of these have preference over ordinary shareholders when it comes to the payment of dividends or distributing the company's assets. Their fixed rate of interest can be a disadvantage when inflation erodes their value, but in times of depression, their guaranteed income is an advantage over ordinary shares, which are unlikely to provide much if any annual dividend.

Sometimes preference shares can be 'cumulative'. This means that if in any one year the profits are insufficient to pay all the preference shares in full, any shortfall in the interest payment is carried forward to be made good in the next year, providing profits improve. Should there still be a shortfall, this will again be carried forward, and so on until the full interest payments have been made. These deferred returns take precedence over all other shares, but not over debentures.

Ordinary shares are often termed 'equities' because these shareholders participate in the equity of the company (*Estates Gazette* 1985: 311). Each share represents a part of the company and thus the shareholder participates in the company's profits to the extent of his or her proportionate ownership of the company.

Should there be a surplus of profits after the prior claims of debenture and preference shareholders have been met, a company may declare a dividend to be distributed amongst the ordinary shareholders. Because of this uncertainty of income, ordinary shares are a riskier investment than debentures and preference shares.

b) Debentures

Debentures are a form of loan to a company and are issued by companies when they wish to borrow money. The assets of the company guarantee the interest payments and capital repayments and therefore, if the company should go bankrupt or go into liquidation, debenture-holders will be repaid before ordinary shareholders.

c) Gilt-edged securities and loan stock

Stock certificates issued by central government are considered a very safe form of investment and are termed 'gilts' or 'gilt-edged securities'. This term originated because the certificates themselves had a gilt-edged border. However, the name nowadays is applied more loosely to indicate a very safe type of financial security investment. Those issued by the UK government are known as Exchequer stocks or Treasury stocks. Some UK and Commonwealth companies and some local and public authorities also issue stocks.

Gilts normally have a nominal or face value of £100. This figure is known as 'Par', and therefore when they are selling on the market at above or below this price, they are selling above or below par respectively. They earn a fixed rate of interest, which is stated when the stock is first issued. It is expressed as a percentage of the nominal value and known as the coupon rate, which is the return per £100 nominal value that will be paid to the holder each year. These interest payments are made half-yearly. Thus for each 5 per cent Treasury gilt held, the holder will receive two payments, each of £2.50 and six months apart, per annum. This nominal rate of interest can have little or no relationship to the 'real' rate of return, or yield, obtained by the owner of a specific stock, which expresses the interest payment as a percentage of the current market value of the stock, not its nominal value. Current yields on UK government gilts are shown in the financial sections of a number of UK daily newspapers, on the DMO website (UK Debt Management Office 2007) and on some other internet sites, such as provided by the stockbrokers Talos Securities Limited, a subsidiary of Boursorama and part of the Société Générale Group (Selftrade Investments 2008).

Gilts may be redeemable, that is, the issuer promises to repay the stockholder the face value of the stock on a specified date. This is also known as 'dated stock'. The redemption dates are usually five, ten, fifteen or more years from their issued date. They are referred to as short-dated or shorts if redeemable in less than five years, mediums if redeemable in five to fifteen years and long-dated if not for fifteen years or more. Alternatively, stock may be irredeemable, or undated. Here the issuer does not guarantee ever to repay the face value to the stockholder. They are known as 'consols', short for consolidated stock or consolidated annuities. Normally, if stockholders do not wish to hold on to the stock any longer, they would sell it back on the Stock Exchange at the then market price.

Yields on gilts are a very good guide to the state of the investment market and the economy generally. As the government guarantees them, gilts represent a very secure form of investment. As explained later in this chapter, other forms of investment must offer sufficiently high yields to offset their increased risks and uncertainty of security of income and capital compared to gilts. Alternatively, they must offer substantially better opportunities of growth in income and/or capital compared to gilts, otherwise there would be no point in investing in them.

10. Real estate

Alternative terms are *real property, property, landed property* or *land and buildings*. It is possible for there to be different interests owned in a single property; all owned by different people at the same time. Each of these interests are likely to have a different value at a given moment in time and so it is important for the valuer to understand and to clearly state what interest is being valued.

In English law, three main types of interest can be acquired in real property:

a)   freehold
b)   leasehold
c)   property unit.

a) Freehold

This is the nearest to absolute ownership of land and buildings that is possible in English law. Unlike other investments, true absolute ownership of landed property is not possible, as in theory the Sovereign retains these rights over all the land in the country. Further restrictions in its usage are imposed by statutory legislation, particularly the Town and Country Planning Acts.

A freehold 'tenancy' is acquired (legally termed a *fee simple absolute in possession*) which gives the freeholder the right to retain possession of the property in perpetuity, in other words forever. Under Section 1 of the Law of Property Act 1925 it is a 'legal estate' and rights of ownership rest in the land itself (this is known legally as rights *in rem*). Ownership is thus retained irrespective of whether any buildings on the land are removed or destroyed. This ownership can be transferred at any time *inter vivos* (during the lifetime of the owner) by deed. However, the ownership will not terminate even on the death of the present owner and may be inherited by his or her successor, or passed in a will to a beneficiary.

Included within the title to the land will be any 'fixtures'. These are chattels, which have been brought onto or attached to the land to improve it or to form part of its scheme of design or layout. In deciding whether an item is a fixture or a 'fitting' (a chattel which is on the land to be enjoyed for its own sake and not intended to form part of the demise), two tests are applied. The first is whether the item is 'annexed' to the land (that is firmly fixed). This test, however, is not conclusive, as some items that rest on the land under their own weight, and are not fixed in any way, have been deemed to be fixtures. The second, and more important test is the purpose of annexation. When it can be shown it was intended to become an integral part of the property, it will be usually be deemed to be a fixture. Each case must be decided on its own particular facts. Fixtures for convenience or ornament cannot be removed where the item may be deemed to be a permanent improvement and not conveniently detachable and is not removable without material injury or damage to the property or the item itself.

As fixtures are considered part of the land, a sale or other disposition of the land (including a lease) will include the fixtures unless the contract of sale (or lease) provides expressly to the contrary. Also if the land is mortgaged any fixtures are part of the mortgagee's security and cannot be removed by the mortgagor without consent.

It is important to establish whether items are fixtures or fittings when undertaking a valuation, as only the fixtures should be reflected in the value of the property itself. The fittings may form a separate valuation and be sold for a 'premium' (cash sum) on the sale of the property, in addition to the main purchase price.

b) Leasehold

Unlike a freehold, leaseholds give their owners the right to exclusive possession of a property for a limited time. This period can be of any duration, providing it has definite and definable start and stop dates. It is thus possible to create a leasehold interest for anything from a weekly tenancy to a tenancy for a term of years, which could last for even as long as up to 999 years. Most UK business tenancies are currently for terms of five to fifteen years, although twenty and twenty-five-year leases have been popular at times.

The leaseholder, also known as the 'lessee' or 'tenant', has a right to occupy and enjoy the property over the lease term. For this privilege, the tenant will pay a rent to the landlord ('lessor') for the benefit of occupation.

The legal title for leasehold is *term of years absolute in possession*. It is the only other 'legal estate' that may exist in English law under Section 1 of the Law of Property Act 1925. As with freeholds, rights exist in the land itself and the lease will not terminate with the death of the lessee or on destruction of any buildings that may have existed on the land at the start of the lease. Ownership may again be transferred *inter vivos* or on death. Under common law, the landlord is generally entitled to the reversion (property free of tenants and their rights) at the end of the lease.

Rent is an essential part of a lease to prevent 'prescription'. This is the acquisition of title to land through a long period of occupation without any formal acknowledgement that another retains ownership of the land. Rent must have some value, but need not be adequate. In other words, rent could be payment of a 'peppercorn' (a very small token amount such as 1 penny per year) or could be payment in kind, rather than in currency. It does not need to bear any relationship to the open market value of the property to prevent prescription. Generally, however, a landowner will seek an open market figure, known as the market rent (MR), when letting property and a valuer will advise what this sum should be.

When considering fixtures present on a property subject to a lease, it is necessary to distinguish between landlord's fixtures and tenant's fixtures. Those considered as tenant's are chattels that are defined as fixtures that would normally form part of the realty, but which are nevertheless legally removable by the tenant on vacation of the premises. As such, the value of tenant's fixtures is usually disregarded and only those owned by the landlord taken into account when assessing the rent.

The following qualify as tenant's fixtures:

- Trade fixtures:
    - These are attached to the land/building for the purpose of the tenant's particular trade, providing it is a chattel in itself, is capable of being removed without being entirely demolished or losing its essential character or value.
    - Any damage caused by removal must be made good by the tenant.
    - Where removal would cause irreparable damage to the premises, it must remain.

- Ornamental and domestic fixtures:
  - These are chattels affixed to a residential property for ornament, domestic convenience or for the better enjoyment of the object itself.
- Agricultural fixtures:
  - These are installed by a tenant farmer for the purpose of his/her agricultural operations.
  - However, tenant must give notice of intended removal to the landlord and allow the landlord an opportunity to purchase the items.

Other fixtures not qualifying under any of these three headings will be landlord's fixtures (even if brought on to the property by the tenant) and may not be removed by the tenant on vacation. Providing the tenant remains in possession, he/she is entitled to remove the above fixtures on final vacation, regardless of whether this is at the end of a single lease term or a consecutive series of lease terms. There is no need to expressly reserve the right to removal on lease renewal – see *New Zealand Government Property Corporation* v. *H M & S Ltd* (1980) 257 EG 606. The cases of *The Ocean Accident & Guarantee Corporation* v. *Next* and *Commercial Union Assurance Company* v. *Next* [1996] 33 EG 91 plus *TSB Bank* v. *Botham* [1996] EGCS 149 also provide relevant precedent on the issues involved with fixtures and fittings.

The two usual ways in which a lease may be created are by: (i) statute; or (ii) express agreement.

i.   *Statute* – leases created by the operation of statutory legislation are termed 'statutory tenancies'. They arise where, in default of agreement between the parties, the statute permits the implication of a lease. An example is under the Landlord and Tenant Act 1954, Part II; where the tenant of a business tenancy has the right to a new tenancy on the expiration of the old, unless the landlord can successfully oppose the grant of the new tenancy under one of the seven grounds listed in Section 30 of the Act. Thus even though landlord and tenant may not be able to agree terms for the new tenancy, the tenant has the right to apply to court, and the court may grant the new lease on terms it specifies, under the authority given by the Act.

ii. *Express agreement* – the most usual way in which a lease is granted, where the parties contractually agree terms for the tenancy, and these are recorded in writing. The Law of Property (Miscellaneous Provisions) Act 1989, Section 2, requires that any agreement for 'the sale or other disposition of land' (which includes granting a lease) of more than three years has to be evidenced (not be) in writing. The minimum four essentials to be recorded to enable such an agreement to be enforceable are:

  - parties – the names of the parties concerned;
  - property – a specific description of the subject property;
  - term – the length of the lease, including its start date;
  - rent – the sum to be paid initially and when payable.

Other terms and details can and usually will be recorded. These *express covenants* are clearly agreed between the parties and stated in writing within the agreement. *Implied covenants* are not stated but can implicitly be considered part of the agreement through sources such as common law or statute where an express covenant does not cover a point at issue at all, or inadequately covers a point.

Express covenants can cover any matter that the parties wish, providing they mutually agree to its inclusion within the lease agreement. It is usual not to rely on implied covenants and thus express covenants will normally be included to cover the terms agreed by the parties, but in specific detail. Some of the main express covenants usually found in leases are as follows.

Rent reviews – to enable landlords to revise the rental they receive from their tenants, and ensure it keeps pace with inflationary increases, it has become usual practice since the 1950s to insert rent review clauses in new leases. These clauses specify at what intervals the rent may be reviewed, together with the legal machinery by which such reviews may be implemented, the basis of valuation and the method of resolving disputes.

The conventional rent review period in the UK nowadays for business tenancies is five-yearly, although three-yearly or even annual reviews are frequently used alternatives. Historically, when inflation was not deemed such an important factor, the preference was for longer review periods of seven or ten years. In the case of ground leases, this could even be extended to 25, 33 or even 50 years. These timings are very unlikely in the modern market.

Most review clauses will specify in detail on what basis the new rent is to be assessed. In some cases, this may require that the property is valued on a hypothetical basis, rather than as it actually stands. For instance, the value of tenant's non-contractual improvements is often specifically to be disregarded. In this case, the building is valued not as it stands, but how it would be if the tenant had not carried out the work. Due to the potential pitfalls, it is essential for the valuer to check carefully the wording of the lease and its rent review clauses whenever instructed to assess the rental value.

In most cases, the landlord and tenant will eventually reach a compromise agreement on the level of the new rental to be paid from the review date. However, to allow for those situations where such an agreement cannot be reached, the review clause normally contains a provision specifying how disputes should be settled. The usual method is by referral to an independent third-party surveyor, who may be jointly appointed by the parties, or appointed or nominated by an agreed independent person, such as the current President of the Royal Institution of Chartered Surveyors.

Assignment or subletting – an 'alienation' lease clause will state whether the lessee may sublet or assign his or her interest, which will usually require the landlord's approval, during the term of the lease. It is unusual for a lease to prohibit absolutely assignment or subletting. It will normally be permitted on the whole of the premises providing the landlord's prior written consent is obtained, such consent not to be unreasonably withheld. However, the right to assign or sublet part only, as opposed to the whole, is more frequently prohibited.

Subletting is occupation granted for a period less than the unexpired period of the tenant's own lease. The party taking a sub-lease is called the sub-lessee or under-lessee and the original lessee becomes the sub-lessor or under-lessor. The holder of the original lease from the freeholder is termed the head lessee and is still responsible to the freeholder for the terms of the head lease, whereas the sub-lessee is responsible to his or her landlord (the head lessee) for the terms of the sub-lease.

Assignment is the transfer by the lessee, with the landlord's approval, of his or her leasehold interest to another party, usually in return for a payment, sometimes termed a 'premium'. Effectively the lease is 'sold' for a capital sum. The acquiring party is the 'assignee', who takes on the obligations and responsibilities under the terms of the lease. With large properties in particular, where there is no right to assign or sublet part (as opposed to the whole) of the premises, the rental value may be adversely affected. Large premises are more marketable if they can be split into smaller units and relet, than if a single tenant must be found to take the entire building.

Repairing and insuring – a lease on 'full repairing and insuring terms' (FRI) implies that the tenant is responsible for all repairs, decorations and insurance. These are the most common terms on which commercial leases are granted. 'Internal repairing terms' (IRT) makes the tenant responsible for internal repairs and decorations only; the landlord for external repairs and decorations, structural repairs and insurance. This is more likely for residential accommodation, but is also occasionally found in commercial leases. An 'internal repairing and insuring' (IRI) lease is similar except the tenant is then responsible for the insurance as well as the internal repairs and decorations.

User – a specific use to which the property may be devoted can be stated, including any prohibition or restrictions on the use. Often this wording is related to uses specified in the Town and Country Planning (Use Classes) Order 1987.

Statutes may modify or restrict the circumstances under which some of the above may operate. In particular, landlord and tenant legislation generally gives tenants security of tenure so that a lease will not normally automatically terminate merely due to effluxion of time. All of the covenants, whether implied or express, must be taken into consideration whenever the property is valued. The exact wording and meaning of each covenant can therefore have a material effect on the value.

c) Property units

The basic philosophy behind property unitisation is that the freehold (or possibly long leasehold) interest in a single property is acquired by a trust which then 'unitises' the interest. This allows the offering of units or shares for sale to investors, each unit representing part ownership of the property.

The principle is similar to the unit trust system of share ownership, whereby it offers opportunities for investors to participate in the commercial property market, without necessarily requiring the financial resources required to purchase outright an interest in a substantial commercial property. This in turn enables them to benefit from the rental and capital growth which such a property could enjoy over a period of time, and yet leave them free to dispose of their interest, in whole or part, at any time and without the need to place the property itself on the market. The investor's liquidity of capital is improved and this could lead to additional long-term funds being made available to the property market.

A real estate investment trust (REIT) 'enables investors to own and transfer shares of an interest in a property or properties' (Parsons 2004: 210). These shares are sold through the Stock Exchange in a similar manner to other investment trust shares.

3.2   Types of investor

The main investors in UK real estate are:

1. Life assurance companies and pension funds:
   - the traditional purchasers of commercial property;
   - both have control of long-term funds, which are acquired without recourse to borrowing;
   - as they are acting on behalf of their savers to provide a good long-term return on their investments, capital appreciation may be considered more attractive than income.
2. Banks: commercial, merchant and investment.
3. Property unit trusts.
4. Property companies and property developers: these buy for investment and/or with the objective of achieving profit through development or redevelopment.
5. Building construction companies:
   - generally buy property for development purposes, where they are able to design-and-build the complete scheme;
   - on completion, they may retain the development for investment purposes, or they may sell it on to the market;
   - the proceeds from these sales will help to finance future purchases and schemes.
6. Other companies: many companies own their own premises, which they view as both an investment and a means of carrying out their business.
7. Landed estates: largely inherited and owned by nobility and aristocracy and with a stately home or mansion at its centre.
8. Crown Estate.
9. Church Commissioners for England and other church estates.
10. Foreign investors: individual, corporate or governmental.
11. Real estate investment trusts (REITs) and property investment trusts.
12. Private individuals: buy for own use to occupy and live in the property or as investment that may be let to a tenant.

3.3   Nominal and real returns on investments

A nominal return is a numerical measure only. For instance, £10,000 invested two years ago has grown with interest to £12,000 today. The nominal return is £2,000 or 20 per cent (£2,000 × 100/£10,000).

The real return is a measure of actual worth or purchasing power. Thus if during the same two years of investment average prices in the economy had risen by 10 per cent; the real return or gain on the investment would have been only £1,000 or 10 per cent; calculated as:

- £10,000 × 1.1 = £11,000 now
- £12,000 – £11,000 = real gain of £1,000
- £1,000 × 100/£10,000 = 10 per cent

Example

The average UK house price in 1973 was £12,500. This sounds ridiculously cheap by today's standards. However, it was not cheap at the time relative to average earnings. The value of the pound then was much higher. But even when that has been taken into account, were houses comparatively cheaper in 1973 than say in 2002, when the average price first reached the £100,000 mark?

- 1973 average house price = £12,500
- 2002 average house price = £100,000
- 1973 average house price-to-earnings ratio = 3.67
- 2002 average house price-to-earnings ratio = 4.47
- Increase in retail prices index from 1973 to 2002 = 7.11

Thus if house prices had just kept pace with inflation, it would be expected that the average price in 2002 would have been 7.11 times higher than in 1973. However, £12,500 x 7.11 = £88,875, not £100,000. In real terms, prices had risen by:

$$\left(£100{,}000 - £88{,}875 \right) \times 100 / £88{,}875 = 12.52\%$$

In terms of affordability, the real cost compared to earnings had increased by:

$$\left(4.47 - 3.67 \right) \times 100 / 3.67 = 21.8\%$$

So it can be seen that over the 29 years between 1973 and 2002, houses in the UK *were* comparatively cheaper in 1973 both in nominal and real terms than they were in 2002, whether measured by average price or price-to-earnings ratio.

Nominal measurement of returns ignores the effects of inflation. It just compares numbers of pounds (or other currency), disregarding their value at the time. Real returns take inflation into account. With inflation, the value of money decreases yearly and prices need to be increased to keep pace. A real return is obtained when the returns over time exceed inflationary increases.

In the UK, the retail prices index (RPI) is often used as a measure of inflation, including by the government in determining pensions and benefits. Full data on it are available from the national statistics web site (http://www.statistics.gov.uk/rpi). This index records monthly changes in the cost, to the average UK consumer, of a specified large selection of goods and services. In its present form the first base of 100 was introduced in 1956. The current index's base of 100 is at January 1987.

3.4   The 'ideal' investment

An 'ideal' investment, if it existed, would possess the following characteristics:

- total security of capital in nominal and real terms;
- absolute assurance of income in nominal and real terms;
- complete liquidity of capital; and
- absolute regularity of income with no inconvenience to the investor.

Is there such an investment? Probably not, but the closer an investment comes to this 'ideal', the better it will be and the more demand there will be for it.

The major way in which the 'quality' of an investment is assessed is through the *yield* it produces. A good investment will attract high demand, which will result in a low yield, and conversely a poor investment will have a high yield.

What is meant by each of the desirable characteristics of the 'ideal' investment?

Total security of capital

This implies that there is no danger of the money that has been invested being lost by the investor. He or she can be certain that the money is safe. In nominal terms it implies that whatever sum of money was initially invested would always be there and never decrease with time. So, if £100,000 were the initial investment, the value of the investment if it was sold for cash would always be at least £100,000.

Real terms, as explained above, are a far more useful and accurate guide to performance. When the value of an investment does not maintain its initial purchasing power there seems little point in investing money into it.

Suppose £100,000 had been the initial investment one year ago, and since then the average increase in prices of other goods and services has been 10 per cent. When sold, if the investment will still only realise a cash sum of £100,000, the investor is 10 per cent worse off now than he or she was a year previously. In which case, why invest? The investors could purchase more with their money twelve months ago than now. In such a case, the investor would clearly prefer an investment that would maintain its value in line with increases in prices of other goods and services. The investor would hope that the value of the investment at the end of the year in question would be at least £110,000, which would be a 10 per cent increase in line with other values. In those circumstances, the investment will have maintained its purchasing power and the investor would be no worse off.

Even more preferable would be if the value of the investment increased by more than the rise in prices of other goods and services, so that the investor could have a greater purchasing power at the end of the investment period than at the start. Clearly this would be an excellent incentive to invest.

Using the retail prices index as a measure of the rate of inflation in the UK economy, any investment that maintains or improves its value compared with this index would be termed a 'good hedge against inflation' (Millington 2000: 15). This is because it would largely protect the investor from the effects of inflation, whereas if the same sum of money were held in cash over the same given period, its purchasing power would decrease in direct proportion to the inflation rate.

Absolute assurance of income

This implies that the income received from the investment will never decrease, and will always be at least what was expected at the time of the initial investment.

For instance, taking the above example of £100,000 invested, if the income receivable was to be based on 5 per cent per annum, then the investor can expect to receive at least £5,000 each year as an income for as long as the £100,000 remains invested. If the annual income were always £5,000 then this would be certainty in nominal terms, as the numerical monetary figure was constant. However, in the ideal investment, the investor would expect the annual income to rise at least in line with inflation, and thus maintain its annual purchasing power. So for example, if the first year's income was £5,000 and over the next year inflation was 10 per cent, then the investor would expect the income in the second year to be at least £5,500 (£5,000 + 10 per cent). This would be an 'assured income in real terms'.

Liquidity of capital

This is a measure of how easily and quickly the investment can be converted back into cash through sale. To be completely liquid, the investor should be able to sell the investment at extremely short notice and have 'cash in hand', paid immediately on sale by the purchaser. 'Cash' in these terms could, and probably would, not be 'hard currency' in terms of bank notes, but a credit paid into a bank account. In realising the cash sum, the investor should be able to sell at little or no cost for complete liquidity. The investment should also be capable of division into smaller lots so that the investor can recover part of his or her investment into cash, again at little or no cost.

Absolute regularity of income with no inconvenience to the investor

This means that the monetary income earned by the investment will be paid in full as and when due, without any demand needing to be made by the investor, and therefore no management charges would be incurred.

As stated at the outset, the above criteria represent the 'ideal'. In practice, it is extremely unlikely that any single investment will possess all these characteristics. Nevertheless, they do represent a useful checklist against which any investment can be judged and analysed. The closer to this ideal that an investment compares, the greater the demand will be for it. This will be reflected in the yield it returns, which will be considered next.

3.5   Risk and yields

The choice of investment will largely depend on how much risk investors are prepared to take and how quickly they wish to reconvert their savings back into cash. The ultimate choice will rest on which of the four 'ideal' factors they are prepared to compromise over most since it is normally impossible to achieve completely all four simultaneously.

The essential nature of any investment is the foregoing of a capital sum now in return for a regular income and/or growth in capital value over a future period. This income can arise from payment of a rate of interest each year or, with property, from receipt of rent.

The annual income return on the investment expressed as a percentage of its market capital value is termed the yield. The percentage that an investor would require from an investment will reflect the characteristics of that particular investment. The basic premise is the greater the risk and trouble involved, the higher the yield required; and the safer the investment, the lower the yield that would be acceptable. The 'ideal' investment would thus represent the lowest yield, as there is no risk, trouble or uncertainty connected with the ownership of such an investment.

Other investments close to the 'ideal' will offer very low yields as there will be high demand for them and low risk attached. Conversely, poor investments that are subject to great risk and uncertainty must offer high yields; otherwise, no investor would purchase them. Such investors recognise that they stand as great a risk of losing money as gaining it, but providing the yield is high enough they would be prepared to take the chance, knowing if the investment 'comes good' they will stand to make an above-average return on their investment.

The yield thus provides a common basis of comparison between different types of investment. Investors can decide which offers the best return, taking into account the risk and uncertainty associated with each. A higher yield, by itself, does not necessarily mean that an investment represents a good proposition. The question that must be asked is does that yield represent 'good value'? Even a very high yield may not represent good value where the investment involves a high degree of risk and uncertainty. It is a balance between risk and reward.

Calculation of investment yield:

$$\frac{\text{Annual financial return from the investment} \times 100}{\text{Capital value of the investment}} = \text{yield per cent}$$

Thus for shares:

$$\frac{\text{Annual dividend per share} \times 100}{\text{Current market value of the share}} = \text{annual yield per cent}$$

In the UK, the yield quoted for shares in the financial section of the newspapers is the latest twelve months' declared net dividend as a percentage of the price. Net in this context is the figure left after deduction of standard rate of Income Tax from the before-tax, or gross dividend. Conversely, for Irish and overseas shares the gross dividend is used to calculate the yield.

For property let at its full rent:

$$\frac{\text{Annual market rent} \times 100}{\text{Market value of property}} = \text{annual yield per cent}$$

With property, actual receipt of a rental income is not necessary to be able to calculate the yield on the investment. An investor that purchases the freehold interest in a property for own occupation will neither pay nor receive rent. In return for the outlay of the capital sum that is the purchase price of the freehold interest, they will occupy rent-free. The benefit to yearly cash flow is the sum saved that would otherwise be the annual rent. The 'annual financial return from the investment' would thus be this saving. Once this has been estimated, the yield obtained by that investor from occupying the property can be calculated.

Alternatively, purchasers of a freehold property that offers 'vacant possession' (VP) could let it to a tenant rather than occupy themselves. The potential annual income would be the current market rent (MR). The capital value, which is the price that could be paid to purchase the freehold interest in the property, would thus depend on the yield required by the investor.

> It is easy to see that every purchase of real property is an investment. A capital sum is given up, in expectation that an income will be received over and above the return of the capital sum. The income is received either in cash or as a saving in rent which would have to be paid. In fact the purchasers of real property are normally divided into the two categories, those who purchase as a pure investment, and those who purchase for occupation and use.
>
> (Lean and Goodall 1966: 10)

3.6   The reverse yield gap

Traditionally the average yields on equities were higher than the yields on gilt-edged or fixed-interest securities. This reflected the additional risk and uncertainty associated with the ownership of equities. The difference between these two sets of yields is known as 'the yield gap'.

The 'reverse yield gap' is where the yields on gilts and fixed-interest securities are higher than those obtained from equities. This has sometimes occurred in recent years and would seem to be at variance with basic theory on yield levels reflecting the degree of risk and uncertainty associated with particular investments. However, considering the yield in isolation can be misleading.

What should be considered is whether the yield offered is a fair return bearing in mind how close to the four characteristics of an 'ideal investment' the asset under consideration comes. The major factor, which is not immediately apparent from an analysis of yields, is security of capital in nominal and real terms. Gilt-edged or fixed-interest securities will seldom offer any appreciable capital growth, particularly in real terms, and although their income is safe, a higher yield is sometimes demanded by the market to offset this lack of growth. Conversely, since the 1960s, equities have generally enjoyed an overall steady growth in capital values. Thus although the income from equities is not guaranteed, and ownership of them involves more risk and uncertainty, the market may at times be prepared to accept lower yields in return for growth. The reverse yield gap is therefore logical, although it may appear not to be at first.

Other variations on the definition of the reverse yield gap are:

- the deficit in the overall investment return from property equities compared to that from gilt-edged and other fixed-interest securities or mortgages; or
- the difference between the borrowing rate and the total return on the investment – for property this may be expressed as the equated yield.

3.7   Real property as an investment

Real property has a number of distinguishing features as a form of investment (Millington 2000: 32, 55–61; Brett 1998: 5–10; Enever and Isaac 2002: 18–22). Real property interests are heterogeneous, in other words each one is unique or different. Although two properties can appear similar, there are usually differences in their size, layout, standard of repair, amenities, etc. which, even if small, will result in their values being different. Even where two buildings are identical in every way, they will not occupy exactly the same plot of land or position within a development and this too can influence value.

Property is relatively durable and so property investment is long-term in nature. Although buildings do depreciate with age, they do have long life spans compared to most other types of capital assets. In addition, the value of the land on which the building stands will often increase in value with time, which largely offsets any decrease in the value of the building.

Costs of buying and selling are relatively high. Estate agents, surveyors and solicitors fees will be incurred, all of which are subject to Value Added Tax (VAT). Stamp Duty Land Tax (SDLT) is an additional burden in the UK.

Should an explicit allowance be made for these costs when carrying out property valuations? Some valuers will do this, adding a percentage at the end of their valuation to indicate the likely expenditures. Others will not, but will mention in their reports that, whilst no allowance is shown, the client should make provision for such costs to be paid in addition to the purchase price or to be deducted from the sale price.

Time involved in buying and selling can be lengthy. The property must be marketed and a suitable purchaser found. Then the terms of the transaction need to be negotiated and agreed and the finance arranged. Finally, the legal documentation must be written, agreed and signed before a deal is completed.

Proof of ownership is sometimes difficult. The exact boundary of the land to be sold or let can be hard to define due to the plans on the deeds being inadequate. Resolving such boundary and title disputes can be costly and time-consuming.

Property is not easily subdivided and thus large amounts of capital are required to purchase. Whereas an investor can usually invest relatively small amounts of cash in other forms of investment, this facility is not available with property investment, where tens or hundreds of thousands, or even millions of pounds, may be required.

There may be substantial management problems. Property needs to be kept insured, repaired, maintained and decorated. Rents must be collected and business rates, Council Tax and other taxes paid, where applicable. Rent reviews need to be implemented and agreed. The investor will therefore generally need to appoint a management surveyor, or agent, to undertake these tasks on his or her behalf. The fees so incurred will of course eat into the financial returns from the investment.

Property is subject to significant amounts of government legislation which can quite frequently be amended, repealed or replaced. The laws affecting landed property are many, complex and varied, and there is seldom a year when modifications or changes do not take place.

In the UK, there is no single national market and property markets can be imperfect. Unlike stocks and shares, where there is a single central market (the Stock Exchange) through which deals are conducted, and which acts as a barometer of market conditions, there is no such system for landed property. Buyers and sellers do not always seek professional advice and may thus lack 'full knowledge of the market'. This, of course, is where professional valuers fulfil a role, obtaining and analysing this information and presenting the results to their clients. Even then, some property transactions are treated confidentially and non-involved parties can find it impossible to obtain information on them.

The supply of property in the short term for any type of use is relatively fixed, or inelastic. An increase in demand will therefore produce larger proportionate changes in price than if the supply was more elastic.

Property is considered a good 'hedge against inflation' (Millington 2000: 32) in the long term as capital and rental incomes historically have a good record of maintaining their value in real terms. Rent reviews help in this respect to maintain the real value of an owner's income flow from a property let to a tenant, in that the income can be increased at regular intervals to keep it in line with current market values. Few, if any, other investments offer this opportunity for regular and sustained income growth.

There can be considerable prestige attached to property ownership, which can lead to 'special purchasers' paying above the normal market rate to acquire specific properties. Such prestige can attach to a particular building due to its address, position, architecture or history. A good example of this is Mayfair in central London, which has traditionally been the centre for headquarters office buildings.

Equated yields on property, which measure the total percentage return making explicit adjustments for the effects of inflationary increases, are generally higher than on many other forms of investment. The differences between the various yields obtained from property are more fully explored in Chapter 11.

3.8   Investment evaluation and selection

The choice of the correct investment for a particular investor at a particular moment in time is a complex business, to which there are few definite answers. Clearly if it was easy to spot the most lucrative investments every time, anybody could make a fortune from investment. However, even experts are not always going to be proved correct in their selection. Invariably they will try to 'spread' their capital outlay across several different types of investments to defray and reduce the risk. If one sector or type of investments performs badly it is hoped any loss will be offset by the above-expected performance of another within the portfolio. It is the well-established principle that it is seldom wise to place all resources into one venture. Some diversification is preferable.

One of the major considerations in deciding on the best investment for particular investors is ascertaining their attitude to risk. The majority of investors are relatively 'risk averse', meaning that they would prefer to minimise the chances of losing some or all of their capital even if this means foregoing extra capital appreciation or income. However, comparatively safe, secure and low-risk investments will provide relatively low returns. More risky investment vehicles must offer some incentive in return for the investor's preparedness to take the additional risk. Thus, these will provide higher returns when they perform well, but may suffer loss when they do not.

In many respects, such investment decisions can be likened to gambling odds offered by 'turf accountants' and 'bookmakers' on the outcome of a sporting or other contest. These in turn are found from an application of probability.

The probability of an event occurring can range from 0 (the outcome is impossible and will never occur) to 1 (the event is completely guaranteed to occur). An alternative way of expressing this is 0 to 100 per cent likely. Thus, if an event is expected to happen 50 per cent of the time, it can be assessed as having a probability of 0.5 (50/100). Such an event would be the likelihood of an evenly balanced coin landing on its 'head' or 'tails' side when flipped and spun in the air. This activity takes place between the captains of a cricket or soccer team before the start of the match with the umpire or referee respectively 'tossing' the coin. The outcome of the spin decides in cricket who may elect to bat or field first and in soccer the selection of whether to take the kick-off or 'choose ends' of the field to defend in the first half.

With spinning a coin, there are only two possible outcomes. Two probabilities, each of 0.5 totals 1. It is 100 per cent certain that either one or the other will happen on each spin. When there are more possible outcomes to an event, the sum total of the probabilities should similarly always equal one. For example, in a soccer match, there are three possible results where team A plays team B. Either team A will win, or team B will win or the game will be a draw. Should all three possible outcomes be considered equally likely to occur then each can be considered to have a probability of 0.3333. This can alternatively be expressed as each having a 33.33 per cent likelihood, which can be quoted as fractional odds of 2 to 1. This is obtained from there being two outcomes that will lose against one that will win.

When a 'book' is compiled, the bookmakers will take into consideration their assessment of the likely probability of each competitor winning. They will frame their odds around this assessment. However, the odds they offer to gamblers will include a 'margin' to provide the bookmaker with a profit. The odds will not be 'fair' and the sum of their implied probabilities will come to more than 100 per cent. It is known as making the book 'over round' and assures a profit margin for the bookmaker.

Nonetheless, assuming 'fair' odds are offered, that reflect the true probability of each outcome with no profit margin incorporated, then each possible result in the equally balanced football match will be offered at fractional odds of 2 to 1 (or 3.0 in decimal odds which are derived from 0.333 = 1/3). Very few contests will be equally weighted in this way though. Certain outcomes will be considered more or less likely to occur than the equal probability position. Based on form, quality of players in the team, injuries, suspensions, advantage of playing at home or disadvantage of travelling to play away and many other factors, a view can be taken on which outcome is the most likely and which the least and the odds will be adjusted to reflect this.

Suppose team A is considered to have a 50 per cent chance of winning the match, team B a 20 per cent chance of winning and the draw is a 30 per cent chance (50 + 20 + 30 = 100 per cent probable). The fractional odds then become:

- Team A to win = 1 to 1 or evens (50 per cent chance of win and 50 per cent of loss = 50/50 =1).
- Team B to win = 4 to 1 (80 per cent chance of loss and 20 per cent chance of win = 80/20 = 4).
- Draw = 7 to 3 (70 per cent chance of loss and 30 per cent chance of win = 70/30).

The most likely winning outcome is offered at shorter or lower odds than the least likely one. This fairly reflects the expected risk versus reward assessment. There is comparatively lower risk attached to betting on the 'favourite', which offers the most likely chance of success. This in turn leads to higher demand and greater volumes of money being wagered on the favourite as it is the most probable winner. Conversely, demand to place bets on team B or the draw will be lower as they have less chance of winning. The lower the probability of winning, the higher the odds so as to entice backers to bet on them. Although such bets stand a lower chance of winning, should they do so, then a larger reward or payout is received. So, if team B do indeed win, every £1 bet on them will provide a return of £4 against only £1 for each £1 if team A had won.

What relevance does this have to investment and real property? The process of assessing the probability of risk versus reward and deciding whether the offered return is 'value' and fairly reflects these factors bears considerable similarity. Instead of a football match, consider the choice between a prime, a secondary and a tertiary property investment opportunity. The prime property is a modern building with full amenities, situated in the best and most prestigious area and let to a 'blue chip' company with vast financial resources, thus assuring that the rental income is very 'secure'. The secondary property is older and less well appointed; it is in a good but not the best area and has a tenant with a reasonably sound financial record. The tertiary property is old and poorly equipped; it is in a quite run-down district and has a small local company with limited resources as a tenant.

Should the all risks yield (ARY) offered on each of these freehold investments be the same, nobody would invest in the secondary or tertiary properties. Why take on additional risk and uncertainty without the potential reward?

However, suppose the all risks yields available were different. The prime property will provide a return of 5 per cent, the secondary an 8 per cent return and the tertiary a 12 per cent one. Now the investor has to assess whether the potentially much larger rewards of investing in the non-prime sector are worthwhile given the additional risks such investments pose. Some will conclude that the risks outweigh the much higher return and even at 12 per cent the tertiary property does not offer 'value'. Others will reason that a return 140 per cent higher than on the prime property (an extra 7 per cent on top of the 5) makes it worth accepting the far higher risks and uncertainties. It is not so different from choosing to back the favourite or the outsider in the football match! For more on the factors involved in the selection and evaluation of the different categories of real property for investment, see Chapter 10.

Yield is also one aspect that will influence the choice of investment when assessing equities (shares). There are a number of other factors as well.

When there is a new share issue, the shares have a nominal face or 'par' value, such as £1 or 25p, which represents the limit of the company's liability to the holders of the shares. This is not necessarily their sale price when first sold on to the market. Once a company is 'listed' and trading in its shares commences on the Stock Exchange, the market price will rise and fall according to supply and demand at the time. This current market price will usually be different from both the face value and the initial sale price. It is the market price at the close of trading (an average between the sell and buy prices at that time) which is reported in the financial columns of the following day's morning daily newspapers.

Market analysts are usually specialised and concentrate on a specific sector of the market, such as banking, property or retail. This enables them to acquire a highly developed information system on the industry in general and the individual companies within it in particular. They are thus best placed to interpret a company's accounts and results in that sector and advise whether its current share price represents good or poor value given the company's present and likely future performance.

The annual financial return is the dividend per share paid for the year, stated as a percentage or as an amount in pence. It represents the nominal return, calculated on the 'face value' per share, not its current market value. To find the true return on the investment, it is therefore necessary to convert this percentage into a yield, which represents the real return. Once this is done, it may show that the dividend on one company's shares is actually less than on another, even though the dividend percent is higher.

Investors would need to know this actual return on the money they have invested, rather than the theoretical dividend. Comparison of yields between different shares allows an assessment of whether the actual return is fair considering the level of risk and uncertainty associated with ownership of each particular company's shares. Consequently, market listings published in newspapers such as the *Financial Times* normally show the yield expressed as a percentage of the current middle market price of the share.

Cover, short for dividend cover, is the ability of a company to pay the dividend it has issued. That is how many times it could have paid out the sum quoted from their net profits after tax. For instance, if a company has made net profits equivalent to £10 per share and paid out £2 of this in dividends per share, its cover is 5 (£10/£2).

Very few companies that make profits will distribute all of them as dividends to their shareholders. It is usually more prudent to retain some of the profits and reinvest the money in the business to enable it to grow and remain efficient and competitive, which will hopefully enable it to continue to make larger profits in the future. Alternatively, the 'surplus money' may be used to repay existing debts. The cover figure will thus indicate what proportion of profits the company has retained.

Example

Company A has announced a dividend of 10 per cent. Its shares have a face value of 25 pence and are currently selling at £1. Company B has a dividend of 7 per cent. Its shares have a face value of 10 pence and a current market value of 20 pence.

The yield on company A is:

$$25p \times 10\% \times 100/100p = 2.5p \times 100/100p = 2.5\%$$

The yield on company B is:

$$10p \times 7\% \times 100/20p = 0.7p \times 100/20p = 3.5\%$$

Thus the yield on company B's shares is actually 1 per cent higher than on company A, even though its percentage dividend is lower.

A cover less than one indicates that the company has used some of its reserves to make the dividend payment to its shareholders. Unless this is a one-off occurrence, it could lead to serious problems for the company as it will eventually use up all its reserves or must cut back or even cease its future dividend payments. Conversely, if the cover is five or more, it shows that the dividend paid was well-secured and future growth of the firm and hopefully its dividends and/or share price may be anticipated.

The price/earnings ratio (P/E ratio) is a comparison between the market price of a company's shares and the amount of money the company is earning. It is a multiple of net earnings used to establish the sale price of the shares. The lower the yield, the higher the P/E ratio will be. It is similar to the years purchase figure used to capitalise income from property to calculate the capital value of that property. As Law (2006: 412) confirms, 'the P/E ratio is one of the main indicators used by fundamental analysts to decide whether the shares in a company are expensive or cheap, relative to the market'.

Different industries can operate on different P/E ratios and yet be equally sound. When comparing P/E numbers it is therefore necessary to make a comparison between similar firms operating in the same industry, and not between those in dissimilar industries.

For companies with a low earnings yield and a high P/E, the market expects future growth and is therefore willing to accept a low yield. However, to obtain a balanced view it would be worthwhile estimating the company's expected growth on a per annum basis and adding this to its yield to find its true total return or equated yield. A similar appraisal undertaken with property will use equated yield explicitly to account for anticipated income and capital growth as well as the existing yield (see Chapter 12).

3.9   Methods of purchase and sale

There are various methods by which an interest in landed property can be acquired or sold:

- private treaty
- auction
- tender
- sale and leaseback
- takeover or merger
- compulsory acquisition.

Private treaty

This is a contract agreed privately between vendor and purchaser or lessor and lessee. The price to be paid to acquire the interest can be agreed through two alternative methods:

a) Stated price

The owners decide what they feel the property is worth or what they are prepared to accept (which may not be the same thing) and (hopefully) act on advice given by the valuer. The decided figure is the 'asking price' and is quoted on the estate agent's details and advertisements. The implication in English law is that the first offer received at the full asking price, and deemed in all others respects satisfactory, will be accepted, subject to contract. No acceptance is binding until a formal legal contract is completed. Should no such offer be made, negotiation may take place, which may result in the eventual selling price being different to the asking price.

b) 'Offers invited'

These can be of three types:

1. Left completely open (that is just 'offers invited for the property'). This implies that any reasonable offer will be considered.
2. Invited in conjunction with a stated price. For example, '£800,000 or near offer' (ONO). This implies the offer can be a figure over, or a little (but not substantially) under the stated price.
3. Invited over a specified minimum. For instance, 'Offers invited over £800,000'. This sets the bottom figure, but there is no limit to how high the offer can be, which is at the offerer's discretion.

Again, any offer that is accepted is usually made subject to contract and not binding until completion and exchange of the formal agreement.

Auction

An auction sale offers the opportunity to achieve the best price in open competition between potential purchasers. As such, it probably comes closest to providing 'open market value' conditions.

A property can be auctioned after interested parties have had an opportunity of receiving the details and inspecting the property. A specified date, time and venue for the auction are given in the particulars and interested parties may then attend to bid for the property.

Properties at auction may either have a reserve price on them (that is a minimum price at which their owners are prepared to sell) or without reserve (that is the highest bid will be accepted). The auctioneer will not disclose in the auction room the method that is being used until either the bidding passes the reserve, when he or she will likely indicate 'it is in the market to be sold today', or he or she withdraws it from sale because the bidding never reached the reserve.

Providing any reserve has been exceeded, the property will be 'knocked down' by the auctioneer (the banging down of the auctioneer's gavel signifies acceptance of the bid) and sold to the highest bidder. The purchaser will be required to complete a binding memorandum of contract and pay a deposit (usually 10 per cent of the purchase price) before leaving the auction. This is a legally binding agreement and auction sales are not 'subject to contract'.

This method of sale is therefore potentially quicker and more certain than a sale by private treaty. However, if bidding fails to reach the reserve price, fees will still be payable to the auctioneer for abortive work.

Tender

This is a formal method of disposal, and is legally 'an invitation to bid'. It requires that full details on the property, including draft leases or deeds, are incorporated in a tender document, identical copies of which are made available to each tenderer, who is invited to submit his or her 'highest price' for the property.

Tenders can be on an 'open' or 'closed' basis. An 'open tender' is where advertisements are published inviting anybody who is interested in acquiring the property to apply for the tender documents and to subsequently submit a tender. A 'closed tender' is where only certain persons or organisations are invited to submit tenders.

Generally, acceptance of a tender forms a binding contract immediately, and therefore the persons inviting tenders usually reserve the right not to accept the highest or indeed any tender submitted. This is to cover the possibility that the highest price submitted may not be high enough, or that the vendor would prefer not to sell to the highest tenderer for personal or business reasons. If no such express statement is made in the documentation, then it is implied that the highest bid must be accepted providing it is submitted on time.

To avoid the possibility of collusion or malpractice, a fixed date, time and place is stated in the documents regarding when the tenders will be opened. All tenders submitted should therefore be in sealed envelopes, which are clearly marked to indicate a tender is enclosed. Any tenders opened or tampered with before the stated date and time are liable to be declared void. As a further safeguard against collusion, many tenderers will hand-deliver their tender very shortly before the due time.

Sale and leaseback

This is where a freeholder or leaseholder sells their present interest and in return takes a lease back on part or whole of the property at an open market rental or a lower rental linked to the sale price. The new owner acquires a property with a tenant and a guaranteed rental income, without incurring letting fees or risking a rental void. The old owner releases capital for alternative investment purposes and yet retains occupation of the property at least until the expiry of the lease.

Takeover or merger

All company assets include property interests, since firms need to occupy property in order to carry out their business, even if they do not trade or invest in property for its own sake, as do property or development companies. When a company is bought out, or 'taken over' by another, the acquiring company acquires all the assets of the acquired company, which will include its property interests. The price paid to acquire the company will depend on the value of its shares, which will be a reflection of the value of its assets, including property interests, in addition to its trading and profit levels and the market's view of its prospects and performance. In this respect, the 'book value' of its property interests will be of great importance.

In some cases, one of the reasons for a takeover is to acquire a company's property, which the purchaser may feel is being under-utilised or incorrectly managed and considers that its capital value could be increased by different management strategies. Another alternative is where a greater capital sum can be realised from the piecemeal sale of the property rather than from its retention in its entirety in the portfolio. However, this can lead to accusations of 'asset stripping', which implies that the only reason for acquiring the company was not as a going concern, but merely to sell off its assets and obtain a quick profit on the purchase price.

On merger, the assets of two or more firms are joined together in one new enterprise. The combining of the former companies' property assets may enable latent value to be realised. For instance, a development project becomes possible when adjoining land is now in single ownership where previously it was held separately by competing organisations.

Compulsory acquisition

In the UK land and buildings can be compulsorily acquired by a public body, such as a local council, development corporation or government department, that possesses the requisite statutory powers provided by an enabling Act of Parliament. A compulsory purchase order (CPO) is served on the existing owners of the property interests, who are obliged to sell to the authority even if they are unwilling. The price paid is normally based on open market value.

Progress check questions

- What are the three major forms of investment?
- What are gilts and why are they considered such a safe form of investment?
- What's the difference between a freehold and a leasehold interest in landed property?
- What is a FRI lease?
- How do rent reviews help to maintain the real value of rental income from business properties let to tenants?
- What is the difference between measuring nominal and real returns from an investment and why is it important?
- What would be the ideal characteristics of an investment?
- How does yield reflect the relative risks and rewards associated with an investment?
- Why do many investors consider real property a good long-term investment?
- In what ways is real property not such a good investment?
- How will investors evaluate and decide which form of investment best suits their requirements?
- What methods are used to acquire or sell land and buildings?

Chapter summary

There are many alternative forms of investment available, although the three categories into which most funds are placed are equities, gilts and real estate. Each type of investment has advantages and disadvantages that investors must carefully consider against their requirements and circumstances when selecting the one or ones into which they will invest their money. Real estate has a number of uncommon and even unique characteristics that make it both an attractive form of investment, especially in the long term, but also can deter investors, particularly those with limited funds.

In English law, the two principal interests in real property that may be acquired are freehold or leasehold. A freehold interest will last forever, or endure 'in perpetuity', whereas a leasehold has a definite duration. Holders of leases are tenants and pay rent to their landlords for the right to occupy the property. This provides the landlord with an annual income from the property, which through rent reviews can often be increased at regular intervals in line with rises in market levels. The investors in real estate range from individual persons to multinational companies and public sector organisations.

When assessing the change in value of an investment over time, it is important to distinguish mere nominal increases from real ones. Just because an investment has increased in value when measured in units of currency, such as pounds sterling or euros, does not necessarily mean the investor is better off than when the initial investment was made. Comparing the change in value of the investment with the changes in the prices of other goods and services is more informative as it shows whether the purchasing power of the invested money has increased and thus its real value has appreciated.

There are four characteristics that an 'ideal investment' would possess and these provide a measure against which any specific form of investment can be assessed. In turn, the yield or annual return from an investment will reflect how close to the ideal it comes, with better investments returning lower yields to poorer ones. Whether a particular yield level represents good value depends on the balance of perceived risks against rewards associated with its ownership.

Further reading

Askham, P. (1990) 'Mainly for students: conventional and contemporary methods of investment valuation', *Estates Gazette*, 9011 (17 Mar.): 84 -6; repr. in P. Askham and L. Blake (eds), *The Best of Mainly for Students* (London: Estates Gazette, 1993), pp. 346–53.

—— (1993) 'Mainly for students: government stock', *Estates Gazette*, 9315 (17 Apr.): 139–40; repr. in P. Askham and L. Blake (eds), *The Best of Mainly for Students* (London: Estates Gazette, 1999), vol. 2, pp. 103–8.

—— (1993) 'Mainly for students: yields and rates of interest', *Estates Gazette* (4 Sep.): 117–18; repr. in P. Askham and L. Blake (eds), *The Best of Mainly for Students* (London: Estates Gazette, 1999), vol. 2, pp. 109–16.

Baum, A. (2002) *Commercial Real Estate Investment*, London: Estates Gazette.

Brett, M. (1998) *Property and Money*, 2nd edn, London: Estates Gazette.

Brown, G.R. and Matysiak, G. (1999) *Real Estate Investment: A Capital Market Approach*, London: Financial Times/Prentice Hall.

Dubben, N. and Sayce, S. (1991) *Property Portfolio Management: An Introduction*, London: Routledge.

Gill, S. (2008) 'Take the measure of the figures', *Estates Gazette*, 0821 (31 May): 133.

Hoesli, M. and MacGregor, B. (2000) *Property Investment: Principles And Practice Of Portfolio Management*, London: Longman.

Investment Property Forum and Imber, A. (2001) 'Mainly for students: making more of money', *Estates Gazette*, 0114 (7 Apr.): 138–41; repr. in L. Blake and A. Imber (eds), *The Best of Mainly for Students* (London: Estates Gazette, 2004), vol. 3, pp. 187–95.

Isaac, D. (1997) *Property Investment*, Basingstoke: Palgrave Macmillan.

United Kingdom Debt Management Office (2007) *Gilt Market*. Online: <http://www.dmo.gov.uk/index.aspx?page=About/About_Gilts> (accessed 19 Nov. 2007).

Valuation mathematics

4

Compounding and discounting

4.1 Simple interest

The *Oxford Dictionary of Business and Management* (Law 2006: 280) defines interest rate as 'the amount charged for a loan, usually expressed as a percentage of the sum borrowed. Conversely, the amount paid by a bank, building society, etc. to a depositor on funds deposited, again expressed as a percentage of the sum deposited'.

Interest rates are an integral function of investment. In return for the forgoing of expenditure now, an investor can place cash into investments that attract interest payments. The interest thus earned will hopefully offset or even surpass the reduction, due to inflation, in real value or purchasing power of the invested cash over time. Interest rates are always quoted as a percentage, and for comparison between investments are normally referred to on an annual or 'per annum' basis.

Simple interest occurs when all interest calculations are solely based upon the initial sum invested, referred to as the 'principal'. The formula to calculate the total value of an investment, at the end of a specified time period, where simple interest has been earned is:

$$P \times (1 + i \times n)$$

where:

P = principal, or initial sum invested
i = interest rate per time period (usually years) expressed as a decimal number. Thus divide the interest rate by 100. So, 5 per cent = 5/100 = 0.05 and 11 per cent = 11/100 = 0.11
n = number of time periods over which interest accrues (usually years)

In strict mathematical notation, the multiplication signs ($\times$) do not need to be shown, thus reducing the formula to:

$$P(1 + in)$$

Example

If £20,000 has been invested for the last 3 years at 6 per cent per annum simple interest, the current amount invested is:

$$£20,000 \times (1 + 0.6 \times 3)$$
$$= £20,000 \times (1 + 0.18)$$
$$= £20,000 \times 1.18$$
$$= £23,600$$

To find the amount of interest that the investment has earned in total, deduct the principal sum from this end-amount: £23,600 − £20,000 = £3,600 interest payments

4.2   Compound interest

However, whenever interest rates are discussed in connection with the economy, property and investment, it must be remembered that these are usually calculated on a 'compound', not 'simple' basis. Compound interest is calculated at the end of each time period over which interest is based (usually yearly) on the accumulated sum of principal plus interest up to that date. In this way 'interest is earned on interest'.

The mathematical formula to calculate how much will be invested over time with the addition of compound interest is:

$$P \times (1 + i)^n$$

where:

P = Principal (original sum invested)

i = interest rate per time period (usually years) divided by 100, so that it is expressed as a decimal number

n = number of time periods (usually years)

Example

£20,000 has been invested at 6 per cent per annum compound interest for the last three years. Current amount invested is:

$$£20,000 \times (1 + 0.06)^3$$
$$= £20,000 \times 1.06^3$$
$$= £20,000 \times 1.191016$$
$$= £23,820.32$$

To break down this calculation into year-by-year figures:

Principal sum = £20,000
× 6% at end of year 1 = £20,000 × 0.06 = £1,200 interest payment for year 1
Total sum invested at end of year 1 is: £20,000 + £1,200 = £21,200
× 6% at end of year 2 = £21,200 × 0.06 = £1,272 interest payment for year 2
Total sum invested at end of year 2 is: £21,200 + £1,272 = £22,472
× 6% at end of year 3 = £22,472 × 0.06 = £1,348.32 interest payment for year 3
Total sum invested at end of year 3 is: £22,472 + £1,348.32 = £23,820.32

It will be seen that the 'compounding' of interest, effectively 'interest being added to interest', has resulted in a higher value of the investment compared to the same sum, invested for the same time period, at the same interest rate when calculated on a simple interest basis. The total compound interest earned is: £23,820.32 – £20,000 = £3,820.32. This compares to the simple interest sum of £3,600 shown in Section 4.1 above. The longer the time period of the investment and/or the higher the rate of interest, the greater the effect of the compounding will be.

As the calculation finds how much a sum of money invested for a specified time period will *amount* to at a stated interest rate, property valuers refer to the compound interest formula as the 'Amount of £1'. This is the term used in *Parry's Valuation and Conversion Tables* (Davidson 2002). The compound interest formula forms the basis of all other property valuation calculation formulae found within the tables: 'In valuation practice and other studies in land use where some aspect of financial analysis is involved, compound interest calculations of a tedious nature are frequently required. Valuation and financial tables may be used to reduce the time-consuming element involved in these calculations' (Davidson 2002: p. xi).

4.3   'Traditional' annually in arrear tables and formulae

Parry's Tables were first published in 1913 and have remained the standard set of property valuation calculation tables in the UK ever since. They provide a 'ready reckoner' source of data, giving the calculated figure for each respective mathematical formula without the need to undertake the calculations manually. In his preface to the ninth edition of the tables in 1969, Alick Davidson stated that 'most of these tables have now been computed to at least four decimal places and in all cases where five or seven decimal places existed this degree of accuracy has been retained'. In practice, most valuers would round off valuation formula figures to no more than five places of decimals. Final monetary values would be shown to only two places of decimals at most. Thus in the UK, values in sterling are quoted in pounds and pence. There is no need to show a final value that includes amounts less than one penny. For instance, a calculation producing a final figure of 521.36218 when expressed in sterling would be £521.36. Even then, it is seldom that such precise monetary sums will be needed. In many cases with property valuations the final figure will be rounded considerably further. Values will be given to the nearest hundred, thousand or even higher amount due to the relatively large sums involved. For instance, rather than state a value as £12,892 it would be commonplace to round to say £13,000 and when the calculation produces a figure of £2,289,245 to round to say £2.3 million.

The first edition of the tables within *Parry's*, and all the mathematical formulae on which they are based, made the assumption that all calculations are undertaken on an 'annually in arrear' basis. This assumes interest is calculated at the end of each year and all incomes are similarly received. The same approach is adopted with most computer spreadsheet financial formulae, such as those within Microsoft Excel.

The majority of the tables within the latest edition of *Parry's* retain this basis, even though this may not be the actual method of payment in the 'real world'. For instance, in UK commercial property leases it is usual for rentals to be paid quarterly in advance and in the equities market dividends may be payable half-yearly in arrear. The need to consider using alternative financial formulae that reflect the actual timing of payments, rather than the notional annually in arrear approach, has been discussed by various writers, notably Rose (1976) and Bowcock (1978) and more recently Creamer (1999). It has led to the inclusion of some tables within *Parry's* that are calculated on a quarterly in advance basis.

Although interest payments and other incomes may not be actually receivable annually in arrear, this approach is still retained in 'traditional' valuation tables and formulae. Why is this? There may be several possible reasons:

- The method has been used before and since the 1913 publication of the first edition of *Parry's* and is well-known and understood.
- The formulae are less mathematically complex than quarterly in advance ones.
- Providing all calculations are undertaken using formulae based on the same premise then the results should be consistent and comparable with each other.

In his introduction to the twelfth edition of *Parry's*, Davidson (2002: xi) explains:

Tables based on the assumption that income is received quarterly in advance have been extended and given more prominence to facilitate their use in response to the increased interest in this concept and recent discussions by investment surveyors concerning their use. However, as many valuers still use the more traditional approach, years purchase figures based on incomes received annually in arrear still form an important section in this volume.

The topic is discussed in more depth in Chapter 7.

4.4 Use of calculators and computers

Using a calculator enables any of the valuation formulae to be found, even when the appropriate number of years or percentage rate required is not included within a book of tables, such as *Parry's*. Although more relevant to when a computer is not available, such as by students in exams, the use of a calculator to find formulae is still a useful skill to develop.

Modern scientific calculators generally have far more functions on them than are required by the average valuer. Apart from the standard +, −, / and × arithmetic operators, the other functions which will be found useful are:

M	A memory for storage of a number while another calculation is undertaken which will in turn need to use the number in memory. Many calculators have more than one memory to enable storage of a series of numbers.
y^x (or x^y)	Enables a number 'y' to be raised to the power of another number 'x' (or 'x' to the power of 'y'). Used in all compound interest calculations.
^	An alternative button for raising a number by a power found on some calculators. Thus 1.2^4 would be entered as $1.2 \char`\^ 4$.
1/x or x^{-1}	Reciprocal which converts the number 'x' into 1 divided by 'x', a function used in many valuation formulae, such as Present Value of £1.
Exp	Exponential function – useful when working with very large figures as it enables a series of noughts to be entered after a number. For example 20,000,000 would be entered as 2 Exp 7.
log and INV log or 10^x	Occasionally useful in deducing an unusual root of a number.
%	Percentage calculator – although often it is just as easy to multiply by the percentage number expressed as a decimal. For example, 10% of 325 = 0.1 × 325 = 32.5.
ANS	Recalls answer to last completed calculation and is an alternative to entering the result into the memory.

Most other functions of a statistical, trigonometric or scientific nature are unlikely to be required often by a valuer, but will be present on calculators that provide the above functions.

It will be noted that the Amount of £1 formula requires the raising of a number by a power. On a calculator this is typically achieved by use of the y^x or x^y button. The sequence of key presses needed to calculate 1.03^5 being:

1.03, y^x, 5, $=$

and giving the result of 1.159274. When using a spreadsheet on a computer, the relevant symbol to raise a number to a power is $\wedge$ (hold down shift key while pressing number 6 in the alpha keyboard top row). Therefore, to input the same equation, the spreadsheet cell would show:

$=1.03 \wedge 5.$

Relevant financial formulae can also be entered into or be available within programmable calculators. With computer spreadsheets, an individual formula can be entered into a cell or the in-built functions used to undertake the calculations. This is considered further in Chapters 11, 12 and 13.

In addition to their printed version of *Parry's Tables*, the *Estates Gazette*, in conjunction with the College of Estate Management, provide subscribers to the online version of their publication with a downloadable computer desktop version of *Parry's* Valuation and Investment Calculator in the 'Extras' section at http://www.egi.co.uk. This provides the following functions that can be calculated at the press of a button:

- Amount of £1
- Amount of £1 per annum
- Annual sinking fund
- Present Value of £1
- Years purchase single rate
- Years purchase dual rate (no tax)
- Years purchase dual rate tax (adjusted)
- Mortgage instalment per £100 per month
- YP perp
- YP in perp years deferred
- YP in perp quarterly in advance
- YP single rate quarterly in advance
- YP dual rate quarterly in advance (no tax).

4.5   The concepts of compounding and discounting

Compounding is the adding of compound interest to an invested sum, so that the total amount of money invested increases over time. The Amount of £1, or compound interest formula, calculates this effect at a constant rate of interest. The Amount of £1 formula will always produce a figure of more than 1 since interest added to the principal increases the total sum.

Discounting is the inverse of the compounding process, whereby the present value of an investment is found allowing for compound interest that would be earned on it over future months or years. The further into the future that a sum of money would be receivable and the higher the interest rate, the greater the discount factor will be, and the lower the present value of the investment.

In valuation terms, discounting is allowing for the purchase now of an interest, which will produce income and/or capital receipts in the future. These future sums are deferred, or discounted, to find out their worth today. For example, would somebody pay £1 today for the right to receive £1 in three years' time? Probably not, but why not, and how much less than £1 would they pay?

If somebody paid £1 today, they would not have allowed for the fact that the £1 outlay will be unproductive for three years and they could have invested it. It would then have earned (compound) interest and would accordingly have amounted to considerably more than £1 at the end of the three-year period, the exact amount dependent on the rate of interest receivable. Inflation does not affect this principle, although it clearly affects the rate of interest that the investor could have, or would have sought, to obtain.

Therefore, it follows that somebody would pay less than £1 today for the right to receive £1 in three years' time. How much, however, should they allow?

To calculate the effect of discounting, the Present Value of £1 formula can be used (explained in Section 4.8 below). The result of this formula will be 1 if there is no delay before receipt (money is paid 'today') or less than 1 if a sum is received at a future date.

It must be remembered when calculating Amount of £1 or Present Value of £1 that the result as applied to £1 (or $1 or €1) is found from the respective formula. The resultant figure is then multiplied by the number of pounds (or dollars, euros, etc.) involved to find the total sum. In other words, to find the Amount of £1 for £100 over three years at 5 per cent, the calculation is £100 × $(1.05)^3$ not 100.05^3.

4.6 All risks yields (ARYs) and implied risk/growth allowances

In property valuation, the all risks yield (ARY) or market yield is the standard comparison measure of the rate of return from the investment. It is defined as: 'The remunerative rate of interest used in the valuation of freehold and leasehold interests, reflecting all the prospects and risks attached to the particular investment, such as the likelihood of future rental and capital growth' (Parsons 2004: 11). It is thus an implicit figure that incorporates all the risks, uncertainties and potential benefits of the investment. As seen in Chapter 3, lower yields are associated with relatively risk-free investments, whereas more uncertain and risky investments will have higher yields. This principle applies equally to land and buildings as other forms of investment and the best or 'prime' properties will attract lower all risks yields than 'secondary' landed investments.

The market yield will vary according to the type and class of property concerned, its location and position and the type of tenant actually in occupation or one who would potentially occupy the premises. Each property must be assessed individually on its merits. All yields adopted within property valuations should be based on comparables from the market; otherwise, the figure taken is merely an unsubstantiated opinion and would be difficult to prove in a negotiation. The market yield will vary according to the type and class of property concerned; the valuer must compare like-with-like properties.

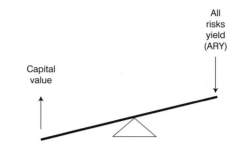

Figure 4.1 'Seesaw' relationship between yield and capital value

Freehold market yields are customarily based on market rent (MR). Where the market rent is not receivable or it is a leasehold valuation, the yield will require adjustment. The choices of all risks yield percentage and adjustments made for leaseholds are considered in Chapter 11.

The price paid by an investor to purchase a property will depend upon the yield he or she wants, or is prepared to accept, from that investment. As the yield goes up, the price that can be paid goes down; and conversely as the yield goes down, the price or capital value goes up (a 'seesaw' effect as illustrated in Figure 4.1). In the same way that prices of freehold property can be derived from the yield, then so can the yields being paid in the market be derived from the prices, or capital values. It is from analysing the prices being paid in the open market by investors, compared to the rental value of the property purchased, that it is possible to establish what constitutes the current market yields. Supplied with this knowledge, valuers can advise investors what return they can reasonably expect from a particular property type in a specific location, and so what prices they can expect to pay or receive for such property.

Property yields are influenced by, though not directly determined by, the level of interest rates in the economy. This is due to the time delay inherent in property development and property transactions, the general imbalance of supply and demand and the overall long-term nature of investment in land and buildings. In turn, general 'rates of interest depend on the money supply, the demand for loans, government policy, the risk of non-repayment as assessed by the lender, the period of the loan, and relative levels of foreign-exchange rates into other currencies' (Law 2006: 280).

As there is no direct relationship between interest rates and property market yields, there is not usually an immediate effect on overall returns from property investment from changes in base lending interest rates. Nevertheless, speculators and developers who borrow large sums and work to very narrow margins will be adversely affected whenever interest rates move significantly upward and will be forced to curtail their activity in the market and become more selective in their choice of investment or risk large financial losses. They will seek out opportunities that offer a sufficiently high return to warrant the risks being taken and cover the higher interest rates they pay. This will tend to force up property yields.

Conversely, when interest rates are falling investors are able to bid higher prices for property as borrowing the purchase funds becomes relatively cheaper. This will result in a decrease in property yields.

Investors will also draw comparisons between the returns available from alternative forms of investment compared to real property and reflect the relative advantages or disadvantages of each in their demand, which will also exert pressure on property yields in either direction.

4.7   Amount of £1 calculations

As explained above, this formula calculates for every £1 (or $1 or €1) invested how much money will accumulate, principal plus compound interest, over a specified period. It is applied to single payments, not regularly recurring ones, and assumes that a constant rate of interest is applied. The formula is:

$$A = (1+i)^n$$

where:

A = Amount of £1

i = interest rate (such as an all risks yield) per time period (usually per annum) expressed as a decimal. So 5 per cent = 5/100 = 0.05.

n = number of time periods (usually years) over which compound interest is to be added to the principal or initial sum invested.

When the interest rate varies then separate calculations are made for each time period at each rate that applies. Where regular investments of money are made over a time period, rather than one single monetary sum, then the Amount of £1 per annum formula or table is used (see Section 4.9 below).

   Although the most common type of calculation is based on annual interest rates and numbers of years, there is no reason why alternative timings cannot be used, such as monthly. One such example of where this is utilised is to show the effective annual rate of interest where calculations are made on a monthly basis. For instance, a monthly interest rate of 1 per cent is charged. Over twelve months the effective interest paid is thus found from 1.01^{12} (being the Amount of £1 at 1 per cent per month for twelve months). This equation produces the result of 1.1268. This is the principal sum of £1 plus interest, which is seen to be 0.1268 of a pound. The effective annual rate of interest is therefore found to be 12.68 per cent per annum.

4.8   Present Value of £1 calculations

'Present Value of £1' finds how much somebody would pay today for each £1 they will receive in the future. It finds the value in today's currency of each future £1 receivable, allowing for the time value of money by discounting the compound interest that could have been earned between now and the date when the money is received. It is the reciprocal, or inverse, of the Amount of £1 formula. The formula is:

$$PV = 1 / (1+i)^n$$

where:

PV = Present Value of £1

i = interest rate per time period (usually per annum) expressed as a decimal.

n = number of time periods (usually years) which will elapse before the sum of £1 is received.

Looking at the Amount of £1 formula and table will show that £1 invested today will amount to £1.1576 in three years at 5 per cent compound interest. Another way of looking at this is to say that somebody would pay £1 today for the right to receive £1.1576 in three years' time if his or her £1 could have been invested elsewhere and earned interest at 5 per cent per annum. Expressed in valuer's terms this becomes:

The Present Value of £1.1576 in 3 years at 5% is £1.
The PV of £1 in 3 years at 5% is therefore £1/1.1576 = £0.8638 or 86.38 pence.

Assuming the sum to be received in the future is other than £1, then the PV number is multiplied by the amount involved. Thus £30,000 receivable in three years' time allowing for 5 per cent per annum interest would have a present value of: £30,000 × 0.8638 = £25,914.

Example

An investment will return the sum of £50,000 in six years' time. What would the present value of this be allowing for discounting at the interest rate of 6 per cent per annum?

PV of £1 in 6 years @ 6% $= 1/(1+0.06)^6$

$= 1/1.418519$

$= 0.70496 \times £50,000$

$=$ Present Value of £35,248

4.9   Amount of £1 per annum calculations

The Amount of £1 calculates the amount that will accumulate at a specific rate of interest if £1 is invested at the *end* of *each* year for a given number of years. To check that the tables are based on this premise, look up the Amount of £1 per annum for one year at any rate per cent. The answer in each case is £1, signifying that no interest has been added.

The formula for this table is:

$$\left((1+i)^n - 1\right)/i$$

where:
i = annual interest rate/100 so that it is expressed as a decimal number
n = number of years over which the annual sums are invested

The Amount of £1 per annum table is not widely used in property valuations, as it is not concerned with a *present* value, but the worth at the end of a time. It can be used to calculate the total effective eventual cost of an annual expenditure allowing for loss of interest on the sums expended yearly. Thus if £1,000 per annum is spent on maintaining a property, the total effective cost over time could be viewed as the sums expended plus the forgone interest these amounts could have earned had they been invested elsewhere. Over say a five year period the total effective cost is not just 5 × £1,000 but also the compound interest that could have accumulated on each sum for four, three, two and one years. This total sum will be found by £1,000 × Amount of £1 per annum for five years at *x* per cent (where the interest rate obtainable on an alternative form of investment would have averaged x per cent per annum).

Example

£1 per annum is invested each year at the end of each year for five years. The investment attracts interest at 5 per cent per annum. What sum would be available at the end of the five-year period?

   The first payment would be invested for four years and would therefore attract four years' interest. That is it would be multiplied by the Amount of £1 for four years at 5 per cent. The second payment would attract three years' interest, and so on, the last payment attracting no interest at all, as in theory it would invested at the end of the fifth year, and immediately withdrawn.

   Thus £1 invested each year for five years will accumulate to the sum of the Amount of £1 in four, three, two and one years' time plus £1, as follows:

Amount of £1 in 1 yr	1.050
Amount of £1 in 2 yrs	1.102
Amount of £1 in 3 yrs	1.158
Amount of £1 in 4 yrs	1.216
Plus £1 in 5 yrs	1.000
Total	£5.526

This same answer could, of course, have been found more quickly and easily by looking up in the tables Amount of £1 per annum for five years at 5 per cent. However, the above example serves to show how the table is constructed and how it aggregates the effect of an *annual*, as opposed to a single or one-off investment of money.

   Using the formula the above answer would have been found as follows:

Amount of £1 per annum for 5 years @ 5% $= \left((1+0.05)^5 - 1\right)/0.05$

$= 0.2762816 / 0.05 = £5.525632$

SAY $= £5.526$

Supposing the sum invested was £2,000 per annum and not just £1, then the total amount invested after five years with compound interest added throughout will be:

£2,000 × 5.525632 = £11,051.26

Progress check questions

- Can you explain the difference between simple interest and compound interest?
- Do you know how to calculate compound interest?
- Do you understand what the 'time value of money' is and why sums receivable in the future are discounted to arrive at their present value?
- Can you understand why property market yields are called all risks yields?
- What are the 'risks' and how does the yield reflect them?
- Can you distinguish between circumstances that require the use of either the Amount of £1 or the Present Value of £1 or the Amount of £1 per annum formula or table?

Chapter summary

Monetary investments usually attract payments of interest to compensate for the foregoing of expenditure now. These interest payments are normally calculated on a compound interest basis, where interest is added to the investment throughout and subsequent interest calculations are based on the total accrued sum. Thus 'interest is earned on interest' as well as the initial principal sum invested. Traditionally, these compound interest sums are calculated on an annually in arrear basis, although other methods of calculation, such as quarterly in advance, are being used to reflect the modern timing of payments.

All financial calculations undertaken for property valuation purposes are based around the compound interest formula. Since 1913, *Parry's Valuation and Conversion Tables* have provided an 'industry standard' reference source for the calculated formulae. Alternatively, valuers can use a pocket calculator or computer to undertake the required calculations.

Discounting compound interest to find the present value of a sum of money is of fundamental importance in property valuation. Rates of interest obtainable on an investment depend on a number of factors. The all risks yield or market yield is the traditional measure of rate of return on property investments. This will vary between property investments depending on the perceived levels of risk and benefits associated with each property and will be influenced by, though not directly determined by the general rates of interest in the economy.

Further reading

Askham, P. (1993) 'Mainly for students: yields and rates of interest', *Estates Gazette* (4 Sep.): 117–18; repr. in P. Askham and L. Blake (eds), *The Best of Mainly for Students* (London: Estates Gazette, 1999), vol. 2, pp. 109–16.

Estates Gazette (2000) 'Mainly for students: times of change' 0027 and 0028 (8 and 15 Jul.): 134–5 and 137.

Ifediora, B.U. (2005) *Valuation Mathematics for Valuers and Other Financial and Investment Analysts*, Enugu: Immaculate Publications.

Imber, A. (1999) 'Mainly for students: valuable lessons in the art of calculation', *Estates Gazette* (18 Sep.): 164–7; repr. in L. Blake and A. Imber (eds), *The Best of Mainly for Students* , (London: Estates Gazette, 2004), vol. 3, pp. 435–43.

—— (2000) 'Mainly for students: time to move forward', *Estates Gazette*, 0023 (10 Jun.): 136-8; reprinted in L. Blake and A. Imber (eds), *The Best of Mainly for Students* (London: Estates Gazette, 2004), vol. 3, pp. 465–72.

Lumby, S. (1994) *Investment Appraisal and Financing Decisions: A First Course in Financial Management*, 5th edn, London: Chapman & Hall.

Millington, A.F. (2000) *An Introduction to Property Valuation*, 5th edn, London: Estates Gazette.

Mott, G. (1997) *Investment Appraisal*, 3rd edn, London: Pitman.

Sinking funds and mortgages

In this chapter …

- What a sinking fund is, why it may be used and how to calculate it.
- The effects of tax on sinking funds and how to adjust for them.
- Why sinking funds are used in conventional appraisals of leasehold interests.
- Types of mortgage, how repayments work on them and how to calculate those repayments.

5.1   Role and relevance of sinking funds

A sinking fund is 'A fund set up to replace a wasting asset at the end of its useful life. Usually a regular annual sum is set aside to enable the fund, taking into account interest at the expected rate, to replace the exhausted asset at a specified date. Some have argued that amounts set aside for depreciation of an asset should be equal to the annual amounts needed to be placed in a notional sinking fund' (Law 2006: 483–4).

Parsons (2004: 235) provides the view that a sinking fund is

A sum of money set aside at regular intervals to earn interest on a compound basis either:

a.   to be set off against the diminution in value of a wasting asset, for example, a lease; or
b.   to meet some future cash liability.

The principle is that a fixed sum of money is invested every year over a specified time period in a form of investment that provides a compound interest payment, so that at the end of the time a pre-determined sum of money will have accumulated. Due to its nature, the interest rate achieved on a sinking fund is sometimes referred to as the 'accumulative' rate of interest.

In this way, future capital liabilities can be met by the provision of a yearly sum out of income. Traditionally, this approach to calculating the yearly cost of capital replacements has been used in connection with leasehold interests. The value of a lease tends to decline as the number of years left to run on it reduces. Once the lease has expired there is no value attributable to it. To overcome this, a sinking fund can be used to replace the original purchase cost of the lease although this view is less favoured in contemporary valuation methodology. The topic is considered further in Section 5.4 below.

The investor will be keen to guarantee that the required sum will actually be achieved over the given time period. Any shortfall would not be ideal when budgeting to cover a liability or to provide sufficient capital replacement. This means the investor would not want to incur unnecessary risk. Thus, a very 'safe' form of investment is chosen in which to invest the sinking fund monies. This could include 'gilt-edged securities', government stock or bonds, an endowment or other insurance or life assurance policy or bank deposit accounts. With the possible exception of some endowment policies in the UK in recent years, all such forms of investment are considered quite low-risk and are thus likely to provide relatively low rates of interest.

5.2   Calculation of sinking funds

The annual sinking fund (asf) formula is:

$$r / \left((1+r)^{n} - 1 \right)$$

where:

r =   net-of-tax annual rate of interest expressed as a decimal (see Section 5.3 below for further explanation of the tax effects). Thus 3.5% = 3.5/100 = 0.035

n =   number of years over which sinking fund is to be invested

The annual sinking fund formula (or table in *Parry's*) will give the equal amount that must be invested at the end of each year for a given number of years in order to accumulate to £1 at the end of the period, allowing for compound interest to accumulate on each payment. The interest sum that each successive annual payment accrues decreases due to the reduced period each payment remains invested in the fund. So a payment made today and left invested for the next five years will accumulate five years' worth of interest, but the payment next year will only attract four years of interest, and so on. The last payment secures no interest at all, as it is deemed it will be paid in and drawn out simultaneously, on the basis that the calculations are made annually in arrear. Thus if one looks at the tables, the annual sinking fund for one year at any rate per cent is always £1. This concept is consistent with the traditional valuation assumption that all incomes are received at the end of each year.

Example

A sum of £80,000 is required in 5 years from now. An annual sinking fund is to be set up that will return an interest rate of 3% over this period. What will be the annual sinking fund instalments?

$$\text{asf for 5 yrs @ 3\%} = 0.03 / \left((1 + 0.03)^5 - 1 \right)$$

$$= 0.03 / (1.159274 - 1)$$

$$= 0.03 / 0.159274$$

$$= 0.1883546$$

$$\times \text{£80,000} = \text{£15,068.37 per annum}$$

5.3   Effects of tax and adjustments made

Income Tax or Corporation Tax affects sinking funds in two ways:

- tax is payable on the interest received on the accumulating sinking fund investment; and
- the sinking fund annual instalments must be met out of taxed income.

It is therefore necessary to allow for the first effect by adopting a net-of-tax sinking fund rate. This is the effective rate of accumulation after tax has been deducted. This net rate may be found in one of two ways, either:

- from the T_N 'Tax Adjustment Factors' table in *Parry's Tables*; or
- by using the formula of $(100 - t) \times$ gross interest rate/100 = net interest rate

where:
t = percentage rate of tax as a *whole* number. Thus 30 per cent = 30 not 0.30

This will give the effective rate of accumulation after tax has been deducted. Failure to use the net rate will result in a shortfall and the capital will not be fully replaced as intended, due to the net interest payments being lower than the expected rate. So if calculations are undertaken using 5 per cent when the true rate of return after tax will be only 3 per cent, the total interest accumulated will be considerably less than the calculations would suggest and the sum required will not be available at the future date as expected.

Having selected the investment vehicle into which the funds will be placed, the investor will ascertain what gross (before tax) rate of interest is likely to be offered. In doing this, the investor must be sure that this rate can actually be consistently achieved in the market over the period required. For example, if the sinking fund is to replace a capital sum over the next five years, is it likely that the required gross interest rate can actually be obtained in the market over each of these years? A rate that may be available today may fall in a future year, which will adversely affect the value of the invested funds. In deciding on the most likely average achievable rate over the time period involved there is thus a tendency to err on the side of caution and adopt a lower rate in case interest rates fall over the period of investment.

Having selected the gross sinking fund rate, tax must be deducted at the rate appropriate to the investor setting up the sinking fund, to arrive at the net rate of interest. Some investors, notably charitable institutions, are called gross funds as they do not pay tax and thus no deduction for tax needs to be made in arriving at their sinking fund accumulation rate. Their annual sinking fund instalments to replace a given figure would therefore be *less* than those by a taxpayer and their return (based on spendable income) would be higher.

This leads to the consideration of what gross income is needed in order to provide the necessary after-tax funds required to be invested in the annual sinking fund. The T_G tax adjustment factors table from *Parry's* or the following formula can be used:

Net annual sinking fund payment $\times\ 100/(100 - t)$ = gross income required

again where:
t = percentage rate of tax as a whole number.

Example

A sinking fund is to be set-up to accumulate to £75,000 in four years' time. The money will be placed in a bank deposit account that is expected to provide an average gross interest rate over this time of 4 per cent. The investor pays tax at 20 per cent. What will be the net annual sinking fund instalments and what deductions from the investor's gross annual income will this require?

Net asf rate $= 4\% \times (100 - 20)/100$
$= 4\% \times 0.8$
$= 3.2\%$

asf for 4 yrs @ 3.2% $= 0.032/\left((1 + 0.032)^4 - 1\right)$
$= 0.032/0.134276$
$= 0.238315$
$\times$ £75,000 = £17,873.62 per annum sinking fund instalments needed

Gross income required to provide an after-tax sum of £17,873.62 per annum

$=$ £17,873.62 $\times 100/(100 - 20)$
$=$ £17,873.62 $\times 1.25$
$=$ £22,342.02 gross income per annum

In other words, the investor will need to earn £22,342.02 each year to be left with £17,873.62 after tax to invest in the sinking fund, which with interest being accumulated at the rate of 3.2 per cent after tax, will amount to £75,000 being saved in the account in four years from now.

5.4   Sinking funds and leaseholds

All leaseholds are depreciating or wasting assets in that their value diminishes as the unexpired term, or length of time left before the lease runs out or 'expires', decreases. To help offset this, a 'sinking fund' can be set up by the leaseholder whereby sums are invested each year to replace the capital sum expended in purchasing the interest. The principle is that when invested these annual sums will accumulate, with compound interest added, over the period of the unexpired term, to the initial monetary outlay. In this way the leaseholder will 'get their money back' at the end of the lease – that is, the sum spent at the start on buying it. In this way, the sum can be reinvested in the purchase of another leasehold, and so on ad infinitum.

Whilst this does not protect the value of the leaseholder's investment in real terms, it does ensure that lease ownership can be continued in perpetuity. In that respect only, the sinking fund thus places the leaseholder in a similar position to a freeholder.

In reality, many leaseholders neglect to set up a sinking fund, even though this will result in their investment having nil value on expiry of the lease. The value is depreciated and 'written off' over the lease duration like other business assets. Nevertheless, it has been traditional valuation practice to allow for the setting up of such a fund, and this is taken into account in the valuation calculations.

A single rate basis sinking fund is when the sinking fund payments are reinvested at the same rate of interest as obtained on the property itself. In other words, the *accumulative* rate of interest (the rate earned by the sinking fund instalments) is the same as the *remunerative* rate of interest (the yield on the property investment). As both rates are the same, the sinking fund is on a 'single rate' basis.

An alternative, and more usual basis, is the dual rate approach. This is where the sinking fund instalments are invested at a different (lower) rate than the property market yield in a relatively 'safe' form of investment, such as an insurance policy or long-term government gilts. The accumulative rate of interest (on the sinking fund) is thus lower than the remunerative rate of interest (the property all risks yield). This gives rise to a 'dual rate' basis being adopted. This approach is considered in detail in Chapter 6.

5.5   Mortgages and their repayment calculations

A mortgage is a loan secured against real property. More than one mortgage can be taken out on the same property. Law (2006: 351) defines a mortgage as

> an interest in property created as a security for a loan or payment of a debt and terminated on payment of the loan or debt. The borrower, who offers the security, is the mortgagor; the lender, who provides the money, is the mortgagee. Building societies and banks are the usual mortgagees for house purchasers. In either case the mortgage is repaid by instalments over a fixed period (often 25 years), either of capital and interest (repayment mortgage) or of interest only, with other arrangements being made to repay the capital, for example by means of an endowment assurance policy (this is known as an endowment mortgage).

   In the UK, mortgages come under the Law of Property Act 1925. Either a legal mortgage or equitable mortgage can be created. The first bestows a legal estate, and the second only an equitable interest in the property on the lender. Most mortgages are of the equitable type and these provide the lender (mortgagee), subject to certain legal limitations, with potential remedies if the borrower (mortgagor) defaults on the loan. These are foreclosure (take away the power of redeeming or repaying the mortgage), repossession (seize the asset and sell it to recover sums due) or, in certain cases, the appointment of a receiver.

   If the mortgagee sells, there will inevitably be arrears of interest that will have accumulated. There will also be expenses incurred in selling, and there may be a need for some work, such as repairs, to be carried out before the property can be put on the market. Therefore, the mortgagee must be able to recover all these costs as well as the outstanding capital on the loan. For this reason, it is unusual for a mortgagee to advance a loan of 100 per cent or more of the valuation of the property. They usually prefer to provide some margin by only advancing a maximum loan-to-value (LTV) sum of 85 to 95 per cent or less of the valuation. The decision will vary from lender to lender and on the circumstances of the borrower. One of the most frequently used approaches is to limit the amount of loan to a specific multiplier of the borrower's annual income.

   The security must be adequate for the whole life of the mortgage, therefore a mortgage should not be granted for a long term if the value of the property cannot be foreseen. With a wasting asset, the level of the mortgage must always be below the lowest expected value of the interest.

   A valuation for mortgage purposes is primarily an assessment of the market value of the property disregarding certain factors such as furniture and fittings (not fixtures) that may be removed by the mortgagor, goodwill and any speculative value, such as development value. The actual sale price of the property should of course be its market value. However, this figure can be suspect as a guide to value if the parties are not acting 'at arms-length', or there are 'special purchaser' circumstances surrounding the transaction. Detailed guidelines on valuing properties for mortgage purposes are set out by the RICS and are covered in more detail in Chapter 9.

   There are two basic types of mortgage:

1. level repayment
2. interest-only.

With a repayment mortgage, the borrower repays a level, constant sum each year made up of the interest on the loan and a partial capital repayment. In the early years, the majority of the annual repayments comprise interest on the outstanding debt, but in the latter stages of the life of the mortgage, this reverses so that relatively larger portions of the payments are repaying the capital. This is due to the reducing outstanding balance so that interest is charged on a smaller amount of capital over time.

   With an interest-only mortgage, the amount of interest repaid each year remains constant (providing the interest rate remains the same), calculated on the initial sum borrowed, not on a reducing balance, as is the case with a repayment mortgage. In addition, an investment vehicle is required to repay the capital at the end of the mortgage period. Traditionally, this is provided by an endowment policy that will mature at the end of the mortgage term and provide a sufficient cash sum to clear the amount borrowed and possibly leave a cash surplus for the benefit of the assured party/parties if the policy is on a 'with profits' basis. This means that the assurance company guarantee to pay a sum to cover the loan plus a basic additional sum, to which bonuses are added each year dependent on the profit levels of the company. Frequently a further terminal bonus is also added to the sum payable. Pension funds and other types of investments can provide alternative methods of providing the necessary cash sum at the end of the mortgage.

In addition to the above-mentioned basic types of mortgage, there are a number of variations available such as low-start mortgages, where the repayments steadily increase in steps over the first few years of the term, which is designed to lessen the initial burden of repayments. There are also variations in how the interest rate is determined, from standard variable rate (SVR) to capped rates, fixed rates, discounted rates and stepped rates.

Mortgage repayment calculations

The annuity that £1 will purchase formula is the basis for the calculation of yearly level repayment mortgage payments. It finds the yearly mortgage repayment per £1 borrowed. The formula is:

$$i + i / \left((1+i)^n - 1 \right)$$

where:

i = interest rate expressed as a decimal number, so that $8.25\% = 8.25/100 = 0.0825$.
n = period of the loan in years

It will be noted that the second half of the above formula is effectively the annual sinking fund calculation. Thus an alternative way of expressing the annuity of £1 is:

$$i + \text{asf}$$

Additionally, it will be seen after reading Chapter 6 that the annuity that £1 will purchase is also the reciprocal of the single rate years purchase formula, so another alternative way of expressing it is:

$$\left((1+i)^n \times i \right) / \left((1+i)^n - 1 \right)$$

or $Ai / (A - 1)$ where A=Amount of £1 or $(1+i)^n$

With interest-only mortgages, the yearly repayments are found by:

Sum borrowed × Annual interest rate/100

As most mortgages require payments to be made on a calendar monthly basis, the necessary repayments can be found by dividing the annual sum by twelve.

Years purchase is the present value of an income flow whereas the annuity is the income that £1 will buy. It must be remembered that the formula and the corresponding tables in *Parry's* are again based on the annually in arrear assumption. With many modern mortgages, this may not hold true. With some mortgages the interest is calculated on a quarterly basis; with others it can be monthly. With 'flexible' mortgages that allow overpayment, payment 'holidays', and other variations, the interest is usually found on a daily basis. This also applies to 'all in one' type of mortgage accounts, which combine bank current account, savings account, and mortgage in one. In each of the cases where interest is not calculated on an annually in arrear basis, the repayments indicated by the 'traditional' formulae will not be correct and will need amendment.

Example

A mortgage of £50,000 is obtained over a five-year term from a building society at an interest rate of 8 per cent. The loan is to be repaid by equal yearly repayments comprising capital and interest. The repayments could be found by any of the following alternative methods:

1. For every £1 borrowed, the repayment each year will be interest of £0.08 and in addition a further sum each year will be required to repay the capital over five years: the sinking fund concept. The payment each year will therefore be: 'interest' plus 'annual sinking fund for 5 years @ 8%'. This could be abbreviated to:

   $i + $ asf 5 years @ 8% $= 0.08 + 0.170456 = 0.250456$

   For £50,000 the yearly repayments will therefore be:

   £50,000 × 0.250456 = £12,522.82

   The annuity that £1 will purchase table (single rate) in *Parry's* is the above formula and from the table it will be seen that 5 yrs @ 8% is 0.250456. Multiply this by £50,000 = £12,522.82 per annum.

2. As the annuity £1 will purchase is the reciprocal of the YP single rate formula (explained in Chapter 6) the sum borrowed can be divided by the YP 5 yrs @ 8% single rate:

   £50,000/3.99271 = £12,522.82 per annum

   The monthly repayments on this loan would be:

   £12,522.82/12 = £1,043.57

   The total amount repayable will be:

   £12,522.82 × 5 = £62,614.10

An illustration of how repayment mortgage payments work

Repayments mortgage payments are based on the single rate table, resulting in equal total yearly repayments of capital and interest (see Chapter 6 for a fuller explanation of single rate basis).

Interest due each year is calculated on the balance of capital outstanding at the commencement of the year. Thus in the early years the amount of interest paid is higher than in the later years of the mortgage.

The following illustrates how repayments on the five-year mortgage given in the example above will enable the loan and interest to be cleared over the repayment period.

As has already been noted, the yearly repayments from the annuity that £1 will purchase table for a loan of £50,000 over five years at 8 per cent are £12,522.82 per annum. Thus, the following cash flow chart in Table 5.1 can be constructed:

The mortgage could be redeemed, or paid off, during the term by paying an amount equivalent to the outstanding capital. Thus, if the property on which the mortgage is secured were sold at the end of the third year, the amount to repay would be £22,331.52. This figure could also be obtained by multiplying the yearly repayments by the years purchase single rate for the number of

Table 5.1   Cash flows for a five-year mortgage

Year 1 start	£50,000.00
Add interest @ 8%	£4,000.00
	£54,000.00
Repayment	− £12,522.82
Year 2 start	£41,477.18
Add interest @ 8%	£3,318.17
	£44,795.35
Repayment	− £12,522.82
Year 3 start	£32,272.53
Add interest @ 8%	£2,581.80
	£34,854.34
Repayment	− £12,522.82
Year 4 start	£22,331.52
Add interest @ 8%	£1,786.52
	£24,118.04
Repayment	− £12,522.82
Year 5 start	£11,595.22
Add interest @ 8%	£927.62
	£12,522.84
Repayment (end of yr5)	− £12,522.82
Balance end of year 5: SAY	zero (mortgage paid off)

The slight error of 2 pence in the final balance is due to 'rounding' the annual repayment sum.

years left of the mortgage. Thus, YP 2 years' single rate @ 8% × £12,522.82 = 1.78326475 × £12,522.82 = £22,331.50 (again just a 2 pence difference between totals due to rounding error). This sum also assumes that the lender does not levy any early redemption penalties or other payments, such as administration charges, in calculating the redemption amount.

Progress check questions

- What is a sinking fund and why might it be used?
- Why are the interest rates used in sinking fund calculations on a net-of-tax basis?
- Do you understand why a purchaser of a leasehold interest may consider setting up a sinking fund to replace the initial purchase price by the end the lease term?
- Can you see why in a repayment mortgage over a 25-year term the borrower will not have reduced the outstanding capital balance by much after making repayments for five years?

Chapter summary

Sinking funds can be used to set aside a regular sum of money each year to cover future capital liabilities or to recover already incurred capital expenditures. These sums are invested annually in a relatively safe form of investment so that the capital sum returned can reasonably be assured. Being quite low risk, these investments are likely to provide lower rates of interest. The effects of taxation on both the income that provides the sinking fund instalments and the interest payments received on the fund must be taken into account to guard against a shortfall in the expected funds available in the future and to ascertain the impact on existing incomes.

Traditionally, one use of sinking funds is to replace the initial capital cost of purchasing a leasehold interest in property. However, in practice, many leaseholders decide not to take out such a fund and accept that the asset will 'waste' with time.

Mortgages are a form of secured lending, using real estate as a means of security for the loan. The mortgagee, or lender, provides capital funds to the mortgagor or borrower, which are repaid with compound interest, through regular (normally monthly) repayments. These repayments can be considered as the annual equivalent of the capital sum involved. This means that a regular series of smaller payments is deemed to equate to the value of a larger, single capital payment over time, allowing for the addition of compound interest. It is the method whereby an annual income can be generated in return for foregoing the use of a capital sum and is the basis of many property valuation calculations, which are covered in Chapter 6.

Further reading

Brett, M. (1998) *Property and Money*, 2nd edn, London: Estates Gazette.

Imber, A. (1998) 'Mainly for students: the loan arrangements', *Estates Gazette*, 9804 (24 Jan.): 144–6; repr. in P. Askham and L. Blake (eds), *The Best of Mainly for Students* (London: Estates Gazette, 1999), vol. 2, pp. 124–9.

Morley, A. (2008) 'It all adds up in the end', *Estates Gazette*, 0806 (9 Feb.): 162–3.

Capitalisation

In this chapter ...

- The difference between income and capital sums.
- What a years purchase is and why it is used.
- The different types of years purchases, how they are calculated and when and why each is employed.
- The difference between remunerative and accumulative rates of interest as applied to valuation of leasehold interests.
- The effects of taxation on leasehold valuations and how these are taken into account.
- How a value may be calculated when the duration of the interest depends on the life expectancy of its owner.

6.1   Income flows and capital sums

It is important to understand the distinction between income flows and capital sums. Law (2006: 265) defines income as 'any sum that a person or organisation receives either as a reward for effort (for example, salary or trading profit) or as a return on investments (for example, rents or interest)'. This is contrasted this with 'capital expenditure (capital costs; capital investment; investment costs; investment expenditure)', which is 'the expenditure by an organization of a significant amount for the purchase or improvement of a fixed asset' (Law 2006: 87). Parsons (2004: 43) states that capital value is 'the value of an asset as distinct from its annual or periodic (rental) value'.

Thus income is receivable on a periodic basis and is recurring. In taxation terminology, income may be 'earned' or 'unearned'. 'Earned' income is that obtained from employment or self-employment. Thus employees expect to receive a wage or salary in return for providing their labour and services to an employer. These payments are normally received on a weekly or monthly basis, but comparisons of earnings are usually expressed 'per annum' – that is, the total income for a whole year is used as a measure of that person's earnings.

'Unearned' income is obtained from investments, including land and buildings. It can be interest payments on money invested in bonds, stocks or bank and building society accounts. It could be dividends paid on equity shares owned in companies. With land and buildings such income will arise where the owner has let the property to a tenant who pays rent. This rent is then the landlord's income from that property.

The timing of receipt of income payments will vary between investments. Some forms of income will be regular and others not. The exact figure paid each time can similarly fluctuate. Whilst interest payments on cash deposits in bank savings accounts will be regularly paid (commonly on a monthly or annual basis) the interest rate may be variable and can be changed at the bank's discretion, meaning the amount of interest paid will vary even when the total deposited has not altered.

Another consideration is whether payments of income are made in advance or in arrear. Does the payment cover what is to come or what has already occurred? Bank interest payments on deposited savings are calculated and paid in arrear. It is only after the cash has resided in the account for a specific time period that the interest is paid. Rental payments on land and buildings, however, are usually made in advance. Rent is paid now to cover the coming month or three months period. Again, to enable ready comparisons to be made between incomes received on different forms of investment, it is usual to quote it on a 'per annum' basis, thus indicating the total income receivable over a twelve-month period. Thus, a tenant paying a rent of £1,000 per calendar month would be providing the landlord with a gross rental income of £12,000 per annum.

Capital sums are payments made on a 'one-off' or 'lump sum' basis to purchase or acquire an investment, good or chattel. For example, an investor might purchase the freehold of a building for £1 million and then let that property to a tenant for a rent of £80,000 per annum. The tenant has the right to occupy and use the premises in return for the rental payments and these provide the owner with an *income* of £80,000 per annum in return for the *capital sum* expended of £1 million.

As previously explained, 'traditional' property valuations and valuation tables, such as *Parry's*, are based on the assumption that incomes and interest are receivable and calculated on an annually in arrear basis. To an investor, the first year's income is therefore not worth £1 but the *Present Value of £1* in 1 year. This means that if an investor had to wait one year to receive £1 they would be prepared to pay something less than £1 today for the right to this sum. How much less than £1 will depend on the interest rate which would be payable on such an investment.

Nowadays, although income is seldom receivable annually in arrear, this standard convention is still commonly used for the calculations. There are alternative sets of tables and valuation formulae (even within *Parry's*) that explicitly calculate income on an in advance and/or paid more frequently than annually basis. These are considered in Chapter 7. In reality, most UK business leases require rents to be paid quarterly in advance. However, these can still be valued using the traditional 'annually in arrear' formulae and tables.

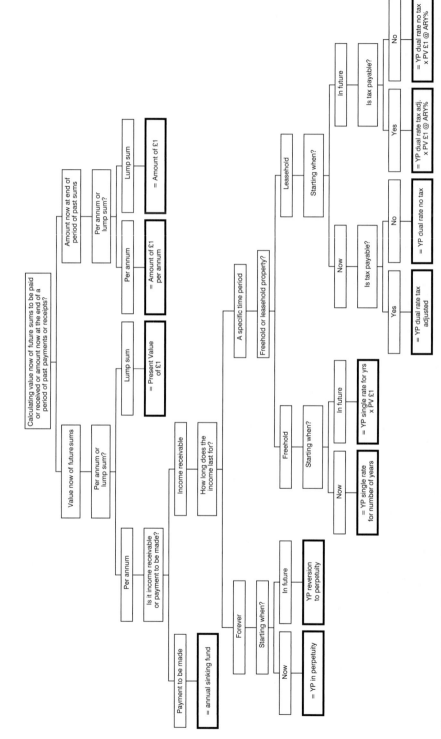

Figure 6.1 Which valuation formula or table to use

Capitalisation of income is achieved by multiplying the annual income flow by a multiplier, known as the years purchase (YP). The YP is a numerical multiplier. It is a number and should never be given a prefix or suffix such as '£' or '%'.

YP is not a value in itself, but merely the mathematical factor that is used to arrive at the capital value, taking account of the duration and timing of the income flow from the investment and the interest rate attributable to that investment. The appropriate years purchase formula to be used to value each 'block' or 'tranche' of rental income will depend on when that rental is first receivable and for how long a period it will continue to be receivable.

To decide which valuation formula or table should be used to undertake the required calculation, the flow chart in Figure 6.1 may assist. The appropriate formula is shown in bold in the box at the end of the relevant flow direction.

6.2 Years purchase in perpetuity

Years purchase in perpetuity ('YP perp') means the *Present Value of £1 per annum forever*. It can also be known as the 'present value of an annuity'. At 5 per cent the YP perp is 20, which is the sum total of all the PVs from year 1 to infinity. A simpler way to calculate this is to use the formula, which is:

$$\text{YP perp} = 1 / i$$

where:
i = interest rate, or yield, expressed as a decimal (rate/100)

Alternatively, the calculation may be easier to undertake mentally if the equation of 100/Yield is used, where the yield is kept as a whole number. For instance, YP perp at 8 per cent can be found from 1/0.08 or 100/8 = 12.5. With time, the relationship between a percentage yield and the corresponding YP perp figure becomes familiar and valuers can refer to the YP number used in a calculation to another valuer without needing to explain the yield used. Thus an investment showing a YP of 20 on analysis will be understood to be an investment providing a 5 per cent yield (1/0.05 or 100/5 = 20).

This multiplier is used to capitalise the income flow from a *freehold* property investment, where the current income is already the full market rent (MR) and thus in present value terms cannot be any higher. This is the conventional approach used in 'traditional' valuations. No attempt is made to forecast what, in nominal terms, the rent may be in the future due to inflationary pressures and growth. However, this is done when using equated yield and a discounted cash flow approach and is explored in Chapter 12.

The formula is:

$$\text{YP perp deferred} = 1\,/\,i(1+i)^n$$

or $1\,/\,iA$

where:
$A =$ Amount of £1 or $(1+i)^n$
$i =$ interest rate expressed as a decimal (ARY/100)
$n =$ number of years before income is first received

Effectively, it is the YP in perp deferred. It can therefore also be arrived at as follows:

YP in perp $\times$ PV for deferment period

that is $1/i \times 1/A = 1/iA$, which of course is the formula given above.

Example

A freehold property is let at a peppercorn (effectively nil rent) for the next four years. Thereafter the market rent will become payable by the tenant. This is currently estimated to be £40,000 per annum. From evidence provided by similar investments, an all risks yield of 5 per cent is considered appropriate for this property.

The present capital value of this freehold investment can thus be assessed as follows:

£40,000 × YP perp deferred 4 years @ 5%
$= £40,000 \times 1\,/\,0.05(1+0.05)^4$
$= £40,000 \times 16.454 = £658,160$
SAY value = £658,000

6.5   Years purchase (single rate) deferred

This calculation is required when an income flow is receivable for a specific number of years, but starting at some future date, not now. It is calculated by finding the YP single rate for the number of years during which the income will be received, and multiplying this by the PV, at the same remunerative rate of interest, for the number of years between now and when the income will commence.

The formula is:

$$\text{YP single rate deferred} = \frac{(1+i)^n - 1}{i(1+i)^n} \times \frac{1}{(1+i)^d}$$

where:
$i =$ all risks yield expressed as a decimal (ARY/100)
$n =$ number of years over which the income will be received
$d =$ number of years before income starts to be received (deferment period)

Or in other words:

> YP single rate for years × Present Value of £1
> (can be abbreviated to: YP for yrs def yrs)

Example

An income of £100,000 per annum will be receivable for a period of five years, the first payment being received at the end of the third year from today. What is the present value of this income flow, valued at a yield of 10 per cent?

Answer:

> YP 5yrs @ 10% × PV 3yrs @ 10% × £100,000
> (which could alternatively be written as YP 5yrs def 3 yrs @ 10% × £100,000)
> = 3.791 × 0.75131 × £100,000
> = £284,800

The YP and PV figures can be obtained directly from the relevant tables within *Parry's*. Alternatively, they can be found by use of the formulae:

$$\frac{(1+0.1)^5 - 1}{0.1(1+0.1)^5} \times \frac{1}{(1+0.1)^3}$$

$$= \frac{1.61051 - 1}{0.1 \times 1.61051} \times \frac{1}{1.331}$$

$$= \frac{0.61051}{0.16051} \times \frac{1}{1.331}$$

$$= 3.790867 \times 0.7513148$$

$$YP = 2.848$$

6.6 Remunerative and accumulative rates of interest in leasehold valuations

Leasehold interests in real property are 'wasting assets' in that their value will eventually diminish to zero. To offset this, the investor in leaseholds could set aside part of the profit rent from the investment each year to be paid into a sinking fund that will recover the initial capital expenditure. In this way, the initial purchase price will be recovered in nominal terms. However, such an approach does not offset the effects of inflation and there will be no capital growth from the initial investment. It is thus only a partial solution and in practice many leasehold investors choose not to use the sinking fund approach and accept the diminution in value of their investment over time in return for a higher net income from that investment.

In Section 5.4 it was seen that sinking fund interest rates tend to be lower than the all risks yield on a leasehold property investment, as the money is placed into a 'safe' form of investment such as government gilts, which are likely to provide a relatively low yield. This gives rise to a 'dual rate' approach. The sinking fund rate used is the net accumulative rate. This rate is low, both due to the tax allowance and as there should be minimal risk attached to the replacement of capital, with all the risk of the investment being reflected in the remunerative rate (all risks yield).

The fact that an asset is depreciating must be allowed for in the price paid. This is achieved through using years purchase dual rate figures that take into account the sinking fund element and the tax effects on it as well as the yield required on the investment. The proposed purchaser's tax rate is thus an important consideration when advising on the price that investor can pay for a leasehold interest.

6.7   Years purchase dual rate for a number of years

This formula or table is used for the valuation of leasehold interests. The formula determines the price which can be paid for a leasehold interest if the prospective purchaser requires a return of x per cent on capital (the all risks yield or remunerative rate) after allowing for the replacement of the purchase price by a sinking fund which will accumulate interest at y per cent, which is normally a lower rate of interest (the net accumulate rate) than the yield; hence dual rate tables.

The formula for years purchase dual rate is:

$1/(i + \text{asf})$

where:

i = remunerative rate or yield expressed as a decimal number (yield/100)

asf = annual sinking fund at the net accumulative rate of interest for the number of years in question

Thus the YP 10 years @ 5% & 3% is:

$1/(0.05 + \text{asf } 10 \text{ yrs } @3\%)$
$\text{asf} = 0.03/(1.03^{10} - 1)$
$\text{asf} = .0872305$
Therefore YP $= 1/(0.05 + 0.0872305)$
$= 1/0.1372305$
$= 7.287$ YP

Example

What price would be paid for an investment providing an income of £80,000 per annum for the next five years, if the required return on capital was 8 per cent and the investor proposed to invest in a sinking fund to accumulate at a net rate of 2.5 per cent?

With reference to the years purchase dual rate tables, the valuation is as follows:

Income   £   8000 pa
× YP for 5 years @ 8% & 2.5%   3.7
      Purchase price to show an 8% return= £29,600

Using the formula the same outcome can be proven:

YP for 5 yrs @ 8% & 2.5% = $1/(0.08 + \text{asf for 5 yrs @ 2.5\%})$
asf = $0.025/(1.025)^5 - 1$
asf = 0.025/0.1314082
asf = 0.19024686
Therefore YP = $1/(0.08 + 0.19024686)$
= 1/0.27024686
= 3.7 YP
3.7 × £8,000 = £29,600

However, although the years purchase dual rate for a number of years formula uses a net rate of interest on the sinking fund, it is on the assumption that the annual instalments into that fund are *not* met out of taxable income. Apart from non-taxpayers, such as gross funds, this will not be the case and thus with most investors an allowance for this additional aspect of taxation needs to be made.

6.8   Dual rate years purchase (tax adjusted)

In Section 5.3 it was seen that tax affected sinking funds in two ways; namely a deduction is made on the gross interest earned by the fund and the monies paid into the fund need to be found out of taxed income. To incorporate both of these aspects into the valuation of leasehold interests, a further refinement of the years purchase dual rate is required. This is the years purchase dual rate tax adjusted formula, which is abbreviated to:

YP for *w* years @ *x*% & *y*% (tax @ *z*%)

where:
w = years that profit rent is receivable
x = all risks yield
y = net-of-tax annual sinking fund rate
z = tax rate paid by investor or purchaser of the interest (or prospective investor)

It is used for the capitalisation of a profit rent from a leasehold interest where the investor is liable to Income Tax or Corporation Tax on the sinking fund instalments and interest.

The formula for years purchase dual rate tax adjusted is:

$$\cfrac{1}{i \ + \ \text{asf}\left(\cfrac{100}{100-t}\right)}$$

where:

i = all risks yield (ARY) applicable to similar leasehold investments

asf = annual sinking fund at net-of-tax interest rate for stated number of years

t = tax rate percentage as a whole number

The current tax rate, appropriate to the investor should be used. In the UK, individuals pay Income Tax and companies pay Corporation Tax. Different rates apply according to the size of income or profits and the percentage rate used can vary from one financial year to another so it is important to use the relevant figure.

Example

A leasehold investment will provide a profit rent income for the next four years of £22,000 per annum. What will an investor be prepared to pay to acquire this interest if an all risks yield of 11 per cent is required; the average gross sinking fund rate of interest over the next four years is expected to be 5 per cent and the investor pays tax at the rate of 30 per cent?

First, the net sinking fund interest rate must be found for use within the years purchase formula. At 5 per cent gross and 30 per cent tax the net rate will be 3.5 per cent. This is calculated as follows:

5% × (100 − tax%)/100 = 5% × 70/100 = 5% × 0.7 = 3.5%.
Annual sinking fund (asf) for 4 years @ 3.5% = $r/((1 + r)^n - 1)$
= $0.035/(1.035^4 - 1)$ = 0.035/0.147523 = 0.2372511
Then multiply this by 100/(100 − tax%) = 0.2372511 × 100/70 = 0.33893
Add 'i' (ARY as a decimal) to this: 0.11 + 0.33893 = 0.44893
Take reciprocal of this to find YP = 1/0.44893 = 2.2275

Then the value can be found by multiplying the relevant YP by the annual income of £22,000:

Capital value = YP 4 yrs @ 11% & 3% (tax 30%) × £22,000 per annum
= 2.2275 × £22,000 = £49,005.39
SAY value = £49,000

Instead of using the formula, the YP could be found by direct reference to the relevant valuation tables. When seeking the relevant figure it is necessary to look-up the information in 'reverse order'. First, find the table that contains the correct tax rate (30 per cent). In *Parry's Tables* a different colour is adopted for groups of pages devoted to the same tax rate. Then check for the pages that contain the required net sinking fund rate (3.5 per cent). Next look along the column headings seeking the one for the all risks yield (11 per cent). Then finally, look down the rows for the number of years sought (four) and where the row and column intersect will be found the required years purchase number (see example below with relevant cell highlighted).

TAX:	30 %		ASF:		3.5 %	YEARS PURCHASE NUMBERS (dual rate, tax-adjusted)				
	ARY Rate Per Cent									
	9	9.5	10	10.5	11	11.5	12	12.5	13	13.5
YEARS										
1	0.6585	0.6564	0.6542	0.6521	0.65	0.6478	0.6458	0.6437	0.6416	0.6396
2	1.2626	1.2547	1.2469	1.2392	1.2315	1.224	1.2165	1.2092	1.2019	1.1947
3	1.8185	1.8021	1.786	1.7702	1.7547	1.7394	1.7244	1.7097	1.6952	1.6809
4	2.3314	2.3045	2.2783	2.2526	2.2275	2.203	2.179	2.1555	2.1325	2.11
5	2.8058	2.767	2.7292	2.6925	2.6567	2.6219	2.588	2.5549	2.5227	2.4913
6	3.2457	3.1939	3.1437	3.095	3.0479	3.0021	2.9577	2.9146	2.8728	2.8321

For convenience when using *Parry's* or similar tables, the net sinking fund rate adopted for the dual rate years purchase is rounded down to the nearest half per cent. This is because the tables do not contain other fractions or decimal parts of a sinking fund interest rate and as the figure is based on a 'safe' form of investment it is better to be cautious and round the number down rather than up. Thus if the gross sinking fund rate was 3.5 per cent and tax is paid at 28 per cent the exact net rate to be adopted would be 3.5% × (100 − 28)/100 = 2.52 per cent. As the next-nearest rate available in the YP dual rate tables when rounding down is 2.5 per cent it is convenient to adopt this figure. The effect on the valuation will be minimal, but if it is preferred, the exact calculated net sinking fund rate can be used when the years purchase is found from formulae rather than tables.

6.9   Deferment of dual rate years purchase

As with deferment of a single rate years purchase, this calculation involves multiplying the YP for the number of years over which the income flow is received by the PV for the number of years between now and when the income is first received.

The YP is calculated on a normal dual rate (tax adjusted if necessary) basis, but the PV is found on a *single rate basis with no tax*, using the *remunerative* rate of interest (the all risks yield). Where capital sums need to be deferred, future assets or sums receivable, such as income, should be deferred at the remunerative rate. Conversely, future liabilities or sums payable should be deferred at the accumulative rate.

Example

A leasehold interest will produce a profit rent of £50,000 per annum for a period of four years commencing in two years from now. Value this interest at a yield of 12 per cent, a net sinking fund rate of 5 per cent and allowing for tax at 40 per cent.

YP 4yrs @ 12% & 5% (tax 40%) × PV 2yrs @ 12% × £50,000
= 1.9736 × 0.7972 × £50,000
= £78,667
Present value SAY = £78,500

6.10 Life tables and years purchases based on them

These tables are used where an income flow needs to be valued which will last for the duration of a person's life, or the longer or shorter duration of two persons' lives. As the precise length of this time cannot be specified in advance, the traditional formulae and tables cannot be used to carry out the valuation. Instead, there are special tables provided in *Parry's* which are derived from the English Life Tables. These are based on statistical evidence of longevity and give the average life expectancy of males and females at any given age. Based on this average, the probable expected time period over which the income flow will be receivable can be estimated, which in turn allows years purchase and present value figures to be calculated.

Thus if on consulting the age and gender on the Life Tables of a person with a life tenancy the average remaining life expectancy is shown as 12.366 years, this will be the period used to calculate the years purchase to capitalise an income only receivable by that person for their lifetime.

Example

What is the present value at a yield of 6 per cent of an income of £20,000 per annum receivable only for the life of an investor, where given their gender and current age, the English Life Tables indicate a remaining life expectancy of 15.67 years?

£20,000 per annum × YP 15.67 years @ 6%
= £20,000 × $(1.06^{15.67} - 1)/ (1.06^{15.67} \times 0.06)$
= £20,000 × 9.9785 = £199,570
Present value SAY = **£200,000**

Whilst insurance companies make considerable use of the Life Tables, valuers seldom need to use them. The only types of valuations of property interests where their use would be appropriate are life interests and entailed interests. By definition, such valuations are not particularly reliable as the subject person(s) can die much younger than expected, or live far longer than anticipated by the average figures in the Life Tables. In either case, the income flow will have been capitalised over the incorrect period and therefore, in retrospect, it will be seen that the capital sum calculated will not have been accurate. However, it is impossible to forecast this accurately in advance.

Progress check questions

- What is the difference between an 'income flow' and a 'capital value' and how is one found from the other?
- What does 'capitalisation' involve?
- What is a years purchase and how is it calculated?
- Why are single rate years purchases used to value freehold interests but dual rate ones for leaseholds?
- What is the difference between a remunerative and an accumulative rate of interest?
- Why in dual rate YP calculations is the sinking fund rate usually lower than the all risks yield adopted?
- How and why are two adjustments made for tax in dual rate years purchase numbers?
- How are YP numbers deferred to allow for a flow of income that will be received in the future, but does not start until after a period of years has elapsed?

Chapter summary

The income flows from an investment can be capitalised, or multiplied, by a mathematical factor known as a years purchase to assess its capital value. This multiplier is based around the annual percentage rate of compound interest that a similar investment to the one being assessed could be expected to earn. The timing and duration of each tranche of income will determine its value in today's market on the basis that the longer into the future the investor must wait before receiving that income, the less its value will be in current, or present value, terms due to the discounting effect.

Freehold property investments have a life 'in perpetuity', or forever. Over the long term, freeholds are generally expected to increase in capital value due to inflationary effects. Leaseholds, on the other hand, are 'wasting assets' with a finite life and with the expectation that their terminal value will be zero. This is reflected in the years purchase formulae and calculations used for each investment type.

Theoretically, investors in leasehold property will set up a sinking fund to recoup the initial purchase cost of the asset over its life span. Many decide not to do so due to the reduction in investment income this will create. Only the original capital sum in nominal terms will be returned with no growth in any case and due to the uncertainties of the interest rate that can be obtained in the longer term on a sinking fund even this may not be certain. Nevertheless, it is customary to assume this capital replacement approach will be adopted and leaseholds are valued on the presumption a sinking fund is used. When calculating the years purchase an allowance is made for the effect of tax on the interest earned on the sinking fund and for the fact that the instalments into the fund need to be met out of taxed income.

Further reading

Askham, P. (1993) 'Mainly for students: government stock', *Estates Gazette*, 9315 (17 Apr.): 139–40; repr. in P. Askham and L. Blake (eds), *The Best of Mainly for Students* (London: Estates Gazette, 1999), vol. 2, pp. 103–8.

Estates Gazette (1992) 'Mainly for students: investing in leasehold property', 9225 (27 Jun.): 129.

Imber, A. (1999) 'Mainly for students: valuable lessons in the art of calculation', *Estates Gazette* (18 Sep.): 164–7; repr. in L. Blake and A. Imber (eds), *The Best of Mainly for Students* (London: Estates Gazette, 2004), vol. 3, pp. 435–43.

7

Alternative valuation tables and formulae

In this chapter ...

- Why different valuation tables and formulae have been devised and can be used instead of the traditional versions.
- How to calculate the value of an investment whatever the actual timing of the financial receipts from it.
- Three specific alternative sets of valuation tables and the theory underpinning each.
- Practical effects of using different tables and formulae compared to the traditional ones, such as *Parry's*.

7.1   Background and historical development

As explained in Section 4.3, the traditional approach to valuing interests in landed property has been on the basis that all rentals are paid annually in arrear. Even if historically some rents were paid in this fashion, once inflation became a regular feature of the economy, landlords would have been less willing to accept payments in arrear, as the purchasing power, or 'real value' of those payments would have decreased the longer one had to wait for their receipt. With annually in arrear a whole twelve months of lost interest would have been incurred. There would also be the insecurity of income associated with the possibility of a tenant defaulting on rental payments after they had already had the benefit of occupation of the property.

Thus, it became normal market practice to require rentals to be paid in advance, rather than in arrear. This considerably increases the real value of the payments to the landlord, on the basis that money 'in the hand' is worth more than 'money to come in the future'. Investments that produced such an income flow would also have become more marketable compared to those investments still based on annually in arrear payments.

The timing of rental payments is linked to a calendar year. With business tenancies in the UK, the usual time period used is quarterly, with payments commonly due on or before each 'Quarter Day'. These are derived from historical custom and tradition and in England are:

Lady Day	25 March
Midsummer	24 June
Michaelmas	29 September
Christmas	25 December

And in Scotland are:

Candlemas	2 February
Whitsunday	15 May
Lammas	1 August
Martinmas	11 November

Alternatively, some other dates can be specified in a lease for rental payments that similarly divide the year into approximately equal quarters of around thirteen weeks each. Rents payable quarterly in advance and due on a Quarter Day would be paid by each of these dates. Each payment would cover the occupation rental for the forthcoming thirteen weeks.

For UK residential lettings, payment of rent on a weekly or monthly in advance basis is more common. There is a growing acceptance from UK retail property owners and landlords that monthly rent payments will become 'the norm' for that sector (Dixon 2008). It may also be possible to encounter different timings of income payments on other investments, such as half-yearly in arrear or advance, quarterly in arrear or annually in advance.

Each of these timings for payments would have a potential effect on the value of the investment and have led to the production of different formulae and tables. The use of these will thus produce a (slightly) different value for the investment, compared to the value produced by the 'traditional' annually in arrear formulae and tables if the *same* yield percentage is used. In practice, all yields should be derived from analysis of market transactions and if a different mathematical base for the calculations is consistently applied to both this analysis and the following valuation, end values should not differ, as different yield percentages will be used according to the timing of income receipts.

However, although the basis for use of these more 'modern' formulae and tables, which reflect actual market timing of receipt of income is perfectly sound, the property valuation profession has been relatively slow to adopt them in practice. Why is this so and should property valuers forever discard the conventional annually in arrear formulae and tables and use quarterly in advance ones?

One possible reason why the alternative formulae and tables have not been universally adopted is that tradition dies hard in British industries and professions. The very fact that a method has been tried, trusted and understood by different generations over many years will lead to a natural reluctance to discard it in favour of 'modern' ideas. To some extent, this factor may also explain why the valuation surveying profession in the UK still mainly works to imperial floor and site measurements, despite metrication being generally introduced into the country since 1971.

Another reason is that as the majority of business properties are let on quarterly in advance terms, the continued usage of annually in arrear formulae and tables may not be too crucial providing everybody works to the same basis. The overriding principle that must be observed is *how you analyse, so should you value*. Following this will effectively cancel out any discrepancies. In other words, if the same yield percentage was used and a valuation carried out using the different-based formulae or tables, then the final value would be different. However, if all analysis of comparable evidence and the valuation of the subject property are done using the same set of formulae or tables, the end value should be the same as that obtained from using a different set. This is because the yields derived from analysis should reflect the different mathematical bases of the formulae or tables in use. The yield found from an annually in arrear analysis will differ from that found by a quarterly in advance appraisal on the same investment.

One occasion when this approach will break down is where two investments that are not like-with-like are being valued and compared. For instance, the rent on one of the investments is actually paid half-yearly in arrear, whereas the other is quarterly in advance. In this case, two different sets of formulae or tables should be used to value each one; otherwise, an error will occur in the final valuation figures.

Apart from this last exception, there is a validity to the 'analyse as you value approach', and together with the possible conservativism of the profession, probably explains why the traditional annually in arrear basis of valuation continues to be popular. Nevertheless, there is increasing pressure on the profession to 'modernise' and this includes 'converting' to using quarterly in advance tables for *all* calculations, especially as this approach is certainly more likely to be acceptable to other non-valuer professionals such as accountants and management consultants. Should this approach be adopted, it is important to remember to use quarterly in advance formulae to analyse all comparables as well as to value the subject property. It is the approach adopted by the influential *Investment Property Databank* when compiling its statistics on market activity.

Having decided to use quarterly in advance calculations, the choice remains of which mathematical basis to use. In *Parry's Tables*, the assumption is that receipts are quarterly in advance with interest added, or reinvested, annually in arrear. This contrasts with *Bowcock's Tables* which assumed quarterly in advance receipts with interest converted half-yearly and *Rose's Tables* that took quarterly in advance receipts and quarterly in advance interest payments. These alternatives to *Parry's* are now dated, but still offer informative views on the basis of valuation calculations and accordingly are worth considering in more detail.

Jack Rose and Philip Bowcock were both instrumental in introducing alternative formulae and tables to the profession and encouraging their use in practice. Each published his own set of tables, which supplemented rather than replaced *Parry's* (Rose 1976 and Bowcock 1978). They also jointly published a succinct booklet that consolidated their earlier work and introduced a number of new valuation formulae (Bowcock and Rose 1979). The authors rightly argued that armed with these formulae and a pocket calculator, the valuer had limitless flexibility in analysing transactions and valuing properties.

A third man who influenced modern thought on valuation practice is Phillip Marshall. His article (Marshall 1976) explained how and why greater use should be made of DCF (discounted cash flow) analysis in property valuation. In connection with, and resulting from this article, he published a set of tables entitled *Equated Yield (Years Purchase – EY) Tables*, which were later republished in conjunction with the surveying firm of Donaldsons, and thereafter became known as *Donaldsons' Tables* (Marshall 1988). Because of this work, greater attention began to be given to the interconnected topics of equated yields, rental growth and DCF analysis in the property world.

The change in attitudes within the profession, mainly initiated by the work of these men, is reflected in the fact that in the current edition of *Parry's Tables* (Davidson 2002), quarterly in advance and equated yield tables are included as well as the traditional annually in arrear tables.

7.2   Annually in advance calculations

The traditional years purchase formulae and tables all assume that £1 is receivable at the end of every year over a set period. The traditional formulae and tables can be used to calculate annually in advance values simply by taking the years purchase factor for one year less than the actual time period involved and adding 1. Thus the annually in advance formulae will be:

The Amount of £1 $= 1 + (1+i)^{n-1}$

Years purchase in perpetuity $= 1 + 1/i$

Years purchase single rate (present Value of £1 per annum for a number of years)

$= 1 + ((1+i)^{n-1} -1)/(i(1+i)^{n-1})$

Years purchase of a reversion to perpetuity $= 1/(i(1+i)^{n-1})$

all above where:
$n =$ number of years
$i =$ annual interest rate, or yield, expressed as a decimal (that is rate/100)

7.3   Income in advance or in arrear and not receivable annually

Bowcock and Rose (1979: 7) provide conversion factors to be applied to 'conventional' years purchases when income is receivable at intervals other than annually in arrear and where income is either *in advance*

$$\dfrac{R}{m\left[1-\left(\dfrac{1}{(1+R)^{\frac{1}{m}}}\right)\right]}$$

or *in arrear*

$$\dfrac{R}{m\left[(1+R)^{\frac{1}{m}} -1\right]}$$

where:
$R =$ all risks yield as a decimal (ARY/100)
$m =$ time period income is received for example, half-yearly $= 2$ and quarterly $= 4$

The following 'conventional' annually in arrear years purchases

- years purchase for a number of years single rate; or
- years purchase of a reversion to perpetuity; or
- years purchase in perpetuity

can be multiplied by the relevant conversion factor above to find the appropriate years purchase on the required timing basis.

Example

Years purchase figures at 9 per cent on an annually in arrear basis are:

YP 5 years single rate @ 9% = 3.88965
YP reversion to perpetuity after 5 years @ 9% = 7.22146
YP perp @ 9% = 11.1111

To convert to quarterly in advance basis, use the first factor shown in Section 7.3, where:

R = 0.09 (that is 9/100)
m = 4 (for quarters of a year)

$(1 + 0.09)^{¼} = 1.09^{0.25} = 1.021778181$
$1/1.021778181 = 0.978685999$
$1 - 0.978685999 = 0.021314$
$4 \times 0.021314 = 0.085256002$
$0.09/0.085256002 = 1.055644143$

Thus conversion factor to convert annually in arrear YP numbers is 1.055644143.

Then multiply each conventional YP figure by this factor:

YP 5 years single rate @ 9% = 3.88965 × 1.055644143 = 4.10608
YP reversion to perpetuity after 5 years @ 9%
= 7.22146 × 1.055644143 = 7.62329
YP perp @ 9% = 11.1111 × 1.055644143 = 11.72938

So on a quarterly in advance basis, the years purchases are:

YP 5 years single rate @ 9% = 4.10608
YP reversion to perpetuity after 5 years @ 9% = 7.62329
YP perp @ 9% = 11.72938

7.4   Quarterly in advance dual rate tax adjusted years purchases

The following formula (derived from Bowcock and Rose 1979: 7) will calculate, on the property market approach, the years purchase dual rate tax adjusted figure on a quarterly in advance basis.

$$\cfrac{1}{4\left[1-\cfrac{1}{(1+R)^{0.25}}\right]+\cfrac{4\left[1-\cfrac{1}{(1+X)^{0.25}}\right]}{(1+X)^n-1}\times\cfrac{100}{(100-t)}}$$

where:
R = all risks yield expressed as a decimal number (yield/100)
X = net annual sinking fund rate expressed as a decimal number (rate/100)
t = tax rate percentage as a whole number
n = number of years

Note that the 'property market basis' assumes that although rentals are paid quarterly in advance, the quoted rate of interest is effective over the period of one year with its reinvestment being at the end of each year.

Example

What is the years purchase for seven years at 10 per cent and 3 per cent with tax at 40 per cent on a quarterly in advance basis?
   Using the formula:

$(1+R)^{0.25} = 1.1^{0.25} = 1.024113689$
$4[1-1/1.024113689]$
$= 4[1-0.976454089] = 4 \times 0.02354591 = 0.094183641$

$(1+X)^{0.25} = 1.03^{0.25} = 1.007417072$
$4[1-1/1.007417072] = 4[1-0.992637536]$
$= 4 \times 0.007362463 = 0.029449855$

$(1+X)^n = (1+X)^7 = 1.03^7 = 1.229873865$
$0.029449855/(1.229873865-1) = 0.128113106$
$100/(100-t) = 100/(100-40) = 100/60 = 1.6666667$
$0.128113106 \times 1.6666667 = 0.213521848$
$0.094183641 + 0.213521848 = 0.307705489$
$1/0.307705489 = 3.24986077$

SAY YP for 7 yrs @ 10% & 3% (tax 40%) quarterly in advance = **3.2499**

7.5   Rose's Tables

Rose's years purchase tables (Rose 1976) were compiled on the same property market assumptions as *Parry's*, but to five decimal places instead of four. In addition, *Rose's* contained 'Effective Rate Tables of Compound Interest' which calculated years purchase on the normal annual effective yield basis (in other words that interest is converted annually), but on the assumption that interest is payable quarterly in advance instead of annually in arrear.

7.6   Bowcock's Tables

Philip Bowcock's *Tables* (Bowcock 1979) were devised to take account of the modifications necessary to reflect the differences between the 'property market assumption' and the 'stock market assumption' in calculating yields.

In the general financial securities market, most dividends are paid half-yearly, but the yield is quoted on an annual basis. Thus, two half-yearly yields of 3 per cent are worth more than an annual yield of 6 per cent, since through reinvestment, six months interest can be obtained on the first payment of 3 per cent. However, most stock market annual yields are quoted on the nominal basis obtained by doubling the half-yearly rate.

Problems can arise when an investor wishes to make a direct comparison between a property investment and a stock market investment. Property yields are generally understood to mean the effective annual rate of interest, whereas stock market yields are the nominal rates; the true effective rate being higher due to the compounding effect.

For example, a property yield of 8 per cent is an effective annual return of 8 per cent on the sum invested. A stock market yield of 8 per cent would probably indicate two half-yearly yields of 4 per cent each. This would give an effective annual return of 8.16 per cent on the sum invested. This sum is calculated as follows.

If the principal sum invested $= P$, then by the end of the first half-year, the sum invested will be:

$$P + (P \times 0.04) = 1.04P$$

At the end of the second half-year, the sum invested will be:

$$1.04P + (1.04P \times 0.04) = 1.0816P$$

Alternatively, the same figure can be found by use of the Amount of £1 formula where half-yearly interest is 4 per cent and thus over a year the principal plus interest will be:

$$(1 + 0.04)^2 = 1.042 = 1.0816$$

Therefore interest earned equals sum at end of year minus the principal invested at start of year. The interest expressed as a decimal is 0.0816. This is converted to a percentage by multiplying by 100:

$$0.0816 \times 100 = 8.16\%.$$

A years purchase table in *Bowcock's* headed 8 per cent would in fact be calculated at an effective annual rate of 8.16 per cent, as shown above, and so the figures are different from those in *Parry's* or *Rose's Tables*.

7.7   Donaldsons' Investment Tables

This publication contained two sets of tables, one for equated yields and the other for equivalent yields to be used with the discounted cash flow method of valuation appraisal.

The equated yield tables set out the equated yield appropriate to given all risks yields, at specified annual rental growth rates and allowing for rent reviews at stated time periods. In this latter respect, review periods of 3, 5, 7, 10, 14, 21, 25 and 33 year intervals were included in the tables.

The equivalent yield tables related to freehold reversionary investments. They provided a quick way of arriving at the single yield to be applied to both the term and reversion calculation to produce the same capital value as that obtained by applying a separate initial yield to the term, and another (higher) reversionary yield to the reversion part of the valuation. Initial yields from 0 to 10 per cent, reversionary yields up to 16 per cent and times to reversion of one to fifteen years were covered by the tables.

7.8   Practical effects of using alternative tables and formulae

It will be seen that using alternative mathematical bases will affect the years purchase multipliers. This in turn could produce a range of different values for the same investment if the same yield is used in each case.

Suppose an income from a freehold property of £100,000 per annum receivable for the next two years is to be valued on the assumption that a market yield of 8 per cent is to be adopted. Then depending on the manner in which the income is received and the assumptions on the reinvestment of the interest, the following calculations can be undertaken (adapted from Mackmin 1981):

Annually in arrear

Using *Parry's Tables*:

	£	
Net income =		100,000 p.a.
YP 2 yrs @ 8%		1.7833
	£	178,330

Annually in advance

Using *Parry's Tables*:

	£	
Net income =		100,000 p.a.
YP 1 yr @ 8% + 1		1.9259
	£	192,590

that is, one immediate payment (£1 receivable today) plus one payment received in one year's time.

Half-yearly in arrear

Using *Parry's Table*:

Net income =	£	50,000 per half year
YP 4 periods @ 4%		3.6299
	£	181,495

that is, half the annual interest rate received over four time periods. The YP figure is found by looking up YP for four years at 4 per cent instead of two years at 8 per cent.

Half-yearly in advance

Using *Parry's Tables*:

Net income =	£	50,000 per half year
YP 3 periods @ 4% (2.7751) + 1		3.7751
	£	188,755

that is, one immediate payment received plus three more taken at half the annual interest rate over three time periods. The YP figure is found by looking up YP for 3 years at 4 per cent.

Quarterly in arrear

Using *Parry's Table*:

Net income =	£	25,000 per quarter
YP 8 periods @ 2%		7.3255
	£	183,138

that is, take a quarter of the annual interest rate and look up the YP for four times the normal time period.

Quarterly in advance

Using *Parry's Table* annually in arrears section:

Net income =	£	25,000 per quarter
YP 7 periods @ 2% (6.472 + 1)		7.4720
	£	186,800

that is, one payment received immediately plus seven time periods at a quarter of the annual interest rate.

Or using *Parry's Tables* quarterly in advance section or *Rose's Tables* (both assume quarterly in advance receipts and interest converted annually):

Net income =	£	100,000 per annum
YP 2 yrs @ 8%		1.8716
	£	187,160

Or using *Bowcock's Tables* (quarterly in advance receipts with interest converted half-yearly):

Net income =	£	100,000 per annum
YP 2 yrs @ 8%		1.8738
	£	187,380

It will be seen that adopting a different approach in each case has produced a diverse range of end values. However, an investment is not worth more or less depending on the mathematics used. Surely there must be a 'correct' value? In a sense, this is shown by the price at which the investment is sold on the open market. This in theory should be its current market value at the time. It is by analysing such information that a more reliable valuation calculation can be undertaken. Doing so, and consistently using only one mathematical base, should remove the inconsistencies present above since the yield used will be market-derived and will differ depending on the valuation methodology employed. This is the basis of the comparison and investment methods of property valuation, which are covered in Chapters 10 and 11.

Progress check questions

- Why are 'conventional' annually in arrear valuation tables still in use when most property rents are not paid in this way?
- Do you understand the meaning and importance of the principle that you should 'analyse and value' using the same methodology and mathematical basis?
- Why are quarterly in advance calculations being used far more nowadays?
- Can you undertake calculations, using the appropriate formulae or tables, to value an income flow whatever its timing (quarterly, half-yearly, annually, in advance or in arrear)?

Chapter summary

Traditionally, valuation tables and formulae have been based on the premise that rental incomes are receivable annually in arrear; that is, once a year at the end of each year. Few, if any rentals are actually paid in this way, although the property market does assume that interest is converted, or reinvested, once a year at the end of each year. To reflect the actual timing of rental payments, however, there has been a steadily increasing call for valuation mathematics to be entirely based on a quarterly in advance basis. 'Alternative' tables and formulae have been produced to enable this to be undertaken. The yield derived from each method of analysis will vary due to the timing of the receipts and it is thus important that the valuer is consistent in applying a chosen methodology when analysing market transactions and undertaking valuations based on that analysis.

Further reading

Askham, P. (1992) 'Mainly for students: valuation tables', *Estates Gazette*, 9205 (8 Feb.): 174–6; repr. in P. Askham and L. Blake (eds), *The Best of Mainly for Students* (London: Estates Gazette, 1993), 364–72.

Creamer, M. (1999) 'Valuations: make them quarterly in advance', *CSM* (Nov./Dec.): 26.

Estates Gazette (2000) 'Mainly for students: times of change', 0027 (8 Jul.): 134–5, and 0028 (15 Jul.): 137.

Imber, A. (2000) 'Mainly for students: time to move forward', *Estates Gazette*, 0023 (10 Jun.): 136–8; repr. in L. Blake and A. Imber (eds), *The Best of Mainly for Students* (London: Estates Gazette, 2004), vol. 3, pp. 465–72.

Valuation methods

8

Methods of measurement

In this chapter ...

- The role of and main definitions in the RICS *Code of Measuring Practice*.
- The principal types of measurement used in property valuation.
- The level of accuracy expected in measuring for valuation.
- How to convert from or to imperial measurements from metric.

8.1 Overview of RICS *Code of Measuring Practice*

The Royal Institution of Chartered Surveyors has set a number of standards and guidelines on property valuation to ensure consistency and the application of 'best practice' within the profession. These are contained in two publications:

- The *Red Book* (RICS 2007d)
- *Code of Measuring Practice* (RICS 2007b)

The purpose of the RICS *Code of Measuring Practice* is:

> to provide succinct, precise definitions to permit the accurate measurement of buildings and land, the calculation of the sizes (areas and volumes) and the description or specification of land and buildings on a common and consistent basis ... The Code is intended for use in the UK only (RICS 2007b: 1).

> The Code deals only with standard measurement practice. Valuation techniques ... do not form part of the Code'. Also 'The Code is distinct from that relating to the Standard Method of Measurement of Building Works (SMM), which is commonly used in the construction industry (RICS 2007b: 2).

Table 8.1   Measurement conversion tables

Metric to Imperial		
1 millimetre (mm)		= 0.03937 inch
1 centimetre (cm)	= 10 mm	= 0.3937 inch
1 metre (m)	= 1,000 mm or 100 cm	= 3.28084 feet
1 kilometre (km)	= 1,000 m	= 0.62139 mile
1 square metre (m² or sq.m)	= 10,000 cm²	= 10.76426 square feet
1 hectare (ha)	= 10,000 m²	= 2.47097 acres
1 cubic metre (m³)	= 1,000,000 cm³	= 35.31448 cubic feet
Imperial to Metric		
1 inch (in or ")		= 2.54 cm
1 foot (ft or ')	= 12 in	= 0.3048 m
1 mile	= 5,280 ft	= 1.6093 km
1 square foot (ft² or sq.ft)	= 144 in²	= 0.0929 m²
1 acre (ac)	= 43,560 ft²	= 0.40469 hectare
1 cubic foot (ft³)	= 1,728 in³	= 0.028317 m³

Apart from the different types of area that can be measured, there can also be differences in opinion on how a specific space measurement should be taken on site or how the *Code* should be interpreted and applied to the subject building. The *Code* helps reduce inconsistencies and disputes but cannot necessarily eliminate them altogether.

This was highlighted in the case of *Kilmartin SCI (Hulton House) Ltd* v. *Safeway Stores plc* [2006] EWHC 60 (Ch); [2006] 09 EG 184 where around £1 million in value depended upon the parties' interpretation of the *Code of Measuring Practice*. The contract in question was only valid if the net internal area (NIA) of the building was more than a specified minimum. 'The defendant disputed the inclusion of certain areas within the claimant's NIA figure and served a notice terminating the agreement' (Murdoch 2006). The key issue on which the case revolved was the definition in the *Code* of the term 'usable'. The judge stated that whether an area fell within this definition was 'often a matter of pure judgment'.

The case also highlighted the level of accuracy question. As Levy and Anderson (2006) state, 'the developer submitted that no measurements are ever 100% accurate and a margin of error of 1% should be permitted. The judge completely rejected this approach. A surveyor asked to measure the NIA must produce a figure, not a range.' These writers thus conclude that 'where measurements matter, lawyers should discuss with their client's surveyor whether any adjustments to the presumptions in the Code should be made. This should reduce the risk of future arguments as to what is or is not included in the measurement.'

Similarly, it is also essential to make it clear when quoting values per square foot or square metre on what space measurement basis the value has been or is to be calculated. Without agreeing the floor area involved and the basis of measurement, it is impossible to finalise the overall rental or capital value for a property.

Valuers must approach the practical task of measuring a building or plot of land in an organised and methodical manner to ensure the task is efficiently undertaken and to provide all necessary documentary evidence in the event of a dispute. Detailed notes of site inspections, with all measurements and dimensions clearly shown on plans or sketches, must be made and retained so that they can be referred back to. Valuers should double-check before leaving the property that they have all the information and measurements needed to enable the calculation of the required floor and site areas.

Taking the measurements on site can be a time-consuming task. Traditionally, measurement tapes and sticks have been used, but increasingly electronic distance measurement (EDM) equipment is now employed. Small, hand-held measurers, utilising infra-red or laser light beams, can enable very accurate measurements to be taken and calculated more swiftly than with the old approach.

There can be practical difficulties encountered on site in gaining access to all the parts of a building required to obtain the necessary measurements. Busy offices, full of workers, furniture and equipment or distribution warehouse buildings stacked with goods on racking systems from floor to ceiling and being transversed by fork-lift trucks can raise many health and safety and risk assessment issues.

Can scaled floor plans be used with a scale ruler to take off dimensions rather than physically measuring the actual building? Possibly, but in most cases this is unreliable. It is vital to be sure the plans are both accurately drawn and an actual representation of the building as it now stands. Architects' drawings, for instance, may show the intended building to be constructed, but the actual finished structure has some variations from the drawn version.

Having obtained all the necessary dimensions, the valuer still needs to calculate the areas and must be conversant with the basic mathematical techniques involved. Not all spaces will neatly divide into squares and rectangles and areas of other shapes, such as triangles, trapeziums, and rhombuses may need to be found. The overall area may need to be subdivided into many such individual shapes, the area for each found separately and then each added together to arrive at the sum total. With considerable scope for errors in such a process, it is essential to check and double-check all calculations before including them in a final report.

8.2   Gross external area (GEA)

This 'is the area of a building measured externally at each floor level' (RICS 2007b: 8). It is mainly used for the computation of plot ratio and other planning matters, and the estimation of building costs for residential buildings. Being an external measurement, it includes all external wall thicknesses and takes each floor into account. Thus it must be remembered that if the building is not single-storey, GEA is not the site area covered by the building. Also check whether each floor is the same shape and size of those above or below it.

8.3   Gross internal area (GIA)

This 'is the area of a building measured to the internal face of the perimeter walls at each floor level' (RICS 2007b: 12). GIA is used for non-residential building costs estimation purposes and for valuation of industrial and warehouse buildings (including ancillary offices), department and variety stores, food superstores, retail warehouses and new homes for development purposes. It is broadly the GEA with all perimeter and party wall thicknesses and external projections and finishes thereto excluded.

8.4   Net internal area (NIA)

This 'is the usable area within a building measured to the internal face of the perimeter walls at each floor level' (RICS 2007b: 16). Mainly recommended for valuation of offices or shops, it excludes 'non-usable' areas that would form part of the GIA. Examples of such exclusions are toilets, toilet lobbies, bathrooms, cleaners' cupboards, lift rooms, plant rooms, stairwells, lift wells, those parts used for the purpose of essential access and internal structural walls, columns and piers.

This is just a brief indication of the meaning of NIA. Remember, it is essential to refer to and apply the exact wording of the *Code of Measuring Practice* for this and all other definitions.

Progress check questions

- Do you understand the distinction between GEA, GIA and NIA methods of measuring a property for valuation purposes?
- Why does the Royal Institution of Chartered Surveyors consider it necessary to publish a *Code of Measuring Practice*?
- How do you decide which method of measurement to use on a specific building?
- For which category of property valuations is there no single accepted practice for measurement?
- Why is it important to keep full and detailed notes of all measurements taken of a building and the calculation of areas based upon them?

Chapter summary

The Royal Institution of Chartered Surveyors has a *Code of Measuring Practice* to provide a consistent basis for the measurement of land and buildings for valuation purposes. This provides detailed definitions and guidance on the types of measurement that can be used and how they are calculated. All valuers working in the UK should adhere to this Code to ensure standardised good practice and to reduce the possibility of disputes.

Although reasonably prescriptive, the *Code* cannot cover all circumstances that may apply to any building such that all disputes are eliminated. Valuers need to exercise judgement when applying the presumptions and requirements of the *Code*. Where measurements are vital to a legal agreement it is important that lawyers fully consider these presumptions and consider whether any adjustments to them are needed in the documentation to clarify what is and is not to be included in the measurement of the subject building.

Traditionally, buildings have been measured by use of tapes, but increasingly electronic means are now used. It is vital to approach the measurement of a property in a systematic and methodical manner to ensure accuracy and all necessary dimensions are taken. Full notes of the measurements and calculations of areas based upon them must be made and retained as evidence of how the figures have been derived. All valuations should indicate the size of the areas on which they are based and the definition of measurement used.

RICS valuation standards
The *Red Book*

In this chapter ...

- What the RICS '*Red Book*' is and why it is needed.
- The difference between valuation standards and guidance notes.
- The range of RICS Valuation Information Papers available.
- Why valuation reports are needed and what they should contain.
- What is expected in valuation reports being used for secured lending purposes.

9.1 *Valuation Standards and Guidance Notes*

The RICS *Valuation Standards and Guidance Notes* are known as the '*Red Book*' due to the colour of the cover used on the printed-paper 'hard copy'. The *Red Book* 'was first published in 1980, and has been updated many times since then' (RICS 2008d). The RICS Valuation Professional Group Board approves changes to the standards. It is mandatory for RICS members worldwide undertaking valuations.

The 'standards are divided into two main parts. The first contains rules and guidance applying to RICS members anywhere in the world and is consistent with the principal rules of International Valuation Standards (IVS). The second contains material that relates specifically to particular countries' (RICS 2008d).

The latest version, the sixth edition, is effective from 1 January 2008 and is the culmination of an evolution of content through the six editions. This content has been influenced by past reports commissioned by the RICS.

The Mallinson Report (Royal Institution of Chartered Surveyors 1994) was undertaken due to:

1. Public disquiet regarding several well-publicised instances of widely varying valuations of specific properties by different valuers. One of the most noted of these was the Queens Moat Houses (QMH) hotels group valuation. In the space of a year, the QMH hotels went from a balance sheet value of £2.03billion (1991 accounts) to £861million (1992 accounts). This huge difference in opinion cast a large doubt at the time over the methodology of such valuations. On further investigation, it was seen there were some good reasons for the discrepancy. The two valuations were not undertaken on identical assumptions concerning a management incentive scheme that theoretically protected the main group's income at a certain level, but in the end was deemed not to work. In addition, between 1990 and 1993, UK hotel values dropped 30–50 per cent. Compounded inflation over the same period was 19 per cent, so making the 'real' decrease in value much higher.
2. Continued criticism of valuation bases and practice from within and outside of the profession.
3. Losses incurred by mortgagees on failed developments and investments resulting from falling values during the 1990s recession. Some lenders started to look critically at valuation advice given at commencement of the loan during the boom period.

The Report contained many recommendations on valuation definitions, methodology and procedures, plus some changes in organisational structure, and most of these were incorporated into the fourth edition of the *Red Book* (Royal Institution of Chartered Surveyors 1997b).

In 2002 a working party chaired by Sir Bryan Carsberg was commissioned to investigate issues relating to valuations for the commercial property investment market in the UK (Royal Institution of Chartered Surveyors 2002b). It made 'eighteen recommendations to RICS on how the valuation process should be tightened up to minimise the risks of valuers' objectivity being compromised and ensure that public confidence in the system was maintained' (RICS 2006c). These recommendations were incorporated into the fifth edition of the *Red Book* (Royal Institution of Chartered Surveyors 2003c).

The *Red Book's* stated principal purpose 'is to ensure that valuations produced by members achieve high standards of integrity, clarity and objectivity, and are reported in accordance with recognised bases that are appropriate for the purpose' (Royal Institution of Chartered Surveyors 2007d: para.1.1, 1).

The Standards define:

- 'criteria used to establish whether members are appropriately qualified;
- the steps necessary to deal with any actual or perceived threat to their independence and objectivity;
- matters to be addressed when agreeing conditions of engagement;
- bases of valuation, assumptions and material considerations that must be taken into account when preparing a valuation;
- minimum reporting standards;
- matters that should be disclosed where valuations may be relied upon by third parties' (Royal Institution of Chartered Surveyors 2007d: para.1.2, 1).

The purpose of the rules and guidance contained in the *Red Book* 'is to ensure that clients receive objective advice, delivered in a professional manner that is consistent with internationally recognised standards. The standards set a framework for best practice in the execution and delivery of valuations for different purposes but do not instruct members how to value, nor do they discuss valuation methodology or techniques' (Royal Institution of Chartered Surveyors 2007d: para.1.3, 1). It has been translated into a number of languages including Chinese, Russian and Arabic and the latest version reflects the new RICS Rules of Conduct, published in 2007.

The contents are arranged into:

- *Practice Statements* (PS): 'National practice statements are denoted by the use of a PS reference number prefixed by letters identifying the Association (for example UKPS)' (Royal Institution of Chartered Surveyors 2007d: para.4.2, 3); and
- *Guidance Notes* (GN): These 'explain how the valuation standards should be applied to certain types of property, or in particular situations, by highlighting issues that are peculiar to the subject of the guidance notes and discussing these in the context of the valuation standards' (RICS 2007d: para.4.4, 3).

The practice statements 'are grouped in the following categories:

PS 1 – Compliance and ethical requirements;
PS 2 – Agreement of terms of engagement;
PS 3 – Bases of value;
PS 4 – Applications;
PS 5 – Investigations;
PS 6 – Valuation reports' (Royal Institution of Chartered Surveyors 2007d: para.4.1, 3).

For each valuation, the 'basis of value that is appropriate to be reported' must be determined. 'It will almost always be necessary to couple a basis of value with appropriate assumptions or special assumptions that describe the assumed status or condition of the asset at the date of valuation. For most valuation purposes it will be appropriate to use one of the bases recognised in the International Valuation Standards and identified in these standards coupled with any necessary assumptions or special assumptions' (Royal Institution of Chartered Surveyors 2007d: PS 3.1, 41).

The bases of value that are recognised in the standards are (Royal Institution of Chartered Surveyors 2007d: PS 3.1, 41):

- market value (see PS 3.2);
- market rent (see PS 3.3);
- worth (investment value) (see PS 3.4); and
- fair value (see PS 3.5).

Further information and explanations on each of these is given in the relevant Standard, but the basic definitions are as follows.

Market value (MV)

'The estimated amount for which a property should exchange on the date of valuation between a willing buyer and a willing seller in an arm's-length transaction after proper marketing wherein the parties had each acted knowledgeably, prudently and without compulsion' (IVSC 2007b, cited in RICS 2007d: PS 3.2, 42).

In relation to this definition, Robert Peto (Chairman of RICS Valuation Faculty) states that, 'Whilst transactional evidence underpins the valuation process, it is not always available or up to date. In such circumstances, and in particular in a fast changing market, the definition of market value places an obligation on the valuer to use his/her market knowledge and professional judgement, in order to make a clear assessment of where the market stands on the date of valuation' (Peto 2007a).

Market rent (MR)

'The estimated amount for which a property, or space within a property, should lease (let) on the date of valuation between a willing lessor and a willing lessee on appropriate lease terms in an arm's-length transaction after proper marketing wherein the parties had acted knowledgeably, prudently and without compulsion. Whenever market rent is provided the "appropriate lease terms" which it reflects should also be stated' (IVSC 2007b: GN 2, para.3.1.9.1, cited in RICS 2007d: PS 3.3, 46).

Worth (or investment value)

'The value of property to a particular owner, investor, or class of investors for identified investment or operational objectives' (IVSC 2007b, cited in RICS 2007d: PS 3.4, 47).

Fair value

'The amount for which an asset could be exchanged, or a liability settled, between knowledgeable, willing parties, in an arm's length transaction' (IVSC 2007b: IVA 2, para.3.2, cited in RICS 2007d: PS 3.5, 48). 'Fair value' is now recognised and defined as 'a price that is fair between two parties acting at arms length for the exchange of an asset' (Thorne 2007).

Fair value differs from market value in that:

- The price must be fair to both parties having regard to all the circumstances – there is no requirement for it to be a price obtainable in the wider market;
- There is no requirement to assume that the asset has been exposed to the whole market, it can reflect a private deal;
- The price should reflect any special advantages or disadvantages to either party – market value requires these to be ignored;
- Market value does not need to be fair to either buyer or seller.

Valuations for financial statements (for example, balance sheets or Stock Exchange flotation) 'prepared under International Financial Reporting Standards (IFRS) shall be in accordance with the IVSC International Valuation Application 1 (IVA 1)' (Royal Institution of Chartered Surveyors 2007d: PS 4.1, 50).

Apart from by purchasing the printed 'hard copy' version of the *Red Book*, its contents may be viewed by subscribers to the online isurv service or by RICS members at the Institution's website www.rics.org.uk. The full wording of the appropriate standard or guidance note should be closely studied and followed when undertaking any valuation.

The standards emphasise the need for the valuer to establish and understand the client's requirements and to confirm the service to be provided in writing. The purpose of the valuation must be given. In essence, valuers should write to their clients stating:

- what they will do;
- what they won't do;
- explain on what conditions they will undertake the above; and
- what their fee will be and what it includes or excludes.

The client is asked to confirm acceptance (in writing) of all these matters before the valuer proceeds further.

Other recommended practices to adopt when undertaking a valuation are:

- prepare in advance of an inspection by checking maps, old files and other information sources so that a knowledge of the location and property is obtained;
- keep full and detailed written notes of everything seen during an inspection of the property;
- make a sketch plan of the property annotated with dimensions and other observations;
- take photographs of the property and comparables;
- double-check before leaving the property that all the information needed had been obtained;
- ensure all the measurements have been taken that are necessary to calculate the required floor and site areas;
- keep full records of all comparables used, including their source, date and analysis;
- retain full notes on how the valuation calculations were undertaken and why;
- check and double-check all calculations before including them in the final report; and
- note all discussions, telephone calls, meetings held, etc. in connection with the case and ensure these are all filed in chronological order along with site notes and calculations.

9.2   Valuation Information Papers

The *Red Book* also has a related collection of Valuation Information Papers that discuss valuation methodology as it relates to specific property types and issues. As at December 2008, the following papers have been published:

No. 1: *Valuation of Owner-Occupied Property for Financial Statements* (RICS 2003f)
No. 2 (2nd edition): *The Capital and Rental Valuation of Restaurants, Bars, Public Houses and Nightclubs in England and Wales* (RICS 2006b)

No. 3: *The Capital and Rental Valuation of Petrol Filling Stations in England, Wales and Scotland* (RICS 2003a)

No. 4: *The Valuation of Surgery Premises Used for Medical or Health Services* (RICS 2003g)

No. 5: *Rural Property Valuation (includes Property Used for Primary Agricultural Production, Leisure/Amenity, Commercial and Dwellings categories)* (RICS 2003d)

No. 6: *The Capital and Rental Valuation of Hotels in the UK* (RICS 2004)

No. 7: *Leasehold Reform in England and Wales* (RICS 2005b)

No. 8: *The Analysis of Commercial Lease Transactions* (RICS 2006a)

No. 9: *Land and Buildings Apportionments for Lease Classification under International Financial Reporting Standards* (RICS 2006c)

No. 10: *The Depreciated Replacement Cost Method of Valuation for Financial Reporting* (RICS 2007e)

No. 11: *The Valuation and Appraisal of Private Care Home Properties in England, Wales and Scotland* (RICS 2007f)

No.12: *The Valuation of Development Land* (RICS 2007g)

Additionally, a consultation draft VIP on Reflecting Sustainability in Commercial Property Valuations has been prepared.

When undertaking a valuation to which any of the papers refer, it is essential to comply with all the recommendations and guidance provided by the paper.

9.3 Form and content of valuation reports

Most valuation work requires a formal written report to be submitted to the client as the end product of the valuer's investigations. Great care must be taken in the presentation and content of such reports, particularly to ensure the legal duty of care under professional negligence is fulfilled. The report must present all relevant information and reasoning in a clear, logical order, using readable and readily understood language. The opinion of the value must be clearly related to the facts and assumptions set out within the report.

Practice statement 6 and Appendix 6.1 of the *Red Book* (Royal Institution of Chartered Surveyors 2007d: 84–6 and 94–7) state what is the *minimum* content of valuation reports:

> The report must clearly and accurately set out the conclusions of the valuation in a manner that is not ambiguous, misleading, or create a false impression. It must also deal with all the matters agreed between the client and the member in the terms of engagement and include the following minimum information, except where the report is to be provided on a form supplied by the client:

> a.   Identification of the client;
> b.   The purpose of the valuation;
> c.   The subject of the valuation;
> d.   The interest to be valued;
> e.   The type of property and how it is used, or classified, by the client;
> f.   The basis, or bases, of the valuation;

g. The date of valuation;
h. Disclosure of any material involvement or a statement that there has not been any previous material involvement;
i. If required, a statement of the status of the valuer;
j. Where appropriate, the currency that has been adopted;
k. Any assumptions, special assumptions, reservations, any special instructions or departures;
l. The extent of the member's investigations;
m. The nature and source of information relied on by the member;
n. Any consent to, or restrictions on, publication;
o. Any limits or exclusion of liability to parties other than the client;
p. Confirmation that the valuation accords with these Standards;
q. A statement of the valuation approach;
r. The opinions of value in figures and words;
s. Signature and date of the report.

Appendix 6.1 provides additional comment and guidance on the text content to be included under each of the above headings (Royal Institution of Chartered Surveyors 2007d: 94–7). The report is conventionally divided into numbered subject headings with sub-numbered subheadings dealing with aspects of that same topic. Paragraphs should be numbered consecutively through to the end and pages also numbered. All this aids cross-referencing. Schedules should normally not be in the body of the report but appended and clearly identified by a letter or number.

In addition to the minimum information requirements, some or all of the following issues are usually addressed or incorporated into a report either under the *Red Book* paragraph headings, as extra topics or in appendices:

- confirmation of original instructions or terms of engagement;
- site plan;
- description of the situation and location of the property;
- description of subject property;
- details of construction;
- services;
- accommodation measurements/size, design, layout and description;
- rating assessments;
- planning consents and policies;
- tenancy details;
- details of any leases;
- explanation of any defects and contamination;
- overview of property market conditions;
- market evidence and comparables;
- valuation methodology, calculations and reasoning;
- conclusion including valuer's recommendations.

The survey by Matysiak *et al.* (1996) identified other information less commonly included in valuation reports related to the wider investment market and the state of the economy.

Who may carry out the valuation? The valuer

> must have sufficient current local, national and international (as appropriate) knowledge of the particular market, and the skills and understanding necessary to undertake the valuation competently … If the valuer does not have the required level of expertise to deal with some aspect of the commission properly then he or she should decide what assistance is needed, assembling and interpreting relevant information from other professionals, such as specialist valuers, environmental surveyors, accountants and lawyers.
>
> (RICS 2007d: PS 1.5, 15)

Regarding the actual inspection of the property, in general it must always be carried out to the extent necessary to produce a 'professionally adequate' valuation (RICS 2007d: PS5.1, 81). Having agreed the instructions and carried out the inspection, the valuer prepares the written report to be submitted to the client. If a valuation is challenged, what the valuer has to show under the precedent set in *Mount Banking Corporation Ltd* v. *Brian Cooper & Co* [1992] 2 EGLR 142 is that the process followed was in accordance with accepted professional practice. Meticulously following all the requirements of the *Red Book* will be good *prima facie* evidence of this.

9.4 Loan valuations

Real estate is often the form of security used for secured lending. Before advancing the loan funds, the lender will need to be assured the value of the property concerned offers sufficient security for the proposed loan and will instruct a valuer to report on the property accordingly. 'Valuations for secured lending shall normally be on the basis of Market value, except when otherwise governed by law or statute' (Royal Institution of Chartered Surveyors 2007d: PS 4.2, 51). In addition, 'Unless valuing an interest in property as part of an operational entity it is usual to exclude trade fixtures, machinery, furnishings and other equipment from the valuation, although this should also be clarified with the client' (Royal Institution of Chartered Surveyors 2007d: Appendix 2.1(c), 30).

'Before accepting instructions, the valuer must disclose to the lender any anticipated, current or recent fee-earning involvement with the property to be valued, with the borrower or prospective borrower, or with any other party connected with a transaction for which the lending is required' (RICS 2007d: UK Appendix 3.1 (2.1), 198).

Especially with residential mortgage valuations, the lender often requires a standardised pro-forma of their own design to be completed rather than a full report. This comprises of a number of boxes to be ticked or completed with additional information appended. Where a full report is required, the RICS have a model report form that should be adopted (Royal Institution of Chartered Surveyors 2003e).

Progress check questions

- What is the *Red Book*?
- Why is the *Red Book* needed?
- What are the differences between Valuation Standards (VS), Guidance Notes (GN) and Appendices in the *Red Book*?
- Distinguish between *worth* and *market value.*
- What kind of special category valuation issues do the Valuation Information Papers cover?
- Where can you find guidance on the content of a valuation report?
- Why must valuation instructions be confirmed in writing?
- Why is it important to clearly state and explain all assumptions to a valuation?

Chapter summary

The RICS have published appraisal and valuation standards since 1980 to ensure that valuations produced by its members achieve high standards of integrity, clarity and objectivity. This publication is colloquially referred to as the *Red Book* and provides a measure of consistency across the profession. It helps to act as a gauge of whether a member has acted with reasonable skill and care and is mandatory worldwide for all members undertaking valuations. It has been accepted internationally and translated into many other languages.

Included in its contents, the *Red Book* provides definitions of the bases of valuation. These are the principal types of value a client may wish to have appraised. There are also a growing number of Valuation Information Papers within the *Red Book* that give additional detailed guidance and comment on specific categories of property or methods of valuation.

The *Red Book* lists the minimum information to be included in a valuation report and emphasises the need whenever a valuation is undertaken to understand the client's requirements and confirm them in writing.

Further reading

Askham, P. (1989) 'Mainly for students: report writing', *Estates Gazette*, 8908 (4 Mar.): 87–8; repr. in Askham and L. Blake (eds), *The Best of Mainly for Students* (London: Estates Gazette, 1993), 154–60.

Brett, M. (2002) *Valuation Standards for the Global Market*, London: RICS.

—— (2004) *Property under IFRS: A Guide to the Effects of the New International Financial Reporting Standards*, London: RICS.

Cherry, A. (2009) *A Valuer's Guide to the RICS Red Book 2009*, Coventry: RICS Books.

Estates Gazette (2004) 'Mainly for students: the value of standards', 0430 (24 Jul.): 104–5.

French, N. (2005) 'The little Red Book', *Estates Gazette*, 0541 (15 Oct.): 200–1.

Royal Institution of Chartered Surveyors Valuation Faculty (2002) *Response to Carsberg Report*, London: RICS.

10

Comparison method

In this chapter ...

- What the comparison method is and why it is the simplest and most reliable conventional approach to valuing property.
- How properties are analysed into value per common unit and what those units are.
- The important considerations when inspecting and valuing any type of property.
- The special characteristics and methods of valuing specific property types, such as the unique approach often adopted to the measurement and analysis of floor areas in retail units.
- The type of market evidence that is required to undertake valuations and where and how it may be found.
- How to adjust values to take account of each property's size and quality.
- How to estimate the market rent for a property or the ground rent for development land.
- The effects of inflation on rental values and how property owners can try and mitigate its adverse effects on the value of their investments.
- How to decapitalise a sum of money to express what it would be worth as a per annum amount rather than a single payment.

10.1 Basis of method

The methods of property valuation used in UK professional practice are:

- comparison
- investment

- profits or receipts and expenditure
- residual
- cost, and
- discounted cash flow (DCF).

The first five are considered 'traditional' or 'conventional' methods, having had a long-established history of use. Discounted cash flow is viewed as a 'modern' technique and is an alternative approach to the investment method. It has increasingly been used from the 1990s onwards to supplement 'traditional' methods. It can help to show potential problems or areas for further investigation or a more complete investment analysis within a conventional calculation.

The comparison method is based on comparisons derived from current market evidence to find rental or capital value directly. The investment method uses some comparison evidence and years purchase formulae or tables to value income flows and find capital value. The 'profits' method can be used as well as the previous two methods or where they cannot be used. It is primarily utilised when property is used for a business purpose, and analysis of the receipts and expenditure accounts can be undertaken to find rental value. Where required, this can then be capitalised using investment method techniques. The residual method is used to analyse development or redevelopment proposals to discover whether they are viable and the price that can be paid to acquire the site or existing premises. The cost method is often considered the 'method of last resort' and used where none of the other approaches can be utilised or to supplement a valuation found from one of the other methods. Value is based around costs of construction of the building on the site plus the land value.

The comparison approach is the simplest and most reliable method of valuation. It is used whenever possible in preference to other 'traditional' methods or used in conjunction with the investment method when final value cannot be found just from comparison. It requires comparison to be made between the property to be valued (the 'subject' property) and other similar properties that have been the object of recent transactions or are currently on the market ('comparables').

Many people in everyday life employ the basic premise of the comparable method. When looking to buy or sell an item they will make comparisons between the item in question and others for sale or that have been sold recently on the 'market'. For instance, a person wishes to sell their car but does not know at what price they should offer it for sale or at which they are likely to have to sell it. They will probably look around to see if there is any 'market evidence' of what may be an appropriate figure. They may look at advertisements in their local newspapers, on the internet or in specialist magazines to see what prices are being asked for similar models of vehicle. They may look at the price guides offered in the monthly automotive journals or in the UK trade *Glass's Guide*. They will compare the engine size, mileage, condition, accessories and features of the model they are selling with the others that are for sale or have been sold. In this way they will adjust what seems a reasonable price for their car up or down depending on its plus and minus points and arrive at a figure which seems a reasonably expected sale price given the current market conditions.

This whole process bears a good deal of similarity to what is required when using the comparable method of property valuation. The property to be valued is compared to others on the market now or that have recently been sold or let on the market. Adjustments are then made to allow for the advantages and disadvantages of the subject property in relation to each comparable to arrive at a figure that can be considered the current market value of the subject.

The three main requirements of property comparables are:

1. similar property type to the subject; and
2. similar location to the subject; and
3. evidence obtained is recent and reflects current market conditions.

How recent will depend on the state of the market. In a bullish market where values are increasing rapidly, comparable evidence will need to be obtained from the past few weeks. In a sluggish or static market, evidence within the last twelve months could still be relevant. It is thus difficult to be too precise over the question of what constitutes 'recent', but as a general rule evidence taken from up to six months prior to the valuation date will normally be sought and considered.

The major problem that faces valuers in the accumulation of comparable evidence is how to obtain all the necessary information. It is essential to obtain as much relevant data as possible on each comparable. The market is not completely 'transparent'. Not all deals are fully reported in the press and some are kept private and confidential. Ascertaining the exact and complete details of a property and the transaction concerning it that has taken place can require extensive research and analysis. The results of this work should be carefully and accurately recorded before the valuation of the subject property is carried out and all such records kept and filed for future reference.

Every item of information on each comparable transaction should be recorded and indexed for ease of reference, and the records constantly updated to ensure the information remains current and relevant. It is always better to have too much than too little information, both in terms of the number of comparables and the details on each comparable, when carrying out a valuation. The valuer also needs to be satisfied that all the available relevant comparable evidence has been found. The information that is deemed to be most relevant and useful can then be extracted from this volume of material. It may well be that the final valuation is principally based on only a few comparables, but it is better to select a small number as being the best evidence from a large selection than not having many items to select from. However, it also depends on current market conditions. Sometimes, when the market is quiet there are very few comparables and every piece of evidence is useful. In all circumstances, it is important that the valuer is confident no relevant material has been missed.

When obtaining comparables, ensure the exact source of the information is noted and referenced, so that further details, clarification or confirmation can be obtained if necessary at a later date. Also note whether the figures mentioned are actually completed transactions, agreed and binding but not yet completed deals, or are merely offers or asking figures. The latter hold little weight in terms of admissible evidence, but may indicate general market trends. Only those figures that are definitely agreed can be considered positive evidence and can thus be relied upon in negotiations or reports.

Having accumulated all the details of each comparable it is then advisable to enter the information onto a chart, schedule, spreadsheet or database, which summarises the main points and makes later reference to the information quicker and easier to comprehend. Suggestions for the basic headings under which this information could be entered for each property type are given later in this chapter under the respective property type. Not all these headings will always be relevant, and there may be additional headings required in specific cases that are not listed in the suggestions. A very full set of categories under which to record comparable information for three types of properties is provided on the RICS isurv online service:

Confirmation of rental evidence – commercial and industrial (RICS 2008a)
Confirmation of rental evidence – offices (RICS 2008b)
Confirmation of rental evidence – shops (RICS 2008c).

Each property is unique, and allowances must be made for the differences between the property being valued and the ones used for comparison, to take into account the various advantages and disadvantages of each. The valuer must use his or her knowledge and experience of the market and the properties involved in deciding what allowances and adjustments must be made in the analysis of the information to reflect these relative plus and minus factors.

All of these adjustments should be undertaken objectively and directly related to evidence whenever possible. However, some aspects of subjectivity can occur in this process of adjustment, depending on the viewpoint and opinion of the valuer concerned. For instance, the valuer acting for the landlord is perhaps more likely to try and place emphasis on those comparables which indicate higher rental figures, whereas the tenant's valuer is more likely to emphasise information which produces a lower rental sum.

Neither approach is, in itself, necessarily incorrect or inaccurate. As has been previously explained, property valuation is an art, not an exact science, and providing each valuer is correctly exercising his/her duty of care to their respective clients, then each of their opinions is valid. The final rental agreed between the parties will usually be determined by negotiation or, in the last resort, by a third-party surveyor acting as an arbitrator or independent expert who will view the same evidence in an entirely impartial, objective manner. In either event, the finally agreed rent is likely to represent a compromise between the initial viewpoints of the two parties' valuers, which will itself form new comparable evidence for subsequent settlements.

The comparison method can be used to calculate freehold capital values, market rent values and freehold all risks yields. Apart from a direct comparison to obtain any one of these three, it is also possible to derive any one of them providing two of the other factors can be found; as follows:

$$MV = MR \times 100/ARY$$
$$MV = MR \times YP \text{ in perpetuity @ } ARY$$
$$ARY = MR \times 100/MV$$
$$MR = ARY \times MV/100$$
$$MR = MV/YP \text{ perp @ } ARY$$
$$YP \text{ perp} = 100/ARY$$
$$YP \text{ perp} = MV/MR$$

where:
ARY =   freehold all risks yield percent
MR =   market rent
MV =   freehold market value with vacant possession or let at current market rent
YP   =   Years Purchase

10.2   Units of comparison and application of the method to the various property types

To enable a comparison to be made between properties of dissimilar sizes, it is necessary to analyse value in terms of a specific unit of area measurement, such as per square metre. The usual units of measurement used for this purpose for each property type in the UK are taken from the *Code of Measuring Practice* (RICS 2007b) and summarised in Section 8.1 above. However, the valuer should be aware that the units selected can sometimes vary from one part of the country to another, between countries and sometimes even between one valuer and another. For consistency where this occurs, select the definition and unit of measurement that most precisely reflects the true market position for that property type in that location, even if this differs from the more commonly accepted method given in the *Code*.

Example

Value the market rent of an industrial unit with a gross internal floor area of 3,250 sq.m. A nearby and very similar comparable building, measuring 4,100 sq.m., has recently been let on the open market for £287,000 per annum.

Comparable = £287,000/4,100 = £70 psm (per square metre)

Thus estimated market rent value of subject property

= £70 × 3,250 sq.m.

= **£227,500 per annum**

Generally, properties with relatively small total areas will produce higher values per unit of area measurement than will substantially larger properties. Thus an allowance for 'quantum' frequently needs to be made to comparables to reflect these differences. In addition to size differentials, there may be other significant qualitative differences between the property to be valued and the comparables. Suitable adjustments in terms of deductions or additions should thus be made to the comparable evidence before applying the figure to the subject property. This is considered in detail in Section 10.4 later in this chapter.

Agricultural property

Valuation of agricultural property is a highly specialised activity. Seldom or never will valuers who normally work with urban property be expected to value farmland.

The rural estate market is highly localised. This can result in special purchasers buying the property for reasons of prestige, or for convenience; where for example the purchaser owns the adjoining land. Intense local rivalry and competition can also help to raise sale prices above the expected 'equilibrium' price. Supply of such land in the UK has decreased considerably over the years, as more land is developed.

The rents obtained from tenanted farmland are very 'secure' in that they are comparatively low (compared to other types of land use) and are usually for long tenancies where investment is also viewed in the long term. Farming requires a heavy investment in capital and time and, having made such an investment, a tenant farmer is unlikely to default on payment of the rent, if at all possible, and place at risk these investments.

Unit of comparison normally used

Capital and rental values are analysed into £ per hectare or acre. This overall figure is usually taken to include all buildings, it being accepted that small farms will tend to have more buildings per hectare or acre than large farms, and this will be reflected in a quantum allowance. However, if the farm possesses any particularly notable buildings it may be preferable to value these separately to the land.

Main factors that will determine or influence value

- Location: as with all properties, the most important factor; nearness to markets, towns and transport routes are the major considerations.
- The lie of the land: height, slope and aspect will affect the use of the land.
- Climate and rainfall: the prevailing conditions in the area will again affect the use of the land.
- Size of the farm:
    - for large-scale mechanisation to be effective and economic, a large farm is required;
    - small farms, however, will generally have higher average values per acre/hectare than large, both on the usual 'quantum' allowance basis and because there will generally be a higher density of buildings per acre/hectare.
- The land itself:
    - in the UK the soil quality is graded by the Department for Environment, Food, & Rural Affairs (DEFRA) from 1 (the best arable land) to 5 (the worst quality, no more than heath land);
    - detailed testing of soil samples would probably be advisable in addition to confirm exact quality.
- Type of farm:
    - will largely depend on some of the above factors, and in particular the soil quality;
    - the major categories are arable, dairy, mixed, market gardening, poultry, pig and sheep farming;
    - each will appeal to a different sector of the market and can be expected to possess its own range of land values in a given location, depending on the other factors.
- Water supply to the fields, for crops and livestock: sometimes fresh water supplies may be available from rivers or streams rather than through mains services.
- Roads and approaches: good hard surfaced roads, both leading from the main transport routes to the farm, and within the farm itself, are necessary to enable access to the farm and for the movement of machinery and produce/livestock around the farm in all weathers.
- Fences and gates:
    - both to keep livestock in and trespassers out;
    - adequate fencing can be expensive to replace and is generally a tenant's responsibility.

- Mains services:
  - few farms will have mains drainage or gas, in both cases due to the expense of laying pipes to a (relatively) remote spot, and also in the case of the former, because the gradient back to the nearest main sewer is unlikely to be sufficient;
  - however, mains water and electricity would normally be considered essential.
- The farm house and cottages:
  - each farm usually has one main house, which will be occupied by the owner or tenant farmer;
  - the size and standard of these buildings can vary enormously, from large period mansions to a cramped cottage;
  - the presence of the house is usually reflected in the overall value of the farm, but it may be worthwhile valuing the house and land separately where the building is particularly large or small or enjoys some additional value due to its age, history, etc.;
  - farm labourers are usually provided with living accommodation as a condition of their employment;
  - these 'tied cottages' are occupied on condition that the worker must vacate the property if he/she leaves the job
- The farm buildings:
  - the number, size, condition and standard of construction of the farm buildings, such as barns, cowsheds, etc. will be reflected in the value per hectare or acre;
  - as mentioned, smaller farms will tend to have higher average values per acre/hectare as they will have proportionately more buildings per acre/hectare than a large farm;
  - for example, a farm of 50 acres (approximately 20 hectares) may have 10 buildings (1 per 5 acres or 2 ha) whereas a farm of 500 acres (around 200 hectares) may have 50 buildings (1 per 10 acres or 4 ha);
  - although the larger farm has more buildings in total, it has less proportionately;
  - repairing and insuring liabilities: the usual statutory responsibilities of landlord and tenant are assumed unless the lease specifies to the contrary.

Special characteristics or considerations of this property type

In addition to the farmland itself, 'sporting rights' may form part of a farm estate. These are shooting, hunting and fishing rights that can be retained by the owner for his/her own exclusive use, or in many cases are let out on a licence or lease basis to individuals, clubs or societies, from anything from a day to a long-term period of some years. The fees that may be charged for the use of the rights will depend on the location and on the quality and quantity of the 'sport'.

Farmland can also be let out for clay pigeon shooting, which obviously incurs far fewer overheads for the farm owner and is an activity without the moral issues connected with the shooting of live targets. Fishing rights in rivers, streams, lakes and reservoirs can also provide additional income for the farm owner.

The yields on sporting rights are normally higher than on the farm itself due to the greater uncertainty of the income, caused by the specialised nature of the market to which they appeal and the possible destruction or disruption of the facility due to pollution, poaching, nearby development, etc.

When a farm contains commercial woodlands, where the timber is grown for sale, these must be valued separately. This value is a combination of the value of the land on which the timber is standing plus the value of the timber itself. Such a valuation is a highly specialised activity and the value will depend on the size, age and type of timber present.

Any sporting rights or commercial woodlands would normally be valued separately to the land, using the comparison method if possible. Specialised buildings on the farmland may have to be valued using the depreciated cost replacement method (see Chapter 15 for more information) rather than by comparison.

RICS guidance or other information on valuing agricultural property

See RICS 2003d and 1998.

Suggested minimum headings for comparables schedule

- Address
- Date of transaction
- Source of information
- Land area (hectares or acres)
- Type of farm
- Description of buildings
- Other information (sporting rights, woodlands, etc.)
- Rent per annum exclusive (agreed or asking)
- Freehold capital value (paid or asking)
- Freehold all risks yield
- Rent per annum exclusive per ha (or acre)
- Capital value per ha (or acre).

Retail property (including zoning/halving back method)

There are a number of different types of retail property:

- in town: prime main street, high street or pedestrianised location; shopping centre; secondary, tertiary or suburban street positions
- out of town: shopping centre; mixed-use retail parks
- department stores
- variety stores
- retail warehouses
- supermarkets
- superstores and hypermarkets.

The measurement and valuation methods used with each can differ. The two basic approaches to valuing retail are:

- zoning and halving back method; and/or
- overall method.

Most in town prime location 'high street' shops ('retail units') are valued using the 'zoning and halving back' method. The alternative 'overall' method can also be employed on main street properties and is used on other retail units, such as department stores, supermarkets and variety stores.

Shop 'zoning and halving back' method

This is based on dividing the ground floor sales area of the shop into 'zones'. Each square metre is allocated a different value depending on which 'zone' it is within. The most valuable space is nearest the shop's entrance frontage and the value per square metre decreases with distance back from the frontage by one half with each 'zone' back from the front area.

There are sound business reasons to support this principle. The glazed shop front provides space for the retailer to place a window display, which is viewable by potential customers from the street outside the shop. The larger the shop front, the more room there is for such a display, which in theory increases the chances of attracting customers into the shop. Once inside, customers are theoretically not likely to venture further into the shop if they do not see something of interest in the front portion of the sales area, and are proportionately even less likely to walk to the rear parts of the shop. According to this theory, goods are therefore increasingly less likely to be sold the further back from the shop front they are placed and therefore retailers will be prepared to pay less for floor space at the back of the shop than that at the front.

It will be appreciated from the above that the major determinants of the value of a retail property valued using this method are therefore the frontage and depth measurements as well as the total floor area.

Different versions of the method are used by various valuers and within different parts of the country, but the most common approach is:

1. Working from main frontage, divide *ground floor sales* NIA into 6 or 6.1 metre-deep 'zones'.
2. Label zones from front to back A, B, C, D (there will only be all these zones if depth of shop sales area is more than 18 metres).
3. If shop unit is more than 24 metres deep, label *all* remaining sales area as R (Remainder).
4. 1 square metre in zone A = 'x' value, then halve-back for each successive zone, that is $0.5x$ for 1 square metre in zone B, $0.25x$ for C, $0.125x$ for D and $0.0625x$ for R (these being the decimal equivalent of the old fractions where zone B value per sq.m. was 1/2 zone A, C was 1/4, D was 1/8 and R was 1/16).
5. Upper and lower floors and non-sales areas are all valued on an 'overall' basis and are *not* zoned but are allocated different values according to their use.
6. There are no standard accepted values for other areas. Sales floor space in the basement and upper floors can be related to the ground floor zone A value or to a proportion of the overall

value of the ground floor. Second floors can be at a lower rate than the first and higher floors even less than the second. Typical figures used are:

a. First floor sales = $0.125x$ per square metre (one eighth) but this can vary depending on unit type and common local practice, so that $0.167x$ (one sixth) may be used in some situations and $0.1x$ (one tenth) in others. In its 2005 revaluation of rateable values, the Valuation Office Agency (2007) adopted the multiplier of 0.1 (one tenth) for first floor or mezzanine floor sales, 0.0667 (one fifteenth) for second floor sales and 0.0333 (one thirtieth) for third floor sales areas.

b. Basement or lower ground sales = $0.1x$ per square metre .

c. Ground floor ancillary or storage areas = $0.025x$ (with first floor or basement storage areas at a lesser value than this). The values of floor space on all floors used for ancillary purposes are valued at a relatively low 'storage rate' of either a fixed £ rental figure per square metre based on similar storage areas or using the proportion of zone A approach indicated. In its Valuation Scale Reference: VSZONEDV1, the Valuation Office Agency (2007) adopted the following multipliers for its rating revaluation in 2005 for internal storage or ancillary office floorspace: ground or lower ground floor 0.1, 1st floor 0.05, 2nd floor 0.04, 3rd floor 0.02, mezzanine 0.05, basement 0.05.

7. Add up all adjusted areas.

8. Result is floor area *ITZA* (in terms of zone A) and this forms base unit for comparison.

Although measurements should be in metric, many in UK professional practice still use the old imperial system. With this, the standard accepted depth of 'zones' was 20 feet. The nearest metric equivalent of 20 feet is 6.1 metres; thus many valuers will use 6.1 metres deep zones rather than round the figure down to 6 metres zones. In central London and in Scotland it is common to use 30 feet zones, which equate to approximately 9.14 metres. Crosby and French (1996) examined the effects of this change from imperial to metric measurement and also the most common methodology used. Their research indicated that 'rounding to 6 metres and 9 metres will have a minor effect' and recommended 'that the RICS and other professional bodies and interested parties encourage their adoption' rather than use 6.1 and 9.14 metres. As well as the variations in zone depth they also found no universal agreement on number of zones. Based on the 6 or 6.1 metres or 20 feet zone depths, they concluded that a minority of valuers use A, B then remainder or A, B, C, D, E then remainder; some 35 per cent use zones A, B, C then remainder whilst the majority (nearly 60 per cent of respondents) favoured the A, B, C, D then remainder approach indicated above.

Chosen zone depths are not altered to 'fit' the shape of the shop. For instance, a shop with a depth of 14 metres and using the 6 metres deep zone approach, would have a zone A of 6 metres deep, a zone B of 6 metres deep and a zone C of just 2 metres deep. The area would not be divided into three equal depth zones.

Example

A shop has recently been let on the open market for £80,000 per annum. It is rectangular in shape and has a clear internal floor area. It has a net frontage of 5 metres and a net depth of 23 metres and is a single-storey structure.

Analyse this market evidence and use it to find the market rent value (MR) of the adjacent shop, which is similar in all respects except its size. It has a net frontage of 8 metres and a net depth of 18 metres.

Comparable property 'zones'

The division of the property into the following zones is illustrated in Figure 10.1. (Assume £x = rental value of 1 sq.m. in zone A.)

Zone A	= 5 × 6 = 30 sq.m. @ £x	= £	30x
Zone B	= 5 × 6 = 30 sq.m. @ £0.5x	= £	15x
Zone C	= 5 × 6 = 30 sq.m. @ £0.25x	= £	7.5x
Zone D	= 5 × 5 = 25 sq.m. @ £0.125x	= £	3.125x
		Total = £	£55.625x

Therefore floor area 'in terms of zone A' (ITZA) = 55.625 sq.m.

Market rent = £55.625x per annum
As actual market rental = £80,000 per annum, the rental value per sq.m. in zone A is:
£80,000 p.a./55.625 = £1,438.20 p.a.
SAY = **£1,440 per annum per sq.m. ITZA**

Subject property 'zones' (illustrated in Figure 10.2):

Zone A	= 8 × 6 = 48 sq.m. @ £1,440	= £	69,120
Zone B	= 8 × 6 = 48 sq.m. @ £720	= £	34,560
Zone C	= 8 × 6 = 48 sq.m. @ £360	= £	17,280
		Market rent = £	120,960
		SAY £	**121,000 p.a.**

Note that for simplicity with both properties it has been assumed NO deductions for 'non useable' floor area needed to be made from any of the 'zones'. Each zone must be measured on NIA (net internal floor area) basis; thus any staircases, lifts, escalators, WCs, etc. are excluded from the measurements as defined in the *Code of Measuring Practice* (RICS 2007b). Therefore, the square metres in each zone could be less than frontage × depth.

Whatever method is used to analyse the comparables must also be utilised in the valuation of the subject property otherwise errors and inconsistencies will occur.

When obtaining comparables from published sources often only *total* floor area is given. In professional practice you would need to confirm exact layout and proportions of property and

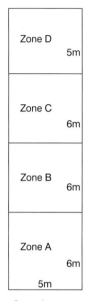

Figure 10.1   Example retail unit zoning valuation: comparable property

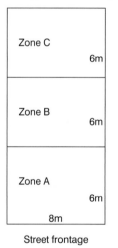

Figure 10.2   Example retail unit zoning valuation: subject property

over how many floors the area is distributed in order to calculate its ITZA. Alternatively, the measurements agreed for the property's rating assessment valuation can generally be viewed on the Valuation Office Agency website (http://www.voa.org.uk) which may give sufficient information to calculate an ITZA area. Note the methodology used by the VOA to calculate the ITZA is based on gross internal area, which may vary from that used by others. Before comparing total ITZA floor area or value per square metre/feet figures from different sources ensure they are on same basis of calculation otherwise a true comparison will not be made.

Disability deductions can be made for 'masked' areas in the property and for those properties of unusual proportions. Thus areas 'tucked away' behind a corner in a L-shaped retail unit would not be considered as good quality space as those parts not 'masked'. Accordingly, the less preferable floor area will be reduced in value. For example, if the area concerned fell within zone C of the shop then instead of taking a multiple of 0.25 times the zone A value per square metre, this may be reduced to 0.167 (one sixth) to provide a mid-value figure between the zone C and zone D multipliers.

There will also be a case for making a quantity allowance for the size of particularly large units and other *end adjustments* can be made to reflect advantages or disadvantages aside from 'quantum' (usually using percentage additions or deductions).

Extra or 'return' frontages can occur with shops on corners or with entrances from two directions (for instance off the main street and through a shopping centre). This can attract additional value, which could be found by a percentage addition to the total value. A typical percentage for this purpose would be 10 per cent, although higher and lower figures can be justified depending on the degree of advantage that the return frontage is deemed to provide the property. Alternative ways of calculating value would be:

- a per metre or foot run of return frontage 'spot' figure; or
- zoning from more than one frontage, using the value attributable to each street frontage and taking the highest rental per square metre for each overlapping zone area (although an 'end deduction' may then be necessary to prevent overvaluing); or
- zoning diagonally, taking an average of the two streets zone A rental or capital value per square metre.

Many smaller shops have residential upper parts that are included in the shop lease although they may be separately sublet. The value of this accommodation is usually assessed alone based on residential property comparable evidence. Alternatively, the value of this upper part could also be expressed in terms of zone A.

Overall valuation method

This can be on the basis of:

- an overall value for each floor, or
- a unit value overall for the whole shop.

It is a method that is commonly adopted for department stores, supermarkets and retail warehouses. The unit value adopted in either method will reflect all the advantages and disadvantages of the property. In the first method the ground floor is valued overall and all other floors valued individually overall related to the unit value used for the ground. In the second approach a single unit value for the whole floor area in the building is used. Again disability or end allowances can be made to the overall figure.

Unit of comparison

For retail property this is normally square metres (or square feet), expressed either in terms of zone A (ITZA) and measured from net internal area (NIA) or in total overall area calculated on NIA or GIA (gross internal area) basis.

Main factors that will determine or influence value

- Location:
    - where in the country;
    - where within a town or city;
    - accessibility and transport links;
    - size and socio-economic grouping of the catchment population.
- Position:
    - in prime shopping area?
    - within shopping centre or on main street?
    - in pedestrianised location?
    - on street corner?
    - near 'magnets' (major retailers)?
    - number and type of complementary other occupiers (cafés, bars, banks, leisure properties, hotels, offices, other retailers, etc.);
    - age, design and condition of adjacent and nearby properties;
    - surrounding infrastructure.
- User: type of shop use and whether limited by planning permission or within the lease user clause.
- The premises style of construction, number of floors and shape.
- Frontage and depth.
- Floor area.
- Floor layout: can affect display area and customer circulation.
- Type of shop front.
- Access for deliveries: front, rear or underground?
- Age and state of repair.
- Sanitary accommodation: adequate for the number of staff and customers using the premises?
- Upper floors and their use: retail, storage, ancillary offices or residential? Each will command a different value.
- Basement and its use.
- Forecourt:
    - is there one and can it be used to display goods or advertisements?
    - examples or uses where this may be relevant are newsagents and greengrocers.
- Structure and whether a 'shell' or fully fitted out: is it adaptable so that future alterations of layout or extensions can be easily undertaken?
- The tenant: size of company, its financial standing and stability plus its type of trade can all influence investment yield.

- Terms of the lease:
  - length;
  - rent reviews frequency and basis of valuation;
  - repairing and insuring responsibilities;
  - assignment and subletting covenants;
  - user clause;
  - are fixtures, fittings and/or improvements included in the rental valuation?
- Are any parts let separately?
  - to other retailers or non-retailers?
  - residential accommodation?
- Service charges: extent, how calculated and when charged?
- Statutes: does property comply with all current legislation?

Special characteristics or considerations of this property type

The major occupiers of prime retail locations, and the most influential, are the large retailing companies, commonly referred to as the 'multiples', due to their having many branches nationally. In a research project carried out by the University of Newcastle and Hillier Parker Research, and funded by the RICS and ESRC (*Chartered Surveyor Weekly* 1985) the definition of a multiple was taken as:

- a retailer with at least ten branches (three in the case of department stores); and
- the branches are located mainly in centres containing clusters of shops, usually in or near the highest rented areas; and
- the branches are nationally distributed.

Parsons (2004: 174) also defines a 'multiple retail outlet' as 'operated by an organisation with 10 or more branches'.

The University of Newcastle project further categorised the multiples according to type of trade as follows:

- stores:
  - department stores
  - variety stores
- apparel:
  - fashion
  - menswear
  - other clothing
  - footwear
- durables:
  - jewellery
  - furniture
  - other durables.

Traditional UK high street shopping historically developed on a linear basis, along the length of the main street. This eventually led to a restriction in the supply of retail floor space in a town or city, as there is only a limited amount of space along that street which can be utilised, irrespective of the amount of growth in the population and its expenditure capacity. Eventually the point is reached where the existing floor space cannot cope with the demand and the town becomes 'under shopped' in that there is spare demand that cannot be met. When this occurs, the solution often taken is to construct a new central development, as close to the prime pitch as possible, which it is hoped will help to regenerate the old 'high street' itself as well as provide additional retail space. Property values can alter as a result of such developments and realignments take place in where is considered the prime pitch.

Out of town and edge of town shopping developments can create difficulties in obtaining comparable evidence since they occupy a unique location and cannot be compared to the traditional in-town units. Shops in non-traditional retail locations, such as airports and railway stations, pose similar valuation problems. To overcome this problem, in many cases the leases of such properties provide for the rent to be assessed on a turnover basis, related to the business accounts. This has similarities to the profits method of valuation and is considered in more detail in Chapter 14.

RICS guidance or other information on valuing retail property

The isurv website has a section on *retail valuations*.

Suggested minimum headings for comparables schedule

- Date of transaction
- Address
- Source of information
- Total floor area and area per floor (NIA or GIA)
- Frontage
- Depth
- Floor area ITZA (in terms of zone A)
- Number of floors and use of each
- Description of building
- Lease terms
- Rent per annum exclusive (paid or asking)
- Freehold capital value (paid or asking)
- Freehold all risks yield
- Rent per annum exclusive per sq.m. (or sq.ft.) ITZA or overall
- Capital value per sq.m. (or sq.ft.) ITZA or overall.

Industrial property

Industrial property is generally understood to include factories, warehouses and distribution centres, workshop units and open storage areas ancillary to industrial processes. Parsons (2004: 139) states that an *industrial building* is

> a property normally defined by the Use Classes Order 1987 as being within Classes B1c, B2 or (possibly) B8. (It was formerly defined by the Town and Country Planning (Use Classes for Third Schedule Purposes) Order 1948 in terms subsequently adopted in the definition of 'industrial process' in the Use Classes Order 1987 (though the latter includes film, video and sound recording)).

He also adds that:

> Briefly, for the purpose of capital allowances for income taxation, 'industrial building or structure' is defined in Section 18 of the Capital Allowances Act 1990, as amended, to mean mills, buildings and structures used for a wide range of activities, including factories; transport, docks, inland navigation, water and electricity undertakings; for the purposes of a 'tunnel undertaking' or a 'bridge undertaking'; for the storage of goods or materials to be used for the manufacture of other goods or materials, finished goods or materials awaiting delivery; mines, oil wells, etc; or for certain agricultural or fishing purposes.

The relevant Town and Country Planning (Use Classes) Order 1987 definitions are:

> Class B1(c) – Any industrial process compatible with a residential area (light industry)
> Class B2 – General industry (not falling within Class B1 or Classes B3 to B7)
> Class B8 – 'Use for storage or as a distribution centre'

Classes B3 to B7 of the Order relate to 'Special' industry. These are the dangerous, obnoxious or 'bad neighbour' type uses and an extensive list of these is specified within the legislation. Many of these specialist types of industrial property are also included in the tax definitions. Often they would be valued using the cost method due to the lack of comparable evidence available on them. This approach is considered in Chapter 15.

Typical modern factory and warehouse 'shed' units are of single-storey construction with a two-storey office element at the front elevation. They are constructed around a portal steel or concrete frame with a low pitch double-skin metal roof. They have good natural lighting from roof lights and windows and a high-quality screeded concrete floor capable of supporting modern racking systems and forklift trucks. The minimum clear headroom under the structure (eaves height) sought is usually 6.1 metres (20 feet) and the electrically operated roller shutter doors are large enough to allow a lorry to back into the unit to load or unload. The buildings are usually on estates developed to a low density (of around 0.5:1 ratio between gross external floor area of all the buildings on the land compared to the site area). This provides good areas around the units for parking of lorries and cars, unloading and loading areas and landscaping.

This style of standard modern industrial buildings can be classified into 'mid-tech' or 'low-tech' categories. Low-tech have office contents comprising 10–20 per cent of the total floor area. The offices are finished to a basic standard with central heating. Mid-tech offers more offices, such as 40 per cent content and has an enhanced building appearance and image and more car parking.

'Workshop' or 'nursery' units are small, usually of no more than 250 square metres each, and are normally used for light industrial purposes. They are typically of similar basic construction to larger factory and warehouse units, but built in terraces, and with eaves heights of some 3.65 metres (12 feet). The office content is single storey only and accounts for a higher percentage of the total floor area than in a standard 'shed'. Possessing small floor areas, they will normally command higher rental rates per square metre or square foot than comparable larger shed units, although in many cases this type of property is sold freehold to owner-occupier small firms.

'Hi-tech' and 'B1' units originated from the demand from the expanding electronics industries in the late 1970s and early 1980s for clean, efficient and easily maintained work units with high office content and which offered good flexibility of use to meet future changes in requirements. Besides production uses, these facilities are used for research and development, storage, design and administration functions. The buildings fall between a description of light industrial and offices, as in many cases the floor area used for each of these functions within the building is equally split between the two. They come within Class B1 of the Use Classes Order 1987.

Unlike during the Industrial Revolution of the nineteenth century, industry is no longer restricted to locating close to its sources of raw materials and power supplies.

Besides the fact that many of these sources are now exhausted, modern transportation and energy systems have reduced the costs of supply to the site of industry. Probably of more importance today is the proximity of desirable living areas to attract and retain highly skilled staff, together with being close to a motorway.

Design materials and styles nowadays display more variety and units can possess a greater individuality, even when built speculatively in grouped developments. The increased concern for the environment is reflected in the design of industrial buildings and in better landscaping. Up to 15 per cent of the site is now usually devoted to soft landscaping, including mature or semi-mature trees and shrubs.

Business park developments, which combine hi-tech industrial units with edge or out-of-town offices, also often provide on-site shopping, eating, hotel, leisure and recreational facilities as well as child care centres or crèches for the children of working parents. This mixed-use style of development aims to provide a 24-hour environment where people work in the day then go out there in the evening.

Unit of comparison

For industrial property the unit of comparison normally used is square metres or square feet of GIA or NIA. With 'low-tech' standard industrial units it is usual to measure the whole building including its office content on a GIA basis, rather than the NIA approach generally used for offices in other situations.

Main factors that will determine or influence value

- Locality:
  - general character of area, including surrounding uses;
  - communications – closeness to roads, motorways, railways, canals, docks, airports, etc.;
  - labour supply – quantity and quality (unskilled, semi-skilled, skilled or professional);
  - public transport – trains, buses, underground, light railways (for workers to reach workplace);
  - mains services – electricity, mains water and mains drainage are essential, gas is more unusual and somewhat of a luxury;
  - access roads – are they adequate in width and load-bearing capacity and devoid of sharp corners and low bridges or other obstructions for large lorries?
  - do overhead or underground cables, drains or pipes impinge upon the use of the property or do they in any way restrict possible future expansion of the buildings?
  - airports – do low-flying aircraft and the associated noise detract in any way from the use of the property?
- Construction:
  - age of building;
  - method of construction and design and flexibility of construction;
  - floor area and method of measurement;
  - eaves height and headroom (a minimum of 6 metres is usually considered desirable in modern buildings);
  - column spacing;
  - what extent of clear floor space is available?
  - access doors (type and width opening);
  - floor loading strength;
  - heating and insulation – type and adequacy;
  - general repair and condition of building;
  - is a sprinkler system fitted?
  - lighting (natural and artificial) – is it adequate?
  - sanitary accommodation – are sufficient washrooms and toilets (both male and female) present for the number of employees within the building?
  - tenant's improvements – have any been carried out; and if so do they add (or subtract) from value of building and can they be included in the valuation?
- General details:
  - overall layout of building, site coverage and site density;
  - loading/unloading space;
  - open storage space;
  - car parking spaces;
  - lorry parking spaces;
  - security;
  - land available for expansion?
  - landscaping;
  - size of development;

- tenant quality;
- availability of government grants or concessions;
- office content percentage and quality.
- Lease terms:
  - length of term – average nowadays is ten or fifteen years, but five is popular with older units;
  - any rent-free periods? (with new property there may be none or up to three months; but with second-hand, older property they could be up to six months or longer);
  - rent reviews – frequency, upwards only or upwards and downwards pattern and basis of valuation to be adopted (for example, standard market rent definition)?
  - any user restrictions?
  - any alienation restrictions?
  - repairing and insuring obligations – usually FRI.
- Statutory legislation:
  - planning permissions, constraints and zoning;
  - compliance with all relevant legislation, which in the UK will include Factory Acts; Offices, Shops and Railway Premises Act; Fire Precautions Act, Health and Safety at Work Act, plus environmental and pollution statutes;
  - valid Fire Certificate for building?

Special characteristics or considerations of this property type

The amount of office floor space in an industrial unit can have a significant effect on its value. To enable comparisons to be made between units it is usual to express the office content as a percentage of the total floor area. For instance, a unit has 800 square metres of production floor space and 200 square metres of offices. Thus, its office content is 200 out of 1,000 square metres, which is 20 per cent of the total. When comparing this with another property with a higher or lower percentage of office space, what adjustments should be made?

The valuation methods that can be used to allow for differences between the office contents of the comparables and the subject property are:

- the overall method; or
- the adjusted floor area method; or
- the excess content method.

The overall method: the valuer uses his or her skill, judgement and experience of the market to make a 'spot' adjustment of the value to take account of the office content percentage.

Example

Overall method

Value the market rent of a unit with total floor area of 2,000 sq.m. that includes 400 sq.m. of offices, i.e. 20 per cent content. There are two relevant comparables as follows:

Unit A has factory floor space of 1,500 sq.m. and offices of 500 sq.m. Office content therefore = 500/2000 × 100 = 25 per cent. It has let at £100,000 per annum (£50 psm overall).

Unit B also has a total floor area of 2,000 sq.m. of which the offices comprise 300 sq.m., i.e. 15 per cent content. It has been let at £92,000 per annum (£46 psm overall).

Using the overall method, the conclusion is:

if 25% content = £50 psm
and 15% content = £46 psm
then 20% content being mid-way between should = £48 psm
Therefore estimated market rent of subject property will be
£48 × 2,000 sq.m. = **£96,000 per annum**

The *adjusted floor area method*: the floor area of the unit is adjusted to allow for the size of the office content and this adjusted area is used to ascertain the rental (or capital) value per square metre or square foot. For this purpose it is usual to say that each square metre or square foot of office floor space is worth 1.5 or 2 times the industrial floor space in the building. The multiple chosen depends on the quality of the offices. Those designed and fitted out to a high standard, similar to purpose-built office properties, will be considered 2 times more valuable and lesser standard accommodation will be multiplied by 1.5 or possibly less

Example

Adjusted floor area method

Using the same subject property and comparables as given in the overall example above and assuming a high standard of office space, the floor areas can be adjusted as follows:

Unit A: (500 × 2) + 1500 = 2500x = MR (where x = value of 1 sq.m. of industrial floor space in the building).
 Thus x = £100,000/2500 = £40
Unit B: (300 × 2) + 1700 = 2300x = MR.
 Thus x = £92,000/2300 = £40

Subject unit: (400 × 2) + 1600 = 2400x = MR
Thus estimated market rent for the subject is £40 × 24,000 = **£96,000 per annum**

The *excess content method*: the unit with the lowest office content is used as a 'base' figure and the 'excess contents' only of the other units are multiplied as in the adjusted floor area method above.

Example

Excess content method

Again, using the previous figures from the examples given above for the overall and adjusted floor area approaches:

Unit B has lowest content at 15% and is valued overall at £46 psm.
Unit A has a 25% content. This is 10% higher than Unit B, so the 'excess content' of A compared to B is 10% of its total floor area.

The adjustment required in the analysis of unit A is as follows:

- Unit has total area of 2,000 sq.m.
- A 15% content would be 300 sq.m. offices.(2,000 × 0.15)
- Actual office space is 500 sq.m., therefore the 'excess' is 500 – 300 = 200 sq.m.
- Multiply this excess by adjustment factor (2) and add to remainder of floor area =
- (200 × 2) + 1800 = 2200x (where x = the value of 1 sq.m. of industrial floor space with a 15% office content).
- Thus x = total rent/adjusted area = £100,000/2,200 = £45.40

Subject property has a 20% content which is 5% in excess of Unit B
A 15% content of 2,000 sq.m. is 300 sq.m.
The 'excess' office space is 100 sq.m. (400 – 300 sq.m.)
Adjusted floor area = 1,900 + (100 × 2) = 2,100 sq.m.
Average adjusted rent from comparables = (£46 + £45.40)/2 = £45.70 per sq.m.
2,100 sq.m. @ £45.70 = £95,970
Therefore estimated market rent for subject property is
SAY = **£96,000 per annum**

RICS guidance or other information on valuing industrial property

Information on isurv website under the heading of *Industrial and Distribution Properties* contains information on valuation methods and practical procedures, including inspection checklists. Three other relevant sections on the site are: *Environmental Appraisal*, *Mineral and Waste Management* and *Plant and Machinery*.

Suggested minimum headings for comparables schedule

- Date of transaction
- Address
- Source of information
- Description of property
- Facilities
- Floor area (GIA or NIA)

- Rent per annum exclusive (paid or asking)
- Freehold capital value (paid or asking)
- Freehold all risks yield
- Rent per annum per sq.m. (or sq.ft.)
- Capital value per sq.m. (or sq.ft.).

Offices

Office properties come in many forms and styles. They may be purpose-built constructions or converted older buildings formerly used for residential or industrial purposes. In the UK many such conversion developments have involved former dockland buildings.

Modern high-rise designs contrast against historic low-rise office buildings in many city centres. Other office buildings are away from the city centre in less expensive secondary positions or in business parks and other locations near to the city or out of town. Occupiers range from huge international companies to small businesses. Some offices are owner-occupied but many are held on leases. Larger buildings can be occupied by one company or by several, each occupying particular floors or suites. This leaves many 'common parts', such as entrance halls, atrium or foyer, lifts, stairs and washrooms, that need to be maintained and cleaned by the landlord, for which a service charge is levied and allocated amongst the occupiers.

Sustainability and environmental awareness issues are growing in importance and are reflected in modern office design, which renders older non-refurbished properties less attractive.

With so many variations in offices, finding good comparable valuation evidence can prove difficult to ensure the comparisons are truly similar in all respects to the building being valued.

Unit of comparison

For offices this is normally square metres or square feet on NIA measurement basis.

Main factors that will determine or influence value

- Location:
    - central business district (CBD), secondary city centre, suburban, edge of town or out of town?
    - accessibility: closeness to main transport routes (both private and public) that is major roads, bus routes, train, tram, light railway and underground stations and availability of car parking space;
    - proximity to similar or complementary uses – other offices, shops, restaurants, public houses, banks, building societies, open spaces, etc.;
    - potential size of catchment area for labour supply;
    - proximity to major clients: with modern communication systems this may be less important than in the past, although personal meetings and discussions are still an essential part of business and proximity can facilitate this;
    - external environment: do the immediate surroundings cause any problems? (for example, it is an advantage, for an office building to be adjacent to a major railway station for

accessibility, but noise and dirt created by the trains must be taken into account in the design of the building if the interior working environment is to be maintained to a high standard).

- Number and types of tenants: a single tenant produces far less management problems, although a higher overall rental income may possibly be derived where the building is subdivided and let to several tenants, which may outweigh the increased costs of management.
- Terms of the tenancy:
  - where buildings are let to more than one tenant, all the leases should be individually checked to ensure the terms are consistent in each;
  - length of term;
  - rent review frequency and basis of valuation;
  - FRI, IRI or other repairing and insuring liability basis; and where building is in multiple occupation, who is responsible for the maintenance of common parts?
  - user clause;
  - ability to sublet or assign;
  - is the rent 'inclusive' or 'exclusive'? Normally exclusive, meaning tenant's rental payments do not include business rates, which are paid separately to the local authority, but occasionally are on inclusive terms and the rates must be deducted from the gross rent received.
- Outgoings:
  - repairs – internal, external and structural;
  - decorations – internal and external;
  - business rates;
  - buildings insurance premiums.
- Services:
  - where these are provided by the landlord, the costs are usually recovered from the tenants by means of a service charge.
  - this is rendered on a quarterly, half-yearly or annual basis, separate to the rental demands, but sometimes it is included in the rent.
  - typical services that may be provided are:
    - air conditioning
    - catering facilities or refreshment/drinks machines
    - cleaning
    - heating
    - lifts
    - maintenance of common parts
    - management charges
    - porterage
    - reception area staff and facilities
    - security
    - telephone exchange
    - washroom/toilet cleaning, towels, soap, etc.
- Facilities:
  - accoustic ceilings;

- air conditioning;;
- car parking;
- carpeting;
- catering facilities;
- central heating;
- double glazing;
- electrical, telephone and computer cable floor and wall trunking;
- lifts and stairs;
- main entrance and reception;
- partitionin;g
- porter/caretaker;
- refreshment/drinks machines;
- security system;
- telephone exchange lines;
- washrooms/toilets.
- Construction:
  - age;
  - floor space, layout and loading;
  - lighting – natural and artificial;
  - mains services;
  - materials and style;
  - number of floors;
  - purpose built or converted;
  - sanitary accommodation;
  - state of repair;
  - lower floors, particularly in office blocks without lifts, are often considered more prestigious, being more accessible, and therefore can command higher rentals than the upper floors in buildings let to different tenants (where one tenant occupies whole building it is more usual to value the total floor space overall at the same rate) – the exception to this rule would be in a high rise block which has lifts, where the top floor is frequently reserved for the boardroom and chairman's office and possibly also a 'penthouse' suite (in these cases, the highest floor may have a considerably higher value than the lowest floors).
- Statutory legislation: compliance with the various laws such as in the UK the Offices, Shops and Railway Premises Act, Health and Safety at Work Act, etc. Does the property hold a valid Fire Certificate?

Special characteristics or considerations of this property type

Most firms seek buildings with large rectangular areas of good clear floor space on each floor, with few columns and good, clear floor heights. Only the largest organisations are likely to require premises of more than 2,000 square metres in total. Good quality heating, ventilation and double-glazing systems are given high priority, as is adequate floor trunking and vertical ducts for computer and electronic communications cabling.

There has been a trend in the last thirty years for the amount of floor space provided per worker in the UK to increase. This 'natural expansion' is largely due to the general improvements in working conditions, both through legislation, and the perceptions, by both employers and employees, of what constitutes an acceptable working environment. Up to 20 square metres per employee is commonplace, although in high-cost city centre locations, this may sometimes be reduced to 10 square metres.

To obtain the accommodation that best suits their requirements, small companies, in particular, tend to move more often than the old standard UK 'institutional lease' term of twenty to twenty-five years. The average period of occupation by small companies is eleven years or less and the trend since the 1990s has been for tenants to seek shorter leases of up to ten to fifteen years, often with a break clause operable around mid-term.

Office users have always required a location that provides good accessibility and complementarity with nearby users, to enable their businesses to operate efficiently. Traditionally this resulted in offices being located in town and city centres. However, with increased traffic congestion, limited car parking facilities and an overloaded transport system, many companies have moved their operations to town and city fringe locations.

Due to the high land values in city centres, office developments there tend to be high-rise to maximise the use of the plot. Conversely, business park offices are low-rise (usually no more than two storeys high) and are located on the fringe of town or in suburban areas amongst landscaped grounds, in mixed-use developments that also have good surface parking provision.

RICS guidance or other information on valuing office property

There is a section under the valuation channel of the www.isurv.co.uk website entitled *office valuations*. This contains information on valuation methods and practical procedures, including inspection checklists.

Suggested minimum headings for comparables schedule

- Date of transaction
- Address
- Source of information
- Number of floors
- Description of building
- Facilities (for example, central heating, air conditioning, lifts,
- carpeting, car parking and double glazing)
- Floor areas (NIA total and per floor)
- Rent per annum exclusive (paid or asking)
- Freehold capital value (paid or asking)
- Freehold all risks yield
- Rent per annum exclusive per sq.m. (or sq.ft.)
- Capital value per sq.m. (or sq.ft.).

Example

Value the freehold capital value of an office building with a net internal floor area of 550 sq.m. and offered with vacant possession. Two very similar buildings have recently been sold for owner occupation nearby. One has a net floor area of 600 sq.m. and sold for £2.4 million. The other measures 500 sq.m. NIA and sold for £2.05 million.

Analysis of comparables

£2.4m/600 = £4,000 per sq.m.
£2.05m/500 = £4,100 per sq.m.

Valuation of subject property

Floor area is mid-way between that of the two comparables, so take average of the two comparable values per square metre, on assumption both comparables are in all other respects apart from size similar to the subject.

550 sq.m. @ SAY £4,050 psf = £2,227,500
SAY = **£2,230,000**

Residential property

The styles of residential property can be categorised as follows:

*   detached houses
*   semi-detached houses
*   end-of-terrace houses
*   terraced houses
*   town houses (modern three-storey dwellings with integral garage to ground floor)
*   maisonettes (units on more than one floor forming part of a larger building)
*   flats (units on one floor forming part of a larger building)
*   bungalows (single-storey detached or semi-detached dwellings).

Each type will tend to command its own market, so that buildings of identical size, location and accommodation will usually achieve different prices if they are of dissimilar styles. Thus two houses, identical in size and in all other respects, except that one is detached and the other mid-terraced, will usually have different values.

Tenure and ownership varies from owner-occupied (either outright or subject to a mortgage), rented from the public sector (such as local authorities and housing associations) or rented from the private sector (individuals or companies). This provides the need for two principal different types of capital valuations. One will be on a 'vacant possession' assumption to find the freehold or leasehold value for owner-occupation. The second will be to find the investment value where the owner does not occupy the property, which is subject to a tenancy and a tenant in occupation that provides a rental income for the owner. Not just market forces will determine the level of this rental, as in some cases UK legislation regulates the figures certain tenants are expected to pay. When valuing tenanted property, as well as considering rental income, the valuer must consider the likely burden of insurance premiums and future repairs, decoration and maintenance costs, as in the majority of tenancies the tenant has very limited responsibility for these 'outgoings'.

As was explained in Chapter 8, 'there is no single accepted practice for measurement of residential property for valuation purposes' (RICS 2007b: 32). Accordingly, there is no single technique of using the comparison method of valuation on this property category.

With capital valuations of owner-occupied residential properties, many experienced valuers who have detailed knowledge of an area and thus of all the comparables can usually value very accurately on the overall impression and 'feel' of a property after a thorough inspection, without specific use of a more mathematical method. They will be able to take into account the overall size, layout and composition of the accommodation, the age, state of repair, position, and so on of the property to be valued and compare it to other properties they know have sold recently in the same location.

However, if a more 'scientific' approach is desired, three main variations of units of comparison can be used in valuing residential properties. Method 1 is the simplest, but not necessarily the most accurate. Method 2 is more complicated but gives more accurate results. Method 3 is relatively quick and provides a more mathematical and reliable basis than the first two. The use of more than one method gives a more balanced view for the final valuation than the use of any one method alone will do.

Method 1: £ per habitable room

Calculate the capital value per habitable room by comparing with other similar properties in the same area that have recently been sold (or less reliably are currently 'under offer' or 'for sale'). A *habitable room* is 'a room used for dwelling purposes but which is not solely a kitchen, utility room, bathroom, cellar or sanitary accommodation' (The Building Regulations 2007). This definition is also taken to exclude landings, halls and stairs.

Exceptionally large or small rooms may be counted as 1.5 or 0.5 habitable rooms, as appropriate. Each habitable room is considered against the 'normal' size for that room in properties of that category. In other words, a third bedroom is usually smaller than the main one, but providing it is a 'normal' size for a third bedroom compared to other similar dwellings, it still counts as one habitable room.

Method 2: £ per square metre or square foot HRF (habitable room floor space)

The floor area of each of the habitable rooms is calculated and totalled and a capital value per square metre or square foot found from comparison with other properties is used to find total value.

A variation on this basic method can be used to take account of the effect of kitchens and utility rooms on the value. As it is generally accepted that such rooms do have some influence on the value, although not as much as the habitable rooms, *half* the floor area of each of these rooms can be included in the habitable room floor space.

Method 3: £ per square metre or square foot based on GEA (gross external floor area), GIA (gross internal floor area) or EFA (effective floor area)

A total overall size measurement is thus needed, with no need to distinguish between number and types of rooms. Limitations of using GEA or GIA is that these do not take into account the amount of floor space occupied by stairs, halls, landings, toilets, etc. which does not 'directly' contribute to available living space. However, both are quicker and easier to measure than EFA.

With each of the three suggested methods adjustments to the £ figure per room, per square metre or square foot will need to be made to allow for the qualitative differences between the subject property and each comparable in the usual way.

Unit of comparison normally used

There is no standard accepted unit. It can be square metre or square foot measured on GEA, GIA or EFA basis; alternatively, a habitable room or square metre or square foot of habitable room floor space.

Example

Value a flat, held on a long ground lease at nominal ground rent, with the following accommodation:

Bedroom 1	11′ × 10′	(3.353 × 3.048m)
Bedroom 2	9′ × 10′	(2.743 × 3.048m)
Living room	20′ × 15′	(6.096 × 4.572m)
Kitchen	12′ × 9′	(3.658 × 2.743m)
plus Bathroom/WC		

The gross internal area (GIA) measures 760 square feet (70.6 square metres).

A flat in similar condition in the same block has just sold for £175,000. It has the following rooms:

Bedroom 1	15′ × 11′	(4.572 × 3.353m)
Bedroom 2	10′ × 11′	(3.048 × 3.353m)
Living room	22′ × 14′	(6.706 × 4.267m)
Kitchen	12′ × 10′	(3.658 × 3.048m)
plus Bathroom/WC		

Its gross internal area (GIA) is 850 square feet (78. 97 square metres).

Analysis of comparable

Number of habitable rooms = 3
Therefore value per habitable room = £175,000/3 = £58,333 per habitable room (phr)
Habitable room floor space (hrf) = areas of:
Bedroom 1 + Bedroom 2 + Living room = 583 sq.ft.
Alternatively, Bedroom 1 + Bedroom 2 + Living room + half Kitchen = 643 sq.ft.
Therefore value per sq.ft. hrf = £175,000/583 = £300.17 per sq.ft. hrf
or = £175,000/643 = £272.16 per sq.ft. hrf
Value per square foot GIA = £175,000/850 = £205.88 per sq.ft.

Valuation of subject flat

Method 1: Rooms are generally smaller than the comparable. Overall area of the main rooms (bedroom 1, bedroom 2 and lounge) are 500 compared to 583 sq.ft. This would represent a 15% reduction (500/583 = 0.85 = 85% of larger property). However, accommodation offered is still broadly similar, although slightly smaller, so say reduce unit value by 10%
> = £58,333 × 90% = £52,500
> 3 habitable rooms @ £52,500 phr = £157,500

Method 2: Habitable room floor space = Bedroom 1 + Bedroom 2 + Lounge = 500 sq.ft.
> @ £300.17 per sq.ft. = £150,085
> Or including half kitchen = Bedroom 1 + Bedroom 2 + Lounge + half kitchen
> = 554 sq.ft. @ £272.16 per sq.ft. = £150,776

Method 3: GIA = 760 sq.ft. @ £205.88 per sq.ft. = £156,469

Conclusion

Value is SAY **£155,000** (see assumptions below).

The above valuation assumes the comparable is similar in all respects, apart from number and size of rooms, to the subject property, and the final valuation required is of a reasonable average figure that may be obtained in the open market. More than one direct comparable would be preferable. In practice, the seller is likely to use the higher rates per habitable room or sq.ft. to obtain an asking price, and the buyer will use the lower rates to arrive at an offer price.

The final valuation suggested assumes reasonably equal negotiating ability on both sides and thus the final agreed figure is likely to be a compromise between both parties' valuations. Remember there is a limited amount of acceptable 'margin of error' between two valuations and that ultimately the price of any property is only what an *actual* buyer is prepared and able to pay.

It will be seen that the use of all the suggested valuation methods leads to a more balanced conclusion, which better reflects both the number and size of rooms in each property than using just one method may have done.

Main factors that will determine or influence value

The general factors that may affect the value of residential property include:

- Size and number of rooms.
- Location; and in particular:
  - surrounding buildings and land uses;
  - shops;
  - schools;
  - open spaces and parks;
  - transport routes and types;
  - recreation facilities;
  - health care facilities.
- Layout of building and plot, including soil type, gradient and risk of flooding, subsidence or heave.
- Mains services – electricity, gas, water, main drainage and telephone lines.
- Heating – water and space.
- Age and condition of building and services.
- Method of construction.
- Planning consents, restrictions and policies.
- Garage – size and position (detached, attached or integral).
- Fencing responsibilities.
- Garden – size, aspect, trees and general features.
- Responsibility for repairs and insurance and any limitations or restrictions on insurable risks
- For tenanted properties, the exact terms of the lease (taking into account terms implied by statute) must be checked, but generally UK residential tenants are only responsible for minor internal repairs and internal decoration.
- For mostly flats and some other properties there may be service charges for services provided by the landlord such as:
  - caretaker;
  - cleaning of common parts;
  - heating and hot water;
  - lifts;
  - lighting of common parts;
  - upkeep of gardens.

Special characteristics or considerations of this property type

Most owner-occupied UK residential property is sold with vacant possession (empty) and for houses the freehold interest is acquired, but flats and maisonettes are usually sold on a (long) leasehold basis. These are ground leases for terms of typically 99 or 125 years and an annual ground rent is payable to the freeholder.

Tenanted property in the UK is subject to much statutory legislation that concerns the tenant's security of tenure, the rights and responsibilities of the landlord and tenant and in some cases the level of rent chargeable. Valuers must ensure they have a full knowledge and understanding of the implications of the lease or tenancy agreement before undertaking a valuation of such property.

RICS guidance or other information on valuing residential property

See *Specification for Residential Mortgage Valuation* (RICS 2003e). Also the isurv website valuation channel contains a section on *student accommodation*.

Suggested minimum headings for comparables schedule

- Date of transaction
- Address
- Source of information
- Type of property
- Number of floors
- Description of accommodation
- Facilities
- Number of habitable rooms
- Habitable room floor space (in sq.m. or sq.ft.)
- Total floor area (GEA, GIA or EFA)
- Freehold capital value (paid,offered or asking)
- Capital value per habitable room
- Capital value per sq.m. (or sq.ft.) habitable room floor space
- Capital value per sq.m. (or sq.ft.) GEA, GIA or EFA.

Development land and property

Development and redevelopment sites or properties are those whose value can be increased by capital expenditure and/or a change of use (which will usually require planning permission). The term development normally implies that a cleared site is available, whereas redevelopment implies that there are buildings or structures already on the land that are to be removed in whole or part and replaced or added to by new buildings.

Unit of comparison normally used

The unit of comparison is hectares (or acres) of GSA (gross site area) or SA (site area). Alternatively, per square metre (or square foot) of GEA or GIA of permitted development or, for residential property, per dwelling or per habitable room of permitted development.

   The basic means of comparison between development sites of £ per acre or hectare does not take into account the density of permitted development on the land. If the various sites being compared can all be developed at identical densities, then this simple method of comparison is adequate. However, if the permitted densities are different, the alternative units given above provide a more accurate basis of comparison of site values.

Main factors that will determine or influence value

There are four main factors that need to be considered in calculating the value of a development property.

1. Establish what is the best (most valuable) use for the site, subject to full planning consent being forthcoming.
2. Estimate the current open market value for the completed development for this use.
3. Estimate what time lapse will be involved to complete the development.
4. Estimate the total costs of the development.

By a process of deduction, the value of the site will therefore be the difference between item 2 above and item 4, allowing for the effects of item 3. If this calculation produces a positive value in excess of the current open market value of the land in its existing use, the site has a development value, and in principle is worth developing. It will be seen that any such value is a residue of the calculation, in that it is the surplus of value over costs. The method of calculating this residue is thus known as the residual method of valuation, and is the traditional approach to valuing development/redevelopment properties. The method is explored in detail in Chapter 13, but in addition to the purely financial considerations, the following main factors will also have an effect on the feasibility and profitability of a development scheme:

1. the local planning authority's requirements
2. supply and demand for the chosen land use, both now and projected at the completion date of the scheme
3. access to the site
4. availability of mains services
5. easements and restrictive covenants which affect the site
6. transportation links
7. soil conditions
8. drainage and water table level
9. topography of the site and surrounding area
10. size, shape and condition of any existing buildings on the site.

Special characteristics or considerations of this property type

Since the Town and Country Planning Act 1947 introduced the Development Plan system in England and Wales, zoning has been used by the local planning authorities as a means of indicating and regulating planning policies. The basis of zoning is that the favoured land use for each area of a borough or town is clearly indicated on published plans. This acts as a statement of the intentions of the local planning authority and gives an indication of the selected character for the area.

The local planning authority must decide which types of land use are appropriate to each area of land within their area. This can depend on a large number of factors, which would include topography, quality of the landscape, transportation links, historical usage of the land, surrounding uses and plans for the future physical, social and economic development and growth of the town. These policies, once formulated, are published in written and diagrammatic form and are kept under regular review so that they may be adjusted or changed to reflect the changing needs and requirements of the local population.

As part of this process of formulating policies, each area of land within the jurisdiction of the local planning authority will be zoned for a specific land use. These uses will be shown on the development plans by specific colours, markings or symbols.

Areas zoned for a particular purpose may contain 'non-conforming' or conflicting uses. A factory in a predominantly residential area would be an example. In these circumstances, the local planning authority would probably seek to limit further incursions of these non-conforming uses and may also encourage such uses within the area to relocate to alternative and more appropriate sites. This could even be in the form of financial assistance or grants and/or by providing suitable alternative accommodation.

Just because a particular land use is indicated on the development plan for an area does not mean, by itself, that any application for planning permission for a similar use would be successful. Nor, for that matter, would a use which conflicts with the prevailing allocated use be automatically rejected. All planning applications must be considered individually on their merits and passed or refused after due consideration of all the relevant factors, including the provisions of the development plan.

Nevertheless, it would be reasonable to assume that applications for uses conflicting with the zoned use are less likely to be considered suitable and appropriate to the area than those for conforming uses, and where permitted, may be subject to various planning conditions to minimise the conflict. Examples of these conditions for an industrial use in a residential district could be landscaping, noise insulation and restrictions on the types of industrial processes that may be carried on from the property.

The local planning authority will also regulate the *density* of developments for which they are prepared to grant full planning permission.

Plot ratio

This is the means by which local planning authorities regulate the density of non-residential development. It is an expression of the ratio between the permitted floor area of the building and the area of the site on which it stands. Floor area in this context is GEA (gross external floor area).

Plot ratio is different from, and should not be confused with, site coverage, which is an expression of the amount of the surface area of the site covered by the building. This will always be different to plot ratio unless the building is single-storey throughout.

Although plot ratio is the normal expression used, it can alternatively be known as the 'floor space index'. There may also be differences in the method of site measurement between local planning authorities. Usually the net site area ('site area' as defined in RICS *Code of Measuring Practice*) is taken, although in some regions the gross site area (GSA) may be adopted. It is therefore necessary to check with the local planning authority on the measurement method used in their locality as considerable variations in the size of permitted development can occur from use of the incorrect area.

It will have been noted from the *Code of Measuring Practice* that the gross site area includes areas of the road onto which the site fronts as well as the site. Where such roadways are adopted public highways, the width of the road is usually taken to the back edge of the pedestrian pavement, not just the road itself. Where the highway so measured is less than 12 metres wide, then half the actual width is used in the calculation, otherwise the 6 metres maximum distance is adopted.

The origin for this somewhat unusual method of calculating site areas probably lies in the fact that at Common Law, the owner of a piece of land has ownership rights in half the width of the (private) roadway onto which the land fronts. Moreover, such owners are also responsible for the costs of repair and upkeep to the road in direct proportion to the length of frontage they own, compared to the total length of the road itself. Once the local highway authority has adopted a private road, these responsibilities for payment pass to the authority, albeit in return for a capital payment from the frontagers. At that point the ownership rights of the frontagers also cease, but nevertheless in planning law, the frontager may still be able to claim some benefit in terms of improved density of development, particularly for corner sites, whose position ensures a higher proportion of the land fronts onto the roadway.

Net site area (SA), on the other hand, is a more straightforward calculation, and is a basic measurement of the actual piece of land on which the development will stand, excluding all roadways.

Plot ratio will therefore normally be:

GEA buildings/Net site area

Although where the density is expressed as the Floor Space Index it will usually be found by:

GEA buildings/Gross site area

In either case, both area measurements should be expressed in a common unit, such as square metres. The result of this simple equation will produce a number that gives the ratio to 1. In other words the first number of the ratio expresses the GEA of the building and 1 is the site area. Thus if the ratio is 0.5:1, this indicates that the GEA of the buildings on the site must not exceed 50 per cent or 0.5 of the site area itself. This low ratio would be appropriate for industrial development, for example, where the buildings would be predominantly single-storey, with some second-floor offices, and adequate open space around the buildings must be provided for loading, unloading, parking and access purposes.

On the other hand, a ratio of 4:1 may be appropriate for a city-centre office development where land availability is very tight and site values are high and thus maximum use must be made of the land available. Such a ratio would enable a four-storey office building to be built, if the building covered 100 per cent of the site; or an eight-storey building if it only covered 50 per cent of the site, and so on. However, as a further restriction to prevent ridiculously tall and thin buildings being constructed, the local planning authority would normally impose a height restriction as well as a plot ratio on the site.

Example

1. The site available for development is 1 acre and the permitted ratio is 2:1. How large can the floor area of the building be?

    1 acre = 43,560 sq.ft. Thus at a ratio of 2:1 the permitted GEA is
    2 × 43,560 = 87,120 sq.ft.

2. A building of 15,000 sq.m. GEA occupies a site of 0.5 ha. What density does this represent?

    0.5 ha = 5,000 sq.m. Therefore, the plot ratio is 15,000/5,000 = 3:1

3. The plot ratio is to be calculated on the gross site area. The piece of land itself has an area of 0.25 ha and has a road frontage of 30m. The road is 15m wide, measured between the back edges of the pavements. How large a development can be constructed on the site if the permitted ratio is 0.6:1 based on a gross site area?

    Gross site area = (30 × 6m) + 2,500 sq.m.
    that is road frontage × maximum permitted width of road (as half total width is 7.5m the maximum distance of 6m must be adopted) plus net site area
    (0.25ha @ 10,000 sq.m. per ha).
    Total gross site area = 2,680 sq.m.
    @ 0.6:1 = 0.6 × 2,860 = 1,608 sq. m. GEA permitted floor area

Residential plot density

This is calculated in a different way to non-residential in that the quantity of permitted floor space is not usually expressed in terms of square metres or square feet but in dwellings or habitable rooms per hectare (or acre). When planning permission is given for a total number of habitable rooms on a site, the number of dwellings that can be built there will therefore depend on the number of such rooms in each dwelling.

The Town And Country Planning (Residential Density) (London, South East England, South West England, East Of England And Northamptonshire) Direction 2005 (ODPM Circular 01/2005) provides the following information on the subject of residential density:

1. Planning Policy Guidance Note 3: Housing (PPG3), issued in March 2000 and as amended in January 2005, sets out a new approach to planning for housing. PPG3 requires new development of the highest quality and for the country's future housing needs to be met in the most sustainable way.
2. Local planning authorities are expected to give priority to re-using previously developed land within urban areas, bringing empty homes back into use and converting existing buildings, in preference to the development of greenfield sites. The presumption is that new development will use land efficiently and be well designed. To avoid the profligate use of land and encourage sustainable environments, PPG3 requires local planning authorities to examine critically the standards applied to new residential development, particularly with regard to roads, layouts and car parking. They are expected to avoid housing developments which make inefficient use

of land (those of less than 30 dwellings per hectare net); encourage developments which make more efficient use of land (between 30 and 50 dwellings per hectare net) and seek greater intensity of development at places with good public transport accessibility.
3. In 2001 the overall density of residential development in England was 25 dwellings per hectare. This had remained unchanged since 1996. Over the period 1997 to 2001, more than half of the land used for housing was built at densities of less than 20 dwellings per hectare and over three quarters at less than 30 dwellings per hectare.

Local authorities are given detailed advice on how site area should be measured for density purposes in PPG3 Annex C, which states that local authorities should use 'net site density' as set out in *The Use of Density in Urban Planning*. It should be noted that, in contrast to the general densities quoted, the average in London itself is around or in excess of 70 dwellings per hectare, reflecting the higher land values there and the greater need to develop available sites more intensively.

Example

The Local Plan indicates that development would be permitted at a density of 250 habitable rooms per hectare. The net site area of a suitable plot is 1.3 hectares. How many semi-detached houses could be constructed on this site if each dwelling comprises 3 bedrooms, 2 reception rooms, kitchen, bathroom and separate WC?

Site area × density = 1.3 × 250 = 325 rooms
Each house comprises 5 habitable rooms (3 bedrooms + 2 reception rooms)
Therefore, number of houses permitted on site = 325/5 = 65 houses
This equates to 65/1.3 = 50 dwellings per hectare.

Ground leases

These are granted by freeholders on building land for long terms. Historically this used to be for a term of 99 years, although 125 or 150 years is often granted. However, it is very occasionally possible for ground leases to be for terms up to 999 years, which is obviously as good as perpetuity.

Ground rent is the rent payable for land let for the purpose of development, or improvement by building on it. The freeholder provides the land and the developer the building. Ground rent is thus charged in respect of the land only and not for the buildings erected or to be erected on it.

Having constructed the buildings on the land, the lessee may retain them for his or her own use and occupation or sublet them to a subtenant for a term not exceeding the unexpired number of years on the ground lease. The freeholder is entitled to the reversionary interest on expiry of the ground lease, which gives the right to possession of the buildings and improvements placed on the land since the start of that lease.

Within the duration of the ground lease, the existence of the buildings provides a measure of security for the rental income receivable by the freeholder, in that the ground lessee is paying a rent that only represents a proportion of the rack rental value of the land and buildings. Thus, where buildings are erected on the land, the ground rent is called 'secured', but if no structures have been erected, it is an 'unsecured' ground rent.

The difference between the market rent value of the land and buildings and the ground rent will constitute the lessee's 'profit rent'. From this sum deductions may need to be made for any outgoings for which the ground lessee is responsible, such as the costs of repairs, insurance and management.

During the nineteenth century and the early part of the twentieth centuries in the UK ground leases were often granted for long terms such as 99 or 100 years at fixed ground rents. Even when it became more commonplace to insert rent reviews into ground lease agreements, these were initially for infrequent periods such as 50, 33 or 25 years. As the rents were considered so secure, and in a low inflation economy did not decrease significantly in value with time, such long review intervals were not considered inappropriate.

Ground rents, particularly when secured, became generally acceptable as good safe forms of investment which in times of nil or low inflation were considered similar to gilt-edged securities. The degree of security of a ground rent was measured by the 'number of times the rent was secured'. For instance, if the ground rent was 20 per cent of the rack rental value, it would be termed as being '5 times secured'. The yields on ground rent thus closely followed those on irredeemable gilt-edged securities.

Nowadays, however, such a lease would be considered a very poor investment, because irrespective of the fact the rent is still 'secure' in terms of being lower than the market rent value of the development, its real value would decrease substantially because of the long intervals between reviews. Such leases are therefore likely to attract very high yields, to offset this disadvantage. The only exception to this would be as the date of reversion nears, when the yield will decrease considerably in expectation of the forthcoming dramatic rise in value, due to the prospect of vacant possession of land and buildings becoming available.

Ground leases now usually conform to institutional requirements in providing security by means of a long term, such as 125 years, but with the rent reviewed at regular periodic intervals. These are usually in line with normal market practice for the type of property constructed on the site. In most cases today in the UK this will result in a five-yearly review pattern, and will allow the freeholder to obtain parity with other open market rental increases in that sector, and thus allow the ground rent to keep pace with inflation.

Often the ground rent is 'geared' to the rack rental value so that a fixed percentage of the market rent value of the land and buildings on the site is taken at each review. This simplifies the valuation at review since, whilst it is frequently difficult to obtain good comparable evidence of ground rents for similar sites at any given time, it is usually rather more straightforward to find such evidence for buildings of the same type as on the site.

The exact percentage adopted will depend on the location of the site, the type of building constructed on it and the negotiating ability of the parties at the commencement of the lease. Typical 'geared' percentages are in the 15 to 40 per cent range, although exceptions to this will be encountered. The lower percentages would be appropriate where land values are comparatively low, and conversely higher percentages would be usual where land has a high value, such as in city centres.

In practice, therefore, if it is stated in a ground lease that the rent payable at review is to be 35 per cent of the open market rent value of the development, the review can be agreed without recourse to finding true ground rent comparables, which are generally far rarer and more difficult to ascertain than rentals on the buildings. Moreover, the freeholders will also know that at each review they will continue to receive 35 per cent of the full market rent and will thus gain proportionately from any rental growth enjoyed by that sector of the market between reviews.

RICS guidance or other information on valuing development land or property

See RICS 2007g.

Suggested minimum headings for comparables schedule

- Date of transaction
- Address
- Source of information
- Type of development proposed
- Site area in hectares (or acres)
- Maximum permitted GEA (gross external floor area) in sq.m. or sq.ft.
- Maximum permitted number of dwellings or habitable rooms (residential developments)
- Ground rent per annum (paid or asking)
- Ground rent per annum per proposed dwelling or habitable room (residential developments)
- Ground rent per annum per sq.m. (or sq.ft.) of proposed GEA or GIA
- Freehold capital value (paid, agreed or asking)
- Freehold all risks yield
- Capital value (site value) per sq.m. (or sq.ft.) of proposed GEA or GIA
- Capital value (site value) per proposed dwelling or habitable room (for residential developments).

Example

Value the freehold interest in a site of 0.8 hectare that has planning consent for warehouse (Class B8) development at a plot ratio of 0.5:1. An adjacent freehold site of 1.1 hectares, with similar planning permission, has recently been sold for £2.6 million.

Analysis

As both sites can be developed at identical densities, a simple comparison can be made. Thus adjacent site has sold for £2,600,000/1.1 = £2,363,636 per hectare

> As subject site is 0.8 ha in size, 0.8 × £2,363,636
> = £1,890,909 site value
> = SAY £1,900,000

Alternatively, suppose the adjacent comparable site had permission to be developed at a density of 0.55:1

> 1.1 ha = 1.1 × 10,000 sq.m. = 11,000 sq.m. site area
> @ 0.55:1 plot ratio = permitted GEA of 6,050 sq.m.

> Therefore site value per sq.m. permitted GEA = £2,600,000/6,050 = £429.75

Valuation of subject site

0.8 ha = 8,000 sq.m. site area
@ 0.5:1 plot ratio = 0.8 × 0.5 = 4,000 sq.m. permitted GEA
@ site value of £429.75 per sq.m. GEA × 4,000 = £1,719,000
SAY site value = **£1,700,000**

Leisure and entertainment properties

The following are examples of the type of property that would fall within this category:

- amusement arcades and centres
- bingo halls
- bowling alleys
- casinos
- cinemas
- golf courses
- holiday camps
- hotels
- leisure and sports centres
- night clubs
- petrol filling stations, garages and other automotive premises
- public houses
- restaurants
- sports stadia
- theatres
- theme parks
- turf accountants (licensed bookmakers).

Where possible, valuation of such properties would be undertaken using a comparison method, but as this relies on recent property market transactions having taken place on similar nearby properties this is often not possible. Many leisure properties enjoy a form of 'monopoly' due to legal or geographical reasons. This makes it difficult to compare one property with another. There may be very few of that type of property in a town or district. A legal consent is often required to carry out the business (a gaming, liquor, fuel or entertainment licence) and the issuing authority carefully regulates the supply of these. This leads to a shortage of market transactions for most of these categories of property. In addition, the buildings, location and style of business conducted can have very individual characteristics that make reliance on comparables difficult and unreliable.

Where the comparison method cannot be used to value the whole interest, it may still be used to provide a base land value for such uses as golf courses and theme parks, to which an extra value may be added to reflect the profitability of the operation carried out on this land. In many cases though, leisure or entertainment properties will need to be valued on a profits method basis that takes account of trading potential (see Chapter 14). In each case, these are 'specialised property' valuations. Each is different and requires specialist knowledge and experience of the property type and of the trades or businesses involved.

In addition to the traditional valuation analysis, the valuer may decide to consult a business management consultant to obtain a professional view on how the business could or should be run to optimise profits and minimise costs. This is known as a 'efficient operator's assessment' (EOA) and is an objective assessment of the business and the market in which it is operating. It is made based on what an efficient operator would assess as being a reasonable expectancy; not the maximum potential. This can be used to inform the profits method calculations.

Unit of comparison normally used

There are no standard units used. These will vary from one property type to another. In many cases it is difficult to find any 'direct comparison' unit that can be used and a form of 'indirect comparison' may be employed. Thus rather than compare value per square metre of floor area, a comparison may be made per cinema seat or per hotel bed space or even expressed as a multiplier of turnover.

Main factors that will determine or influence value

There are three elements that comprise the total 'going concern value' of this category of property:

- land and buildings;
- trade fixtures, fittings, furniture, furnishings and equipment; and
- 'inherent goodwill' – the market's perception of the trading potential excluding personal goodwill.

This assumes the property is sold as a fully fitted and functioning premises for the continued conductance of the existing profitable business activity taking place there. The purchaser is acquiring not just the premises but an established business as well. This will not always be the basis of valuation required though. For mortgage lending purposes financial institutions are more likely to only require a 'bricks and mortar' valuation, seeing only the value of the land and buildings as being relevant as security for the proposed loan advance. It may also be difficult to clearly distinguish the individual value of each element in a purchase price.

Chapter 14 examines the use of the profits or receipts and expenditure method of valuation likely to be employed for leisure and entertainment properties.

Many of the main factors influencing value for these property types will be similar to those outlined already for other business properties. These will include:

- location and position
- accessibility
- proximity to similar or complementary uses
- potential size of catchment area for customers/visitors
- external environment

- type of tenant
- terms of the tenancy
- outgoings
- services and service charges
- facilities
- construction style, age and condition
- mains services
- number of floors
- floor areas
- compliance with statutory legislation
- possession of licence(s) essential to conduct business activity
- planning permissions
- reputation and goodwill of the business
- the accounts of the business and trading history.

Special characteristics or considerations of hotels

Large or city centre hotels are usually sold as part of a group rather than as individual properties. The majority of UK hotels are held freehold although ground leases are used for some large modern properties. The ground rent can be geared to a percentage of the takings, which either may be the gross takings of the whole hotel or the bedrooms only, or may sometimes be the takings for all the rooms, with takings for the restaurant and conference rooms at different percentages to the bedrooms. A minimum rent is usually reserved where takings fall below a certain specified level. In some cases, bands of income may be used and different percentages used dependent on the income range. Alternatively, percentages related to net profit of the hotel can be used, which allows the hotelier's costs to be reflected in the figure paid.

Factors that will influence the valuation include:

- occupancy rate percentage: measure of how many of a hotel's rooms or bed spaces are occupied by paying guests at a given time (per month or per annum), expressed as a percentage of the total number of rooms or bed spaces available within the establishment
- average room rate: £ per day
- average turnover per bedroom
- gross operating profit percentage
- grade or rating of hotel.

Special characteristics or considerations of automotive properties and petrol filling stations

These comprise a number of different property types within the one property and as such present an unusual and challenging valuation problem. In addition to the sale of fuel and oil, a garage business can also include the sale, service and repair of motor vehicles, together with the sale of motor accessories and spare parts. Most modern filling stations have also diversified to provide a retail outlet for general goods such as food, confectionery, drinks, toys, DVDs, CDs, flowers, plants, maps, books, etc.

As far as possible, valuations should be made by direct comparison with other garage properties. Straight comparison could be based on £ per square metre or £ per litre of fuel throughput. However, as the average automotive business includes elements of retail, industrial/workshop and warehouse uses, some indirect relationship between the local values of these types of property uses and the corresponding use within the garage can also be made. The values are not directly comparable but can influence the figures adopted to value each part of the garage separately depending on its usage.

Special characteristics or considerations of public houses, bars and other licensed premises

In England and Wales, with certain limited exceptions, it is a criminal offence to sell, or to expose for sale, any intoxicating liquor without the possession of a 'liquor licence' (issued by the local licensing authority). In this respect, 'intoxicating liquor' is defined in Section 201(1) of the Licensing Act 1964 as: 'spirits, wine, beer cider and any other fermented, distilled or spirituous liquor'. All liquor licences are granted personally to the applicants, and as such do not actually form part or the corporal hereditament. Nevertheless, where the existence of such a licence is considered to enhance the value of the property occupied by the licensee, this is usually reflected in the valuation of that property. With public houses, the level of trade is the biggest determinant of value.

Trade is regarded in most cases as being an inherent benefit of the licence, and thus goodwill is not normally valued separately, but is integral to the valuation of the property. Trade is largely determined by the barrelage of beer and gallonage (or litreage) of wine and spirits sold on the premises. Apart from drink, the other elements of a pub's turnover are food sales, trade from letting rooms and machine income. The latter is usually separated into AWP (amusement with prizes games machine) and SWP (skill with prizes games machine).

Comparison method valuations can be based on £ per square metre or a multiple of gross receipts or net profits.

Special characteristics or considerations of restaurants

Restaurants can pose unique valuation problems due to the fact they are frequently found in retail locations and yet the nature of their trade is different to normal retail usage. This diminishes the relevance of valuation by direct comparison with nearby retail property transactions. This method can be adopted where the property is a restaurant at present, but there are no restrictions on its use for alternative commercial purposes.

The number of seats (covers) and an estimate of average expenditure per customer can be used as a basis for valuation, but the difficulty with using the comparison method is that two restaurants are seldom directly comparable. Different locations can affect trade. The diverse style, nature and layout of the buildings and their types of operation or operator can render each establishment unique. For instance a branch of a 'fast food' chain compared to a traditional waited table establishment is likely to have considerably different customer throughput and monetary turnover.

RICS guidance or other information on valuing leisure and entertainment property

See RICS 2006b, 2003a, 2003g, 2004.

Relevant sections on isurv website valuation channel are: Care sector, Cinemas, Doctors' surgeries, Garden centres, Golf courses, Hotels, Nightclubs, Petrol filling stations, Public houses, Restaurants, Telecoms, Vehicle dealerships, Vineyards.

Suggested minimum headings for comparables schedule

Each property category would require a unique set of headings, as all are different. The following factors could be common to most and extra headings added to reflect the property type concerned:

- Date of transaction
- Address
- Source of information
- Description of property
- Facilities
- Floor or land area
- Rent per annum exclusive (paid or asking)
- Freehold capital value (paid or asking)
- Freehold all risks yield
- Rent per annum per chosen unit(s) of comparison
- Capital value per chosen unit(s) of comparison.

10.3 Property market evidence and information sources

To carry out the comparison method of valuation property valuers constantly need to obtain reliable, accurate and current market evidence and information. The major problem is where and how to accumulate the required relevant comparables.

The principal sources of information are:

- lettings or sales personally undertaken by the valuer him or herself;
- lettings or sales undertaken by others in the same company or organisation for which the valuer works;
- direct information on sales and lettings undertaken by other surveyors or agents, obtained from the persons themselves who completed the transaction;
- indirect information from industry 'contacts' on transactions they are aware of;
- deals reported in the professional property journals;
- property auction results;

- property deals reported in the national and local press;
- published indices, data, statistics and research reports by:
  - the Investment Property Databank (IPD);
  - the Royal Institution of Chartered Surveyors (RICS);
  - surveying and property companies;
  - professional journals;
  - financial institutions;
  - the Valuation Office Agency (VOA); and
  - the Land Registry.

Increasingly, large amounts of the above information can be accessed through the internet. It is important for valuers to properly reference all sources when obtaining comparables, so that further details or clarification, or even written confirmation, can be obtained if necessary at a later date.

The best and most reliable information is obtained when the valuer personally has dealt with the transaction or can speak directly to the person who did. The full details and facts of the case can then be known, otherwise there is the possibility that some of the information is mere 'hearsay' and cannot be fully proven or has been misreported. Often though, due to a shortage of 'primary' data, there is no option other than to place a greater reliance on 'secondary' sources.

The marketing particulars produced by agents for the properties on which they act, or have acted, can provide many useful details, which will save the valuer considerable research time when compiling a schedule of comparable evidence. Speaking to the agents concerned will usually enable the valuer to obtain remaining information on asking prices or rents and the figures at which the transaction was completed. Much useful background information concerning the circumstances surrounding the transaction can also be obtained in this way.

Providing the agents are made aware that the information is being sought by a valuer for a bona fide valuation, most will be cooperative in furnishing details of their transactions or other relevant information. In some cases, their clients will have specifically requested that the details remain confidential and so they will be unable to discuss it, but the majority of transactions can be made public, since if they were not, the comparable method of valuation would break down through a lack of market evidence.

All valuers are particularly aware of this latter point, and therefore the trading of market knowledge is usually freely asked for and given by members within the profession, since it is only through such reciprocal cooperation that the comparable method can operate. However, it is implicit that, in providing details to another valuer, a similar request can be made of the other party at some future date. Valuers who constantly seek comparable information from other people, but who are not prepared to spend the time and trouble to return the compliment, will create bad-feeling and are likely to encounter an obstructive attitude from others in future.

The property valuation profession is very much a 'people' business, in that it relies on personal contact and the good professional relationships between people engaged in the work, to function efficiently. Through the continual process of seeking and providing comparable evidence, the valuer will build up a 'network' of a large number of personal 'contacts', both in private and public practice. These are other people in the profession who are known and can be approached on a personal level when required to assist in the gathering of comparable information. In many cases a valuer will decide to specialise in one area of valuation work and over time will come to know a large number of other such specialists within other organisations. Knowing whom to contact when necessary can considerably ease the task of gathering comparables.

As was explained in Chapter 1 (Section 1.6), the property market is an imperfect one, and as such both parties involved in a transaction usually do not have full knowledge of it. However, by assiduous research and through the mutual cooperation of others, valuers can and should acquire this knowledge to be able to properly advise their clients.

Some examples of secondary data sources

- The *Estates Gazette* (online version www.egi.co.uk): this is the UK property world's oldest journal, first being published in 1858. Although primarily UK centred, it does cover Europe and other global locations. Published weekly it contains substantial sections of advertisements for properties on the market, together with summaries of deals completed, auction results, financial news, professional opinion articles, summaries of property legal cases, etc. The centre section focuses on a particular region or subject, as do the occasional separate supplements. In addition, its bi-weekly 'Mainly for Students' articles contain invaluable information on a variety of subjects suggested by readers, which is of benefit not just to students, but many qualified practitioners as well.
- *Property Week* (online version www.propertyweek.com): this is the other main UK weekly journal that similarly has advertisements of properties, deals summaries, news items, events, professional opinion, financial news, regional or overseas reports, etc.

Careful perusal of current and back copies of both these journals will enable considerable amounts of comparable and general market information to be gleaned. It is therefore wise to maintain collections of back issues in the office for later reference. They can provide details of deals or properties on the market and the agents concerned, which can then be contacted for further information.

Property auction results

In principle, public property auctions offer excellent comparable evidence, in that the prices achieved are deemed to be the best possible on the open market. There is open competition, and bidders are fully aware of other bids in the market and the level of interest shown in the property from the activity in the auction room. The downside is that many properties are only auctioned because of actual or expected difficulties in finding a buyer and concluding a deal by agreement on the market. Examples are forced sales or properties in need of considerable refurbishment. This may render some of the transactional evidence less useful when valuing more prime property.

The auctioneers publish full particulars of the lots to be offered for sale in advance. When seeking comparable evidence from these sales, it is therefore worth the valuer obtaining copies of these details as well as the final selling price of the lots. The auctioneers themselves usually publish the results of the sale on a separate leaflet, although the property press also provides a full weekly summary of the results of UK property auctions. However, personal attendance at the auction is often worthwhile, not only to see if a lot sells and at what price, but to judge the level of interest shown and the pace of the bidding.

Also see the website: www.propertyauctions.com

The national and local press

Apart from the specialised property journals, the national and local press often devote weekly sections of editorial to property matters and also contain advertisements of available properties. Occasionally snippets of comparable information can be gained from these sources. Local newspapers can be especially useful for small secondary properties for sale or rent.

Websites

Here are the websites of some of the major companies of valuers, surveyors and property consultants:

> www.cbre.com/International/Sites/UK/Our+Research/
> www.colliers.com/Markets/UnitedKingdom/
> www.joneslanglasalle.co.uk/en-gb/research/
> www.kingsturge.co.uk/commercial/research/ (rental data)

The Investment Property Databank (www.ipdglobal.com) has huge volumes of statistical market data for worldwide locations. The UK government's Valuation Office Agency (www.voa.gov.uk) property market research reports cover whole of UK and most major property types (VOA 2008b).

Research publications

Examples of overall market data for UK commercial and industrial properties are the publications by Cushman & Wakefield, such as *Marketbeat UK*, 'a quarterly review of market trends and price movements in the UK property market' (Cushman Wakefield Healey & Baker 2008).

Examples of UK residential valuation data sources

- www.rightmove.co.uk: claim to hold 'largest database' of sales and properties for sale or to let
- www.mouseprice.com: similar database of sales, for sale and to let and also includes 'automated valuation model'
- www.nationwide.co.uk: Nationwide Building Society house price quarterly data
- www.hbospic.com/view/housepriceindex/nationalcommentary.asp: Halifax national commentary and analysis
- www.landreg.gov.uk/ppr: Land Registry residential property price reports.

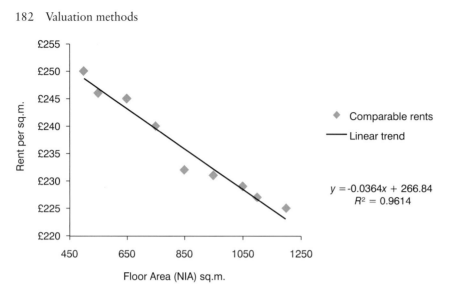

Figure 10.3   Comparables quantum adjustment analysis

- rent-free periods or other form of landlord's 'incentives'
- repairing covenants
- user restrictions.

The exact choice of figures in the adjustment should be based on market evidence, but where this is lacking it may have to be on a more subjective basis derived from the valuer's knowledge and experience of the market. The basic principle is that any features that will add to costs of construction and will meet tenants' or occupiers' needs or will enhance the building's usability or marketability should lead to higher rentals. Restrictions or encumbrances should similarly result in lower figures. When selecting adjustment figures the overriding question is would a potential tenant in the open market expect to pay more or less for the feature in question and by how much?

The VOA, for the revaluation 2005 of rateable values, provides an example of how adjustments are made. These are detailed in its general 'valuation scales' and the 'valuation scheme' applied to individual properties. For instance, Valuation Scale (Reference VOOVERALLV1), which applies to offices valued on an overall basis, assumes that a basic standard of accommodation comprises central heating and lifts, but no air conditioning nor raised floors. 'If central heating is not present the main space price applied is reduced by 5%. If a property is air conditioned the main space price is increased by 5%. Where raised floors are present the price is increased by 2.5%' (Valuation Office Agency 2007).

Example 3

The rental value is required of a warehouse of 800 sq.m. This floor area includes 160 sq.m of purpose-built offices on two floors on the frontage. The lease is for 15 years on full repairing and insuring terms with five year reviews and has a break-clause option at mid-term.

Comparables are around £80 per sq.m for similar, but less modern buildings, on full repairing and insuring leases without a break clause; that have a 10 percent office content and less parking and unloading space and for unit sizes of up to 500 sq.m.

	£
Base rent (from comparables)	80
Add for extra office space	7.27 overall (+9.1%)
for better standard of parking and access	2 (+2.5%)
for inclusion of break clause in lease	4 (+5%)
Less for size ('quantum')	(2.40) (– 3%)
User: no restrictions	nil
Repairs: no adjustment as both FRI	nil
Total adjustments =	£10.87 (+13.6%)

Estimated market rent for 'subject' £90.87 per sq.m per annum
× 800 square metres = £72,696
Market rent of subject SAY = **£72,700 per annum**

It is assumed that all the above adjustments, made on a percentage or value per square metre basis, have been derived from analysis of market evidence, which indicates similar amendments to value have been made on comparable properties.

The adjusted floor area method has been used to allow for the additional office space in the subject property as follows:

- Typical comparable = 10 per cent office content in 500 sq.m. total size.
- Therefore, offices = 50 sq.m. and industrial space = 450 sq.m. in typical comparable premises.
- Using office space = 2 × value of industrial, adjusted floor area = (50 × 2) + 450 = 550 sq.m.
- Rent @ £80 × 500 sq.m. = £40,000 per annum
- Divided by adjusted floor area = £40,000/550 = £72.73 per sq.m. p.a.
- Subject adjusted floor area = (160 × 2) + 640 = 960 sq.m.
- @ £72.73 per sq.m. = £69,818 estimated rental value
- Unadjusted overall figure would have been 800 sq.m. × £80 = £64,000
- ERV compared to this is £69,818/£64,000 = 9.1 per cent higher or £87.27 per sq.m. overall, which is an increase of £7.27 per sq.m.

In compiling and analysing comparison evidence, the valuer must be careful it is indeed comparable and can be used as evidence when required. In *English Exporters Ltd* v. *Eldonwall* [1973] 1 All ER 726 it was stated that submitting a list of comparables that are not clearly documented in writing and confirmed by the actual parties concerned will only be considered 'hearsay' evidence (and thus carry far less weight) in a formal hearing, such as an arbitration or court case. Conversely, the *Land Securities* v. *Westminster City Council* [1992] 44 EG 153 case established that an independent expert's decisions on another property may be considered inadmissible evidence.

National Car Parks v. *Colebrook Estates* (1983) 266 EG 810 concerned a car park in W1 area of London. In this case the judge stated that 'the comparables produced by both experts are, in my judgment, in no way comparable. There seems to be a ridiculous idea that in W1 one is in central London and the conditions are the same throughout. Nothing could be more untrue.'

In *Newey & Eyre Ltd* v. *J. Curtis & Son Ltd* (1984) 271 EG 891, the comparable used was the property next-door to the subject premises. Nevertheless, in determining the rent under Section 34 of the Landlord and Tenant Act 1954 (Part II), the court stated that the comparable was similar to the subject only in respect of position, access, floor area and rent review period; *but* it was different because of its age and layout, the repairing covenants in its lease, it was subject to a restrictive user clause and its rent had been agreed with a sitting tenant rather than on the open market. From this it can be concluded that a wide range of factors will be considered in ascertaining true comparability of comparables and even a seemingly similar property can be quite dissimilar on closer inspection.

An interesting view on the analysis of comparable evidence was provided in the Court of Appeal case of *Oriani* v. *Dorita Properties Ltd* (1987) 282 EG 1001. This involved retail premises in East Precinct of St George's Walk, Croydon. The tenant provided nine lease renewal and two rent review comparables from the West Precinct (a more favourable area). The landlord produced two comparables of open market lettings from East Precinct. It was held that a fair and reasonable assessment of the open market rent was:

- sum of zone A equivalent figures of the nine lease renewals in West Precinct;
- less 20% (to allow for difference between locations);
- plus adjustments for increase in values from date of comparables to date of hearing;
- plus the sum of the two zone A equivalents from the East Precinct;
- divide this total sum by 11 to give an *average* figure per square metre, which was applied to the subject property.

10.5 Methods of assessing market rent

As explained in Section 9.1 above, the RICS *Red Book* defines *market rent* (MR) (RICS 2007d: PS 3.3, 46). Historical terms used to describe what is now known as *market rent* are sometimes still used or found in articles and books and include:

- full rental value (FRV)
- 'rack rent'
- estimated rental value (ERV), or
- open market rental value (OMRV) – as used in the fourth edition of the RICS *Red Book* (1997b).

As rent is a payment by a tenant for the use of land or land and buildings, the amount of rent depends on the use or potential use to which the land and buildings can be put and the supply and demand for that use. In *declining reliability* (first method is the best, the second next best and so on) the methods that can be used to calculate the *market rent* of a property are:

- actual rent paid for subject property
- derived from comparable rents
- decapitalised comparable sales
- a percentage of profit or turnover
- a percentage of construction costs
- a percentage of occupier's income.

Actual rent paid for subject property

This can be used, providing it has recently been agreed in the open market 'at arms length' between a willing lessor and a willing lessee. In many instances, for example at rent review, the rent has to be calculated on a hypothetical basis. A typical situation at review is that the property is to be assumed as vacant and to be let, even though it is actually occupied by a tenant under an existing lease. Speaking in relation to a rating case, the words of Lord Buckmaster, in *Poplar AC v. Roberts* [1922] 1 K.B. 25, are still relevant: 'the actual rent paid is no criterion, unless, indeed, it happens to be the rent that the imaginary tenant might be reasonably expected to pay'.

Grounds for challenging that the actual rent is not the best evidence of current market rent value include:

- the rent was not fixed by the 'higgling of the market' (for instance, it was agreed between related companies or even members of the same family); or
- the rent includes business goodwill; or
- the rent is part of a larger transaction, such as a sale and leaseback arrangement; or
- the rent is out of date; or
- a premium has been paid and this is reflected in the current rent payable.

Derived from comparable rents

This relates to rent paid for identical or similar properties, again, providing it has recently been agreed in the open market 'at arms length' between a willing lessor and a willing lessee.

Decapitalised comparable sales

Here, recent capital transactions on comparable properties are decapitalised to find the rental involved. The price must have been agreed in the open market 'at arms length' between a willing seller and a willing buyer. Decapitalisation should be at an all risks market yield appropriate to the property. The basic equations needed to find the rent by this method were considered in Section 10.1 above.

A percentage of profit or turnover

This is where a business is conducted on and from the property and relates to the profit or turnover from that activity. This method is considered in more detail in the sections on turnover rents and the profits method of valuation in Chapter 14.

A percentage of construction costs

The *current* costs of construction of the property standing on the site are estimated less, for older properties, a deduction for age and obsolescence. This method is considered in Chapter 15 on the cost method of valuation.

A percentage of occupier's income

The ultimate last resort method, it is a 'means test' that has no direct relevance to the market value of the property. It comes down to what the occupier can afford to pay, which may or may not be what the property is actually worth. Fortunately it is extremely seldom that this method has to be used as invariably one of the other five methods of assessment can be utilised.

10.6 Assessing ground rents

Ground leases and ground rents were considered in Section 10.2. The methods of assessing a ground rent for a piece of land, in *declining reliability*, are as follows:

- direct comparison with other ground rents;
- a percentage of the market rent;
- decapitalisation of the site value;
- decapitalisation of the completed development value; or
- annual equivalent of the total costs of the completed development.

Direct comparison with other ground rents

For similar properties and providing as always these rents have recently been agreed 'at arms length' in the open market between a willing lessor and a willing lessee.

A percentage of the market rent

The market rent value of the completed development that stands on the plot of land. As previously explained, the principle is that the ground rent can be 'geared' to the rack rental value of the development, which is assessed on land plus buildings. This method has the advantage that it is usually easier at rent reviews to obtain relevant comparable evidence for buildings than for bare sites. When the rent reviews on the ground lease are timed to coincide with those on the head lease for the completed buildings, this makes the valuation even easier.

The percentage of market rent used depends on the type of property and its location. In areas of high land values, the site on which a building stands will constitute a far higher proportion of the total value than will a site in a poorer location. Similarly, two buildings of identical size and standard of construction, and standing on similar sized plots, will command different values according to their location. It is the value of the land that makes the difference.

Decapitalisation of the site value

Here the site capital value is either known from a recent sale or is itself estimated from comparables. The all risks yield chosen for the decapitalisation will be influenced by, although not necessarily the same as, the yield appropriate to the buildings that stand on the land.

Decapitalisation of the completed development value

The capital value of the development is either known from its recent sale or is estimated from comparable evidence and then decapitalised at an appropriate all risks yield.

Annual equivalent of the total costs of the completed development

This is a percentage of the actual costs incurred. Being last in the list, this is the method of last resort. It attempts to relate costs to open market value, which is always a tenuous link and is seldom used as invariably one of the other methods can be utilised.

10.7 Indexed rents

Not a method that has proven popular in the UK but is the standard provision in most commercial leases in a number of other countries in the European Union. Indexation provides that the rental shall be increased in line with a specific index number at regular intervals. This therefore largely removes the need for the usual rent review machinery, which has become a standard part of UK commercial leases. In turn this also effectively removes the need for rent review valuers to advise the landlord and tenant on such occasions, since a standard national index number is easily obtainable and understood by both parties and can be applied without the need for negotiation.

In choosing which indexation to use, the index of retail prices (RPI) or its equivalent could be selected as a measure of inflation. Rents indexed to this number would maintain their value in line with inflation and so the 'real' value of the rent would not decline. However, since the 1960s, UK property values have frequently risen faster than retail prices and rents linked to the retail prices index would have lagged behind open market rent values. Conversely, in times of property recession, such as in the UK in the early 1990s and in 2008, property values can fall while prices of other goods and services can continue to rise. Linking rents to an inflation-measuring index would result in rent demands rising when the values of property were falling and businesses were failing. There would be little likelihood of these increased rental demands being met and the owners would suffer income voids. These contributing reasons are why this system has proven unpopular so far in the UK.

Example

Current rent payable, fixed three years ago, is £50,000 per annum. The lease provides that rent will be reviewed every three years in line with the change in a specified national index number. At the time the rent was last fixed this index number was 220 and now it is 253. What will be the new rent payable from this year?

New index number/old index number = 253/220 = 1.15
New rent = old rent × change in index number = £50,000 × 1.15
New rent = **£57,500 per annum**

10.8 Annual equivalents

An annual equivalent is a way of expressing a capital sum in terms of an equivalent figure per annum. It is found by the inverse of the process used in capitalisation. It can thus be termed 'decapitalisation'.

Parsons (2004: 14) defines annual equivalent as 'Having regard to the period (being either a term of years or a freehold), the annual income calculated to correspond to a capital sum, such as a premium, paid for an interest in land'.

Example 1

A freeholder expends £500,000 on a property for which the all risks yield is 8 per cent.
  The annual equivalent of this expenditure in perpetuity is thus:

8% × £500,000 = 0.08 × £500,000 = **£40,000 per annum**

or found by:

8 × £500,000/100

Example 2

An improvement costing £200,000 and with an expected useful life of 20 years is undertaken on a freehold property. What is the annual equivalent of this expenditure allowing for an all risks yield of 6%?

Annual equivalent = £200,000/YP 20 years @ 6% = £200,000/11.46992
= **£17,437 per annum**

Example 3

A leaseholder has acquired her interest on the basis of a dual rate YP of 12% + 4% (tax 40%). She now expends £25,000 on the property. What is the annual equivalent of this sum if the lease has an unexpired term of five years?

AE = £25,000/ YP 5yrs @ 12%+4%(tax 40%)
= £25,000/2.338
= **£10,692 per annum**

Example 4

The leasehold interest in an office let on a fifteen-year FRI lease (with five-year reviews) at a rent of £30,000 two years ago has just been assigned for £25,000. The price paid did not include any other factors. What is the estimated current market rent of the property, found in terms of rent paid plus annual equivalent of the purchase price if a single rate all risks yield approach is used at 8 per cent?

Valuation

There are three years left until the next rent review, at which time the rent will be increased to the full market rent. The rent payable was agreed two years ago and is now adjudged to be below current market rent value, which is why the capital sum has been paid on assignment. The estimated current market rent value can be found as the sum of the rent received by the freeholder plus the annual equivalent of the price paid as follows:

Rent paid under lease =		£ 30,000
Capital sum paid =	£ 25,000	
divide by YP 3 yrs @ 8%	2.5771	
Annual equivalent of price paid =	£ 9,701	
Estimated current market rent =	£39,701	
SAY =	**£39,700 p.a.**	

Progress check questions

- What are the six different methods of valuation that can be used?
- Why is the comparison method of valuation the best?
- What is the basis of the comparison method?
- What are the three essential requirements of any comparable evidence?
- Why is it important to record full details on each comparable found including its source?
- What is the 'zoning' method and when and why is it used?
- How is the density of development of development on a site measured and controlled?
- What is a quantum adjustment and why may it be made to a comparable?
- How can market rent be estimated?
- What approaches can be used to find a ground rent payable under a ground lease?
- What is an annual equivalent and how and why is it calculated?

Chapter summary

The comparison method of valuation is one of six that can be used. It is the 'first choice' method, being the simplest and most reliable and is used whenever possible. It requires full details of comparable transactions to be found and analysed. These comparables must be as similar as possible to the property being valued and have been involved in a recent transaction.

To compare one property with another a common unit of measurement is used. This varies from one property type to another. Retail property often uses a unique approach to assessing the floor area, known as zoning and halving back, with the floor area expressed in terms of zone A. The definitions of how to measure each property type are mostly to be found in the RICS *Code of Measuring Practice*.

The investment method of valuation also utilises aspects of comparison as well as mathematical formulae or tables. With some categories of property it is difficult to use the comparison method due to the lack of transactions taking place or because the properties are unusual or used for specialised purposes. With these it may be possible to use a form of indirect rather than direct comparison.

Each property type has its own characteristics and factors that influence value. In many instances the RICS provides guidance on their valuation in Information Papers or expert opinion on the isurv website.

Obtaining all the information on each comparable can be time-consuming and require considerable research. Details can be obtained in person, from networking with others in the industry, at auctions, through the internet, in publications and from research work undertaken by major organisations and others. Having obtained information on all comparables, the subject property is compared to each one and adjustments made to reflect any differences in size and quality. Through comparing with as many relevant comparables as possible, the value of the subject property can be estimated.

A number of different approaches can be used to assess the market rent of a property or the ground rent payable for a ground lease. These vary from a comparison approach through to using costs and reflect the overall methods of valuation.

Further reading

Askham, P. (1993) 'Mainly for students: the valuation of agricultural property', *Estates Gazette* (2 Oct.): 121–2 and (30 Oct.): 107–8; repr. in P. Askham and L. Blake (eds), *The Best of Mainly for Students* (London: Estates Gazette, 1999), vol. 2, pp. 281–94.

—— (2003) *Valuation: Special Properties and Purposes*, London: Estates Gazette.

—— (2008) 'Mainly for students: it doesn't have to be a drama', *Estates Gazette*, 0829 (26 Jul.): 86–7.

Estates Gazette (1984) 'Mainly for students: analysis of shop rents', 269 (28 Jan.): 333–6.

—— (1995) 'Mainly for students: agricultural lettings', 9525 (24 Jun.): 153–4.

—— (1997) 'Mainly for students: inns and outs', 9724 (14 Jun.): 117–20.

—— (2000) 'Mainly for students: how to reflect price paid', 0019 (13 May): 136–8.

—— (2000) 'Mainly for students: search for the elusive figure', 0043 (28 Oct.): 172–3.

—— (2005) 'Mainly for students: pricing adjustments', 0523 (11 Jun.): 136–7.

French, N. (2005) 'Compare notes', *Estates Gazette*, 0543 (29 Oct.): 184–5.

Grenfall, W. (2003) 'Perfect pricing', *Estates Gazette*, 0348 (29 Nov.): 122–3.

Harper, D. (2007) *Valuation of Hotels for Investors*, London: Estates Gazette.

—— (2008) 'Mainly for students: an opinion of balance', *Estates Gazette*, 0811: 88—89.

Harris, R. (2005) *Property and the Office Economy*, London: Estates Gazette.

Hayward, R. (ed.) (2008) *Valuation: Principles into Practice*, 6th edn, London: Estates Gazette.

Hattersley, M. (1981) 'How to value existing and new hotels', *Chartered Surveyor* (Jul.): 784–5.

Hillman, P. and Dalby. R. (2000) 'Stranger than fiction', *Estates Gazette*, 0039 (30 Sep.): 136.

Lynch, T. and Clark, K. (eds) (2006) *Real Estate Transparency Index*, Chicago: Jones Lang LaSalle.

Mackmin, D. (2007). *The Valuation and Sale of Residential Property*, 3rd edn, London: Estates Gazette.

Marshall, H. and Williamson, H. (1997) *Law and Valuation of Leisure Property*, 2nd edn, London: Estates Gazette.

Mendell, S. (1994) 'Hotel valuations: a modern approach', *Estates Gazette*, 9406 (12 Feb.): 126–7.

Murphy, G. (1999) 'Valuation of farm buildings', *CSM* (Jul./Aug.): 26–7.

Parnham, P. and Rispin, C. (2000) *Residential Property Appraisal*, London: Spon Press.

Plimmer, F.A. (1998) *Rating Law and Valuation*, London: Longman.

Saunders, O. (2001) 'Earth, wind and water', *Estates Gazette*, 0106 (10 Feb.): 152–4.

Scarrett, D. and Smith, M. (2007) *Property Valuation: The Five Methods*, 2nd edn, London: Routledge.

Valuation Office Agency (2006) *Rating manual*, vol. 5, *Rating Valuation Practice, All Classes of Hereditament*. Online: <http://www.voa.gov.uk/instructions/chapters/rating_manual/vol5/frame.htm> (accessed 25 June 2007).

Williams, R.G. (2008) *Agricultural Valuations: A Practical Guide*, 4th edn, London: Estates Gazette.

11

Investment method
Traditional

In this chapter ...

- The conventional method of valuation used when the required value cannot be found just from comparison.
- The basis of valuing many freehold and leasehold interests.
- How to select the correct level of all risks yield to use.
- The other different types of yield that may be used to calculate the value and how to choose between them.
- The valuation of reversionary interests.
- The types of tax and their effects on property valuation.
- The annual expenses incurred in the ownership and occupation of land and buildings and how to estimate them.
- Insurance of buildings and how this differs from valuation.

11.1 Basis of method

Using the investment method, the capital value of a property investment is found by multiplying the annual income flow by a multiplier, which is the cumulative present value factor and more usually termed the *years purchase* (YP). This process is known as 'capitalisation'. It relies on use of all risks yields, obtained from comparables, to calculate the YP number. Freeholds are usually valued on a *single rate* basis and leaseholds usually on a *dual rate tax-adjusted* basis.

The basis of the method is thus:

- For freeholds: Net income × Years purchase single rate = Capital value
- For leaseholds: Net profit rent × Years purchase dual rate tax adjusted = Capital value (also known as lease value on assignment).

For a freehold, when the current net income is the present market rent (MR), the calculation requires the YP in perpetuity factor (YP perp), which was covered in Section 6.2 above.

The current net income from the property could be an actual sum, in that the premises are let to tenants paying a rent deemed equivalent to the current market rent. Alternatively, it could be notional in that the property is owner-occupied and thus the owner has the use of a building that would command the market rent if let. Through ownership, the occupiers are able to use a building that would otherwise have cost them this rent. The 'saving' of this sum is the same as an equivalent addition to income of that amount. The same premise is true of a building being offered for sale with 'vacant possession' as the purchasers have the option to occupy themselves or let to tenants at the market rent.

Where current net rental income from a freehold investment is the same as the present market rent figure capitalisation can be undertaken using the years purchase in perpetuity factor. When the rental income from a freehold investment is not its current market rent, other years purchase numbers will be employed. The appropriate years purchase formula to be used to value each block or 'tranche' of rental income will depend on when that rental is first receivable and for how long a period it will then be receivable (covered in Chapter 6).

By necessity, the investment method relies on the comparison method to find the appropriate yield and/or the market rent to use in the calculation. As this can introduce an aspect of subjectivity in the selection of figures to be adopted, it is the method of second choice, where a direct comparison of capital values cannot be made. Nevertheless, it is the most reliable method of valuation other than the comparison method.

What if there is a lack of comparable evidence on which to base the selection of the yield or market rent? The chairman of the RICS Valuation Faculty has addressed this issue: 'Transactional evidence underpins valuation process ... but what if not available or up to date? Especially in a fast changing market, the obligation is on the valuer to use market knowledge and professional judgement, to make assessment of where the market stands on the date of valuation' (Peto 2007b).

Example

Investment method

Value the freehold interest in a factory that has just been let on the open market at the full market rent of £100,000 per annum. Comparables indicate an all risks yield of 8% will apply.

Valuation

	£	
Net income =	100,000 p.a.	
× YP perp @ 8%	12.5	
	Market value =	£1,250,000

Notes

- The all risks yield has been selected after careful analysis of comparable market evidence and taking account of the location and nature of the property; the strength of the tenant covenant and all other relevant factors.
- As explained in Section 3.7 an addition can be made to the final value to indicate the anticipated costs of purchase, for instance: estimated purchase price of £1,250,000 plus costs at 5 per cent = £1,312,500 total expenditure by purchaser.
- Alternatively, no explicit addition is made to the calculation in respect of costs, but the client is informed in the report that although the valuation does not reflect them, due allowance should be made for costs to be added to the recommended purchase price, or to be deducted from the sale price, depending on whether the client is the purchaser or vendor respectively.

Where a lease is granted for occupation and the lessee (tenant) pays the full market rent to the lessor (landlord), there is no 'profit' and the lease by itself has no value on the market. When the lessee pays a rent to the lessor that is less than the market rent, then the lessee potentially enjoys a 'profit rent' and the lease does have a saleable value. In this case, the rent paid to the landlord will be termed the 'rent reserved' or 'rent payable' under the lease, or if the lease is a head leasehold interest, it will be termed the 'head rent'. The profit rent is then the market rent of the demised premises minus the rent payable to the landlord.

This profit rent could be notional, in that the lessee actually occupies the premises and is therefore enjoying the use of a property that is worth more rent in the present market than he/she is actually paying to his/her landlord. Alternatively, the property could be sublet and therefore the head lessee actually receives a rental from his/her tenant (which may still be less than the market rent), and providing he/she pays a lower rental to his/her landlord, will enjoy the 'profit' in terms of a cash surplus from the transaction. The phrase 'rent receivable' will be equally applicable, irrespective of whether it is a notional market rent that could be received if the property was to be sub-let or if it is the rent actually being received from a sublessee.

In setting out leasehold valuations, it is therefore customary to calculate the profit rent as follows:

Rent Receivable =	£	p.a.
Less Rent Payable=	£	p.a.
Profit Rent =	£	p.a.

Providing the rents receivable and payable are both on full repairing and insuring (FRI) terms, the resultant profit rent will be the *net* sum direct. Otherwise, it will be the *gross profit rent*, from which an adjustment for outgoings will need to be made to arrive at the net sum (this is considered in Section 11.14 below).

The net profit rent is capitalised by multiplying by the appropriate years purchase number to find the capital value of the lease. Where a lease does have a capital value, this would represent the sum or 'price' that the current tenant would seek if he/she were to 'sell' their lease to another party. In legal terms, this would amount to an 'assignment' of their lease and the 'purchaser' would acquire the right to take over the lease and occupy the property under the same lease terms as the 'outgoing' tenant.

Where there is nil profit rent then there is no value to the lease itself, although if the lease is assigned there may be value associated with the fixtures and fittings present on the premises that will pass on assignment. Additionally, further sums may be paid for the value of the 'goodwill' associated with use and occupation of the premises for business purposes and/or 'key money'.

Key money is defined as: 'That part of the capital price obtained for the freehold or lease of a shop excluding the value of any goodwill but which is additional to the sum of the capitalized value of the estimated profit rental and of the value of any shop fittings of use to the ingoing occupier' (Parsons 2004: 151).

When the rent payable is greater than the rent receivable a negative or 'loss' rent arises which when multiplied by the appropriate YP gives the 'reverse premium' needed to be paid on assignment. This is how much the present tenant would need to pay somebody to take an assignment of the lease, as it has become a liability, not an asset (see Chapter 16).

Example

Leasehold valuation

The lessees (tenants) hold a twenty-year FRI (full repairing and insuring) lease, with five-yearly rent reviews, at a current rent payable of £40,000 per annum. The lease has thirteen years unexpired and the present market rent of the property is estimated at £50,000 per annum. The lessees seek an all risks yield of 13 per cent and pay tax at 40 per cent.

Notes and assumptions

*   The all risks yield of 13 per cent has been derived from comparable market transactions of similar leasehold properties.
*   The lessees have ascertained they can obtain 5 per cent gross on their sinking fund instalments over the next three years in a 'safe' low-risk investment, such as short-term government gilts.
*   After deduction of tax at 40 per cent, this gross annual sinking fund rate equates to a net rate of 3 per cent:

    100 – tax = 100 – 40 = 60
    60/100 = 0.6
    × 5% = 3%

The 'rent receivable' is the current market rent (MR). This sum is notionally receivable by the lessees as they occupy property of this annual value. That is they would have to pay this rent to occupy an identical property.

The term to be valued is only three years, which is until the next rent review. At that time, the rent payable will be increased to market rent and there will be no profit rent.

The remaining ten years of the lease have no effect on the assessment now of the present value as no profit rent = no value assuming there are no other factors such as goodwill, fixtures and fittings or 'key money' to be included.

Valuation

Rent receivable=   £   50,000 p. a.
Less rent payable=   £   40,000 p. a.
                                       Net profit rent=   £   10,000 p.a.
              × YP 3yrs @ 13% + 3% (tax @ 40%)   1.49435
                                       Capital value = £ 14,943
                                                       SAY= £ 15,000

Example

Analysis of a leasehold valuation

A head leasehold interest of a property available with vacant possession has just sold for £160,000. The head lease has 50 years to run at a fixed ground rent of £3,000 per annum. It is known that the purchaser used the 11 per cent and 3 per cent table with tax at 40 per cent. Estimate the value of the market rent for the property.

This will not be very reliable evidence of rental value due to the many uncertainties involved. Was the sale price just based on the value of the lease and no other factors? Is an assumption that the same net sinking fund rate is likely to be receivable for the next 50 years realistic? Indeed, would the purchaser really set up a sinking fund over this long a period? How can it be ascertained that the purchaser used this valuation table? Leaving aside these questions, an estimate of the market rent could be obtained as follows.

Set out what is known and work backwards to the market rent.

MR say =   £ x p.a.
LESS ground rent payable =   £   3,000 p.a.
                       Net profit rent =£ x – £3,000   p.a.
       × YP 50 years @ 11% & 3% (tax 40%)   8.0144   (from tables)
                       Capital value =   £160,000 (known sale price)

To obtain the net profit rent you thus need to divide the YP into £160,000. Add £3,000 to the result and the full rental value is obtained.

that is £160,000/8.0144 = £19,964
£19,964 + £3,000 = FRV of £22,964 per annum
SAY market rent estimated at **£23,000** per annum

11.2 Valuation layout

To minimise the potential for errors and provide clearly presented figures, it is important to adopt a consistent and logical format for all valuation calculations. This will make the valuation more easily understood by third parties and assist the valuer to check the figures and the arithmetic used. The various worked examples given throughout this book provide some guidelines on this aspect, but points to note in particular on layout are:

1. Always give the valuation a heading, describing which type of value is being calculated, and for what property. As was explained in Chapter 1 there are many different types of value, and the lack of a clear statement at the outset of what is being found will lead to possible confusion and mistake later, particularly if the valuation is consulted by a third party not familiar with the case. In terms of specifying the property in the heading, be as precise as possible, giving the address including post or zip code and building name and floor numbers, where appropriate.

2. It is customary to refer to the rental sums collected by freeholders from their tenants as an 'income'. If the lease is on full repairing and insuring (FRI) terms, this sum received will be a 'net income' as no deductions will need to be made from it. Conversely, if the lease is on terms other than FRI, the sum received will be a 'gross income' from which 'outgoings' will need to be deducted to arrive at the net income. Only the *net income* is capitalised through multiplying by a years purchase.

3. With leaseholds, only the *net profit rent* is capitalised by a years purchase to find the capital value.

4. Any 'outgoings' are listed separately to the left hand side of the main valuation, and the total carried forward to be deducted from the gross income or gross profit rent figure.

5. With dual rate years purchases the property all risks leasehold yield is stated first and the sinking fund interest rate second. That is YP for 5yrs @ 12% + 3% means the years purchase for five years at an all risks yield of 12 per cent and a net sinking fund interest rate of 3 per cent.

6. Single rate years purchases are used to value freehold interests. Traditionally, dual rate years purchases (with or without adjustment for tax) are used for the valuation of leaseholds, although modern practice questions the validity of this approach, particularly for longer leases. This aspect is considered further in Section 16.11. The appropriate years purchase formula to be used to value each 'block' or 'tranche' of rental income will depend on when that rental is first receivable and for how long a period it will continue to be receivable and may require using both a YP and a PV number. For instance, an income lasting for three years and starting in two years' time will be capitalised through multiplying by the YP for three years × PV for two years (see Chapter 6).

7. Any capital deductions or additions are usually made at the end of the valuation, after the capitalisation of the income flows has been completed.

8. The layout looks neater and is easier to follow and check, if all income flows are in the same column, and all capital figures are in a separate column, to the right of the incomes column. Reference to the examples will make this principle clearer.

9. Although unnecessary for the purposes of giving examples in this book, in professional practice all valuations should be dated and signed. Without a date, the figures are meaningless and could be quoted out of context. In a market that is moving rapidly upward or downward, even a valuation carried out last month could now be out of date. A signature prevents any uncertainty as to who carried out the work and to whom all queries should be addressed.

11.3 Choice of all risks yield

As was seen in Section 4.6, the all risks or market yield (ARY) reflects all the risks, uncertainties and likely beneficial prospects of a property investment. All yields adopted in valuations should be based on market comparables; otherwise, the figure taken is merely a guess and cannot be substantiated in a negotiation. The importance of relevant evidence to support an opinion cannot be overstressed.

However, in what way could an investor reason whether the yields being accepted by others in the market are a fair reflection of the level of risk associated with a specific property investment? IPD (Investment Property Databank) have found that, in comparison to long-dated gilt yields, historically the addition of a 2 per cent 'buffer' has been made by the market to provide a risk premium for property. This figure is also suggested by Askham (1989c: 53) to 'reflect the particular risk and liquidity problems associated with property'. This is in relation to *average* all risks yields on UK prime property, but each individual investment must be considered on its own merits and many properties will be seen as requiring a considerably larger premium than this, with a few viewed as such good prospects that slightly lower than a 2 per cent 'buffer' may be acceptable. This is confirmed by Imber (1997: 198) who suggests that 'the risk premium for prime property could be 1–2%' whereas for secondary property it 'would be significantly higher than for prime property'.

The market all risks yield for any specific property will vary according to:

• type and class of property concerned;
• its location and position; and
• type of tenant in occupation or type who would potentially occupy the property.

All risks yields are based upon the risks and benefits of a property being let at its full market rent. This is particularly important when calculating 'term and reversion' valuations (see Section 11.10 below). In deciding a rate at which to capitalise an income, the principle is that the lower the rent being paid as a proportion of the full market rent, the more 'secure' that income is and the *lower* the yield that will be taken. Effectively, as the tenants are paying less than the full rent, they are less likely to default on payment of that rental. Conversely, if the rent paid by the tenants is higher than the market rent, a *higher* yield is expected to be taken as the income is 'less secure' and implies taking more risk.

However, no matter how secure it may appear, any income becomes insecure in *real* terms in times of inflation if it is fixed for too long a period. The longer the period an income is fixed, the less desirable that investment is as its real value will steadily decrease with each passing year. Thus a very low rent, such as a ground rent fixed for long periods, such as 25 or 33 years before review, would not attract high demand in the investment market and would only sell providing the yield was sufficiently high to offset its disadvantages.

Comparables will clearly indicate that freehold interests are more attractive to investors than leasehold interests are, even after allowance is fully made by provision of a sinking fund for the fact that leaseholds are, by their nature, depreciating assets. This is understandable when it is realised that a lessee is bound by covenants to the lessor and therefore is not as free to do as he or she wishes, as is a freeholder. For example, the freeholder may impose restrictions on the type and manner of use to which the property can be put, which can restrict its marketability as an investment.

Higher yields are required on leasehold investments compared with freeholds because there is more risk and uncertainty associated with leaseholds. To obtain an income from a leasehold, for example, the property must be sublet and rent collected from the sublessee. This involves management costs and the risk of rent not being paid on time, even though rent under the head lease will still be due to the freeholder. Thus if the sublessees are in arrears with their rental payments, the head leaseholder can suffer a negative cash flow from the property. Although freeholders are equally at risk from their tenants being in arrear with their rental payments, they are not at least obliged to make any payments themselves in return for the ownership of their interest in the property (unless of course the property is mortgaged).

Another reason for higher yields on leaseholds is that increases in inflation or decreases in savings interest rates can seriously erode the real value of the sinking fund payments, making it difficult for lessees to purchase a similar interest on expiry of their present leasehold.

11.4 Differences between freehold and leasehold all risks yields

Given that all risks yields on leasehold property investments are always higher than on directly comparable freehold property, where both are let at the full market rent value, what size of adjustment is appropriate? An addition of 1 or 2 per cent is frequently adopted. For more consistency through the yield range, the following additions are suggested as a guide to obtain a leasehold yield given the corresponding freehold yield:

Freehold yield range	*Leasehold yield*
up to 4%	+ 0.5%
>4 to <10%	+ 1%
10 to <15%	+ 1.5%
15 to 18%	+ 2%
over 18%	+ 2.5 to 3%

Should the valuer prefer to adopt a more proportionally consistent approach to adjusting the leasehold yield, a 'sliding scale' could be adopted on the basis that, say, an uplift of 20 per cent is used (that is leasehold ARY = 1.2 × freehold ARY) and rounded to nearest 0.25 per cent. This would provide the adjustments shown in Table 11.1. Nevertheless, good comparable evidence is *always* the best way of arriving at the correct leasehold yield to be adopted rather than making a subjective adjustment such as suggested above. However, obtaining such evidence for leasehold transactions can prove difficult making the use of an assumption unavoidable. Any such assumption must take into account the property type, location, lease length and tenant quality.

11.5 Initial yields

Initial yield or initial return is defined as: 'In an investment analysis, the initial income at the date of purchase expressed as a percentage of the purchase price including the costs of purchase' (Parsons 2004: 141).

Table 11.1   Suggested corresponding leasehold all risks yields to freehold allowing for an uplift of 20% and rounded to nearest 0.25%

Freehold ARY%	Corresponding leasehold ARY%	Freehold ARY%	Corresponding leasehold ARY%
2.00	2.25	11.25	13.50
2.25	2.50	11.50	13.75
2.50	3.00	11.75	14.00
2.75	3.25	12.00	14.25
3.00	3.50	12.25	14.50
3.25	3.75	12.50	15.00
3.50	4.00	12.75	15.25
3.75	4.50	13.00	15.50
4.00	4.75	13.25	15.75
4.25	5.00	13.50	16.00
4.50	5.25	13.75	16.50
4.75	5.50	14.00	16.75
5.00	6.00	14.25	17.00
5.25	6.25	14.50	17.25
5.50	6.50	14.75	17.50
5.75	6.75	15.00	18.00
6.00	7.00	15.25	18.25
6.25	7.50	15.50	18.50
6.50	7.75	15.75	18.75
6.75	8.00	16.00	19.00
7.00	8.25	16.25	19.50
7.25	8.50	16.50	19.75
7.50	9.00	16.75	20.00
7.75	9.25	17.00	20.25
8.00	9.50	17.25	20.50
8.25	9.75	17.50	21.00
8.50	10.00	17.75	21.25
8.75	10.50	18.00	21.50
9.00	10.75	18.25	21.75
9.25	11.00	18.50	22.00
9.50	11.25	18.75	22.50
9.75	11.50	19.00	22.75
10.00	12.00	19.25	23.00
10.25	12.25	19.50	23.25
10.50	12.50	19.75	23.50
10.75	12.75	20.00	24.00
11.00	13.00		

11.6 Reversionary yields

A reversionary investment is one where a property is let at less (or more) than open market rental value but there is the expectation that the rent will be adjusted to the market rent in the future, either at rent review or on reletting at the expiry of the existing lease. A reversionary yield or reversionary return is: 'The income on reversion from a property expressed as a percentage of the purchase price' (Parsons 2004: 222).

With such investments, there are two alternative approaches to valuing:

- 'term and reversion', also known as the 'block income' approach; or
- 'hardcore' or 'layer' approach.

11.7 Equivalent yields

Equivalent yield is: 'The constant capitalisation rate applied to all cash flows, that is, the internal rate of return from an investment property reflecting reversions to current market rent, and such items as voids and expenditures but disregarding potential changes in market rents. Conventionally calculated on a nominal basis, assuming that rents receivable/payable annually in arrear' (Parsons 2004: 95). It is thus 'the discount rate applied to the cash flows so that the capitalised sum of all the incomes discounted at this rate equals the capital outlay and where all rents, now or in the future at review, lease renewal or reletting are taken at present values current at the date of valuation'.

An alternative view is that it is the single yield to be used in both the term and reversion of a 'block income approach' valuation rather than using differential yields to adjust between the perceived relative securities of the term income relative to the income on reversion.

The *true equivalent yield* (TEY) is defined as: 'The internal rate of return from an investment property reflecting reversions to current market rent, and such items as voids and expenditures but disregarding potential changes in market rents and reflecting the actual timing of cash flows. In the UK rents receivable quarterly in advance' (Parsons 2004: 257). This gives a more accurate assessment of the overall return from the investment as it takes into account actual market practice.

To calculate the equivalent yield received on an investment requires interpolation of the discounted cash flow gross present value to find the internal rate of return (IRR). This is considered in Section 11.11 below.

11.8 Equated yields

Equated yield is defined as: 'In valuing an investment property, the internal rate of return, being the discount rate which needs to be applied to the flow of income expected during the life of the investment (reflecting reversion to market rent and such items as voids and expenditures) so that the net income discounted at this rate equals the capital outlay. Rents at review, lease renewal or reletting take account of expected future rental changes' (Parsons 2004: 92).

In other words, it is the internal rate of return (IRR) found from a discounted cash flow (DCF) analysis, where on every occasion when the rent can be adjusted or a capital sum received or paid, the figure used is one that allows for expected compounded rental or capital growth between now and then. It thus requires explicit allowances to be made for anticipated future growth, whereas in 'traditional' valuations such factors must be implied in the choice of the all risks yield. In this way, equated yields reflect the true overall return from the property, after taking into account the effects of inflationary increases. This approach is considered in Chapter 12.

11.9 Analysis of a property yield

Investors have to assess their attitude to risk and the main objective of their chosen form of investment when deciding on what constitutes an acceptable and appropriate yield. Risk has to be balanced against security. The more risk an investor is prepared to take, then by implication, the less secure will be their investment and the higher the level of yield that will be sought to make it worthwhile accepting the greater risk. There has to be a balance between risk and reward. Should the rewards not be sufficiently high to tempt an investor to take greater risks then there is no incentive to do so. It would be much better to seek a safer, more secure form of investment that offers the same return. On the other hand, even if the investment was considered absolutely safe and risk free, there would be a minimal level of yield the investor would seek in return for being unable to use the invested capital during the period of investment. This is illustrated in Figure 11.1.

As mentioned in Section 3.8, the majority of investors tend to be relatively risk-averse. They would prefer to forgo some yield return in place of the greater safety of keeping the capital value of their investment intact or growing at a slower rate, than to take additional risk with the possibility of this negatively affecting the capital. It is important for any investment adviser to formally ascertain the investor's risk attitude and the primary goals of the investment strategy, which is whether higher income or greater likelihood of capital growth is required. The choice between them can often be a taxation consideration, where if capital gain is to be taxed at a lower effective rate than additional income, then the growth in capital is considered preferable and vice versa.

Figure 11.1   Relationship between all risks yield (ARY) and risk level of investment

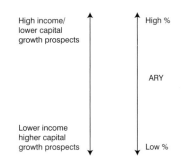

Figure 11.2   Relationship between all risks yield (ARY) and income and capital growth prospects of an investment

These objectives can be mutually exclusive, in that high-yielding investments provide higher annual income but are less likely to provide such good capital growth, as shown in Figure 11.2.

In arriving at an acceptable yield on a property, the investor will also have regard to available rates of return on other forms of investment and the level of inflation within an economy. As was seen with the 'ideal investment', maintaining the real as well as nominal value of the capital is desirable and for this to occur the value of the capital invested should at least rise in line with other inflationary increases. The average return on gilts and other 'safe' forms of investment, such as bank deposit accounts, together with those on equities, will also strongly influence the choice of yield.

Isaac and Steley (1999: 45) showed that, in essence, three main components to the total yield would be sought:

- compensation for not being able to use the invested money for another purpose, which can be termed a 'return for time preference' (that is forgoing use of the money today in preference for using it at a future date by which time it will hopefully have grown to a larger sum); plus
- the prevailing average rate of inflation present in the economy, as otherwise the purchasing power of the income received and the value of the invested capital will be eroded; plus
- reward for taking risk, which will depend on the extent of the risk and uncertainty involved.

The sum total of these components should provide the equated yield that will be sought by the investor, as this provides the total return and incorporates the initial yield plus anticipated growth.

11.10 Term and reversion valuations using the 'block income' approach

This is the 'traditional' approach to valuing a reversionary investment, which occurs when, in present value terms, the current rental income or profit rent will rise or fall at some future date.

Freehold term and reversion valuation

As the current income from a reversionary property is not the estimated present market rent, then a term and reversion valuation approach can be adopted. This means there will be (at least) two different blocks of income receivable over different times. One income will be receivable over the term and this will be followed by another (usually higher) rental, on the reversion, at some future date.

In the conventional investment method, no adjustment is made to the present market rent to reflect rental growth or anticipated inflationary increases between now and the review date. All figures are computed at present value, in other words their value today, not what they may be in the future. Once the market rent is reached, in present value terms, it cannot be increased any higher and the income is deemed receivable at that level in perpetuity. The implicit assumption is that the property will be continually relet at market rent thereafter.

The market all risks yield, as obtained from comparable evidence, is usually a reflection of the risk rate when the market rent is being received. As the rental during the term is not the market rent, it can be judged a lower risk (when substantially below the market rent) or higher risk (when above the market rent) and this should be reflected in the choice of yield adopted for the term calculation. There is no standard rule on what percentage adjustment should be made to reflect this different level of security of income. It will be a subjective assessment of the perceived degree of risk involved. Where the difference between the rents is reasonably substantial, for example the term rent is less than 80 per cent of the reversion rent, most common practice will be to lower the yield by 1 per cent. However, smaller or larger adjustments may be justifiable, depending on the opinion of the valuer and the facts of the case.

Where the term rental is a low figure and/or is fixed for a long period, such as ten years or more, it is arguable whether enhanced security of income would be more than offset by loss of value in real terms. In such cases, a term yield equal to, or even higher than, the reversion yield may be appropriate.

To avoid the subjectivity that selection of the term yield may involve, the equivalent yield approach is preferred where the same yield is used throughout the calculation. This is a simpler and more consistent approach (see Section 11.11 below).

The capital value of each block is separately calculated and the sum of the value of the blocks represents the capital value of the freehold investment. Figure 11.3 is a worksheet that provides the standard format and layout of the valuation. This can be used as a template for calculations with the relevant figures inserted in the blank spaces.

VALUATION ANALYSIS OF:

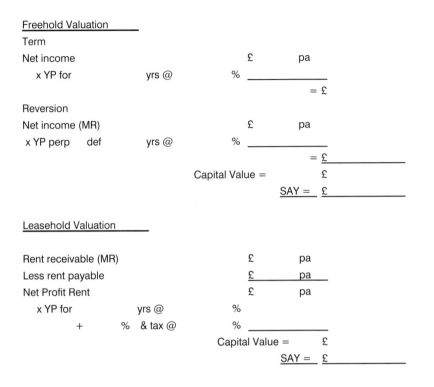

Figure 11.3 Worksheet for standard freehold term and reversion valuation and a leasehold valuation

Example

Freehold term and reversion example 1

A freeholder currently receives a rental income of £100,000 per annum under a full repairing and insuring (FRI) lease granted three years ago for a term of 25 years with five-yearly rent reviews. The present estimated market rent (MR) is £150,000 per annum. In other words, in present value terms, the rent can be increased to £150,000 per annum at the next review in two years time.

There are thus two blocks of income: £100,000 per annum for the next two years (the term), followed by £150,000 per annum in perpetuity, and commencing in two years' time (the reversion). The present capital value of the freehold interest will be the sum of the value of the term + reversion blocks. This is illustrated in Figure 11.3.

There is no need to take account of the additional rent reviews which will be due in 7, 12, 17 and 22 years from now under this lease. Once the rent has been increased to MR, it (usually) cannot go any higher, and thus the interest can be valued in perpetuity from that point on.

A valuation of the interest, assuming an all risks yield of 8 per cent is:

Term

Net income =	£	100,000 p.a.
× YP 2 yrs @ 7%		1.808
	= £	180,800

Reversion to MR

Net income =	£	150,000 p.a.
× YP perp def 2 yrs @ 8%		10.717
	=	£ 1,607,500
	Capital value (MV) =	£ 1,788,300
	SAY =	£ 1,788,000

- The all risks market yield is applied *to the reversion* not the term.
- It is a yield derived from market analysis of lettings at MR.
- The term yield is then adjusted to reflect the additional 'security' of the term income that is if it is *significantly* lower than the MR, as in this instance; it is therefore deemed 'safer' than the MR in that the tenant is less likely to default on the payment of a lesser sum than the present full rental value.
- In these circumstances it is traditional to reduce the ARY in the term by 1 per cent.
- Using the conventional layout, the written descriptions are in the 'left-hand column', rental income and YP numbers in the 'middle column' and capital values in the 'right-hand column'.
- The market value is then found be adding together the capital value sums.
- This final value is always likely to be an odd amount so it is usual to round it off – this is indicated by 'SAY £'.

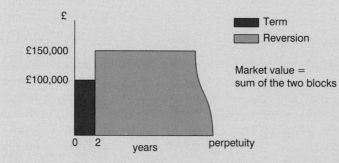

Figure 11.4   Diagrammatic representation of freehold term and reversion: example 1

Example

Freehold term and reversion example 2

A freehold property was let two years ago at £60,000 per annum net on a twenty-year lease with provision for upward rent revisions at the end of every five years. Value the freehold if the estimated MR is £80,000 and an all risks yield of 7 per cent is required.

Term

Net income =	£	60,000 p.a.
× YP 3 yrs @ 6%		2.673
		= £ 160,380

Reversion

Net income =	£	80,000 p.a.
× YP perp def 3 yrs @ 7%		11.6614
		= £ 932,912
	Capital value =	£ 1,093,292
	SAY =	£ 1,093,000

- 1 per cent lower yield used in term as term income considered 'more secure'.
- Term is only until next rent review when rent will be increased to market rent.

Freehold with more than one reversion

This will arise when the property is occupied by a tenant paying a rent below the market rent and subject to a forthcoming rent review, lease renewal or contractual increase that will still result in a sum payable below the full market figure.

Example

A lease is granted with a stepped rent arrangement whereby the rent for the first three years is agreed at £50,000 per annum with a prearranged further increase to £60,000 per annum for the next two years and otherwise subject to rent reviews every five years to the market rent. Suppose there is a significant rise in market rental levels over the first two years of the term, so that by the end of year two the estimated market rent is £70,000 per annum. There will now be a term of one year at £50,000 per annum, followed by an initial reversion to £60,000 per annum for two years (deferred one year) and then a final reversion to £70,000 per annum in perpetuity deferred three years.

Assuming a market all risks yield of 8 per cent is deemed to apply to this investment. Each lower 'tranche' of income is only slightly 'more secure' than this risk rate and so a 0.5 per cent reduction in yield for each step down may be considered appropriate. The resulting valuation of the freehold interest will be:

Term

Net income =	£	50,000 p.a.
× YP for 1 yr @ 7%		0.935
		= £ 46,729

Reversion to intermediary rent

Net income =	£	60,000 p.a.
× YP 2yrs @ 7.50%		1.796
× PV 1yrs @ 7.50%		0.930
		= £ 100,218

Reversion to market rent

Net income =	£	70,000 p.a.
× YP perp def 3yrs @ 8%		9.923
		= £ 694,603
	Market value =	£ 841,550
	SAY =	£ 841,500

Leasehold term and reversion valuation

Valuing leasehold interests on a term and reversion approach will arise when the 'profit rent' which the leaseholder enjoys varies. An example of this could be with a head lease where the ground rent is fixed or reviewed at less frequent intervals to the occupational lease given to a subtenant.

Example

A property company is interested in purchasing the head lease of an office building.

The net internal floor area of the building is 200 sq.m. per floor on each of five floors. The building is at present occupied by a sublessee who holds a lease for ten years, granted six years ago, without review. The rent payable under this sublease is £135,000 per annum on full repairing and insuring terms. The full rental value is estimated at £200 per sq.m. The head lease has an unexpired term of nineteen years at a fixed rent of £20,000 per annum.

Allowing for payment of tax at 30 per cent, net annual sinking fund rate of 3 per cent and for an all risks yield of 8 per cent, advise the company what premium they should pay to purchase the head leasehold interest.

Term

Rent receivable =	£	135,000 p.a.
Less rent Payable =	£	20,000 p.a.
Net profit rent =	£	115,000 p.a.
× YP for 4 yrs @ 7%		
+ 3% & tax @ 30%		2.430
	= £	279,488

Reversion

Rent receivable =	£	200,000 p.a.
Less rent Payable =	£	20,000 p.a.
Net profit rent =	£	180,000 p.a.
× YP for 15 yrs @ 8%		
+ 3% & tax @ 30%		6.377
× PV 4 yrs @ 8%		0.735
	= £	843,734

Market value (assignment price) = £ 1,123,221

SAY = **£ 1,123,000**

- The term yield is lower to reflect the fact that the rent receivable in the term is more secure than in the reversion. The level of risk is assessed on the rent to be collected not the net profit rent.
- The estimated market rent has been applied equally to all floors. Alternatively, if the building lacked a lift, it could be reasonably argued that slightly different rates should be applied to each floor to reflect their position in the building.

11.11 From term and reversion with differential yields to equivalent yield approach

Present-day term and reversion methods were devised in the 1950s when inflation was negligible and fixed incomes below market rent levels were more secure. In most economies nowadays, they are far less secure in real terms due to the effects of inflation, so reducing the yield to reflect increased security on below market rents; particularly where they are fixed for more than five years this can be an arguable practice and open to too much subjective variation.

A much simpler and more consistent approach is to adopt the equivalent yield approach. In this, the same yield is used throughout the appraisal to capitalise all incomes, whatever their level and for however long they are fixed. Providing such yields are derived from an equivalent yield analysis of market transactions, then the figures adopted are reliable.

When the equivalent yield is found from an appraisal using conventional annually in arrear formulae or tables, it can be converted into its quarterly in advance 'true equivalent yield' (TEY) by use of the conversion formula provided in Section 7.3 above. Tables 11.2 and 11.3 provide a

Table 11.2 Quarterly in Advance Conversion Table: nominal equivalent to true equivalent yield

Quarterly in Advance Conversion Table

Nominal Equivalent Yield (annually in arrear) to True Equivalent Yield (quarterly in advance)

Nominal Equivalent	True Equivalent	Nominal Equivalent	True Equivalent	Nominal Equivalent	True Equivalent
2.00%	2.0253%	7.50%	7.8652%	13.00%	14.1290%
2.10%	2.1279%	7.60%	7.9752%	13.10%	14.2471%
2.20%	2.2306%	7.70%	8.0853%	13.20%	14.3653%
2.30%	2.3334%	7.80%	8.1956%	13.30%	14.4836%
2.40%	2.4364%	7.90%	8.3060%	13.40%	14.6021%
2.50%	2.5396%	8.00%	8.4166%	13.50%	14.7207%
2.60%	2.6428%	8.10%	8.5273%	13.60%	14.8396%
2.70%	2.7462%	8.20%	8.6381%	13.70%	14.9585%
2.80%	2.8497%	8.30%	8.7491%	13.80%	15.0776%
2.90%	2.9533%	8.40%	8.8602%	13.90%	15.1969%
3.00%	3.0571%	8.50%	8.9715%	14.00%	15.3163%
3.10%	3.1610%	8.60%	9.0829%	14.10%	15.4359%
3.20%	3.2650%	8.70%	9.1945%	14.20%	15.5556%
3.30%	3.3692%	8.80%	9.3061%	14.30%	15.6755%
3.40%	3.4735%	8.90%	9.4180%	14.40%	15.7955%
3.50%	3.5779%	9.00%	9.5300%	14.50%	15.9157%
3.60%	3.6825%	9.10%	9.6421%	14.60%	16.0361%
3.70%	3.7872%	9.20%	9.7544%	14.70%	16.1566%
3.80%	3.8920%	9.30%	9.8668%	14.80%	16.2773%
3.90%	3.9969%	9.40%	9.9793%	14.90%	16.3981%
4.00%	4.1020%	9.50%	10.0920%	15.00%	16.5191%
4.10%	4.2073%	9.60%	10.2049%	15.10%	16.6402%
4.20%	4.3126%	9.70%	10.3178%	15.20%	16.7615%
4.30%	4.4181%	9.80%	10.4310%	15.30%	16.8830%
4.40%	4.5237%	9.90%	10.5443%	15.40%	17.0046%
4.50%	4.6295%	10.00%	10.6577%	15.50%	17.1263%
4.60%	4.7354%	10.10%	10.7712%	15.60%	17.2483%
4.70%	4.8414%	10.20%	10.8850%	15.70%	17.3704%
4.80%	4.9475%	10.30%	10.9988%	15.80%	17.4926%
4.90%	5.0538%	10.40%	11.1128%	15.90%	17.6150%
5.00%	5.1602%	10.50%	11.2270%	16.00%	17.7376%
5.10%	5.2668%	10.60%	11.3413%	16.10%	17.8603%
5.20%	5.3735%	10.70%	11.4557%	16.20%	17.9832%
5.30%	5.4803%	10.80%	11.5703%	16.30%	18.1062%
5.40%	5.5873%	10.90%	11.6850%	16.40%	18.2294%
5.50%	5.6944%	11.00%	11.7999%	16.50%	18.3528%
5.60%	5.8016%	11.10%	11.9150%	16.60%	18.4763%
5.70%	5.9090%	11.20%	12.0302%	16.70%	18.6000%
5.80%	6.0165%	11.30%	12.1455%	16.80%	18.7238%
5.90%	6.1242%	11.40%	12.2610%	16.90%	18.8479%
6.00%	6.2319%	11.50%	12.3766%	17.00%	18.9720%
6.10%	6.3398%	11.60%	12.4924%	17.10%	19.0964%
6.20%	6.4479%	11.70%	12.6083%	17.20%	19.2209%
6.30%	6.5561%	11.80%	12.7244%	17.30%	19.3455%
6.40%	6.6644%	11.90%	12.8406%	17.40%	19.4703%
6.50%	6.7729%	12.00%	12.9570%	17.50%	19.5953%
6.60%	6.8815%	12.10%	13.0735%	17.60%	19.7205%
6.70%	6.9902%	12.20%	13.1902%	17.70%	19.8458%
6.80%	7.0991%	12.30%	13.3070%	17.80%	19.9713%
6.90%	7.2081%	12.40%	13.4240%	17.90%	20.0969%
7.00%	7.3173%	12.50%	13.5411%	18.00%	20.2227%
7.10%	7.4266%	12.60%	13.6584%	18.10%	20.3487%

Table 11.3   Quarterly in Advance Conversion Table: true equivalent to nominal equivalent yield

Quarterly in Advance Conversion Table

True Equivalent Yield (quarterly in advance) to Nominal Equivalent Yield (annually in arrear)

True Equivalent	Nominal Equivalent	True Equivalent	Nominal Equivalent	True Equivalent	Nominal Equivalent
2.00%	1.9754%	7.50%	7.1671%	13.00%	12.0369%
2.10%	2.0729%	7.60%	7.2584%	13.10%	12.1227%
2.20%	2.1702%	7.70%	7.3496%	13.20%	12.2084%
2.30%	2.2675%	7.80%	7.4407%	13.30%	12.2940%
2.40%	2.3646%	7.90%	7.5317%	13.40%	12.3795%
2.50%	2.4617%	8.00%	7.6225%	13.50%	12.4649%
2.60%	2.5586%	8.10%	7.7133%	13.60%	12.5502%
2.70%	2.6553%	8.20%	7.8040%	13.70%	12.6354%
2.80%	2.7520%	8.30%	7.8946%	13.80%	12.7206%
2.90%	2.8486%	8.40%	7.9850%	13.90%	12.8056%
3.00%	2.9450%	8.50%	8.0754%	14.00%	12.8905%
3.10%	3.0413%	8.60%	8.1656%	14.10%	12.9754%
3.20%	3.1375%	8.70%	8.2558%	14.20%	13.0601%
3.30%	3.2336%	8.80%	8.3458%	14.30%	13.1448%
3.40%	3.3295%	8.90%	8.4358%	14.40%	13.2294%
3.50%	3.4254%	9.00%	8.5256%	14.50%	13.3138%
3.60%	3.5211%	9.10%	8.6153%	14.60%	13.3982%
3.70%	3.6167%	9.20%	8.7050%	14.70%	13.4825%
3.80%	3.7122%	9.30%	8.7945%	14.80%	13.5667%
3.90%	3.8076%	9.40%	8.8839%	14.90%	13.6508%
4.00%	3.9029%	9.50%	8.9733%	15.00%	13.7348%
4.10%	3.9981%	9.60%	9.0625%	15.10%	13.8188%
4.20%	4.0931%	9.70%	9.1516%	15.20%	13.9026%
4.30%	4.1880%	9.80%	9.2406%	15.30%	13.9863%
4.40%	4.2829%	9.90%	9.3295%	15.40%	14.0700%
4.50%	4.3776%	10.00%	9.4184%	15.50%	14.1536%
4.60%	4.4721%	10.10%	9.5071%	15.60%	14.2370%
4.70%	4.5666%	10.20%	9.5957%	15.70%	14.3204%
4.80%	4.6610%	10.30%	9.6842%	15.80%	14.4037%
4.90%	4.7552%	10.40%	9.7726%	15.90%	14.4869%
5.00%	4.8494%	10.50%	9.8610%	16.00%	14.5700%
5.10%	4.9434%	10.60%	9.9492%	16.10%	14.6530%
5.20%	5.0373%	10.70%	10.0373%	16.20%	14.7360%
5.30%	5.1311%	10.80%	10.1253%	16.30%	14.8188%
5.40%	5.2248%	10.90%	10.2132%	16.40%	14.9016%
5.50%	5.3184%	11.00%	10.3010%	16.50%	14.9842%
5.60%	5.4119%	11.10%	10.3888%	16.60%	15.0668%
5.70%	5.5052%	11.20%	10.4764%	16.70%	15.1493%
5.80%	5.5985%	11.30%	10.5639%	16.80%	15.2317%
5.90%	5.6916%	11.40%	10.6513%	16.90%	15.3140%
6.00%	5.7847%	11.50%	10.7387%	17.00%	15.3962%
6.10%	5.8776%	11.60%	10.8259%	17.10%	15.4784%
6.20%	5.9704%	11.70%	10.9130%	17.20%	15.5604%
6.30%	6.0631%	11.80%	11.0001%	17.30%	15.6424%
6.40%	6.1557%	11.90%	11.0870%	17.40%	15.7243%
6.50%	6.2482%	12.00%	11.1738%	17.50%	15.8060%
6.60%	6.3405%	12.10%	11.2606%	17.60%	15.8877%
6.70%	6.4328%	12.20%	11.3472%	17.70%	15.9694%
6.80%	6.5250%	12.30%	11.4338%	17.80%	16.0509%
6.90%	6.6170%	12.40%	11.5202%	17.90%	16.1323%
7.00%	6.7090%	12.50%	11.6066%	18.00%	16.2137%
7.10%	6.8008%	12.60%	11.6928%	18.10%	16.2949%
7.20%	6.8925%	12.70%	11.7790%	18.20%	16.3761%
7.30%	6.9842%	12.80%	11.8651%	18.30%	16.4572%
7.40%	7.0757%	12.90%	11.9511%	18.40%	16.5382%

quick and readily available look-up method of converting either from nominal (annually in arrear) to true (quarterly in advance) equivalent yield or vice versa.

To find the nominal equivalent yield from a conventional term and reversion appraisal, it is necessary to find the internal rate of return from the cash flows, where no explicit addition for estimated future capital or rental growth is made and all figures are based on present values. When the cash flows are entered into a spreadsheet, the software's rate of return solver can be used. A less sophisticated, but time-honoured approach is trial and error followed by interpolation. In this two trial rates are used, as close as possible to either side of the expected outcome and then the outcomes interpolated to find the actual yield returned. This technique is considered again and in further detail in Chapter 12 in relation to finding the internal rate of return from a cash flow.

Example

Equivalent yield approach

Value the freehold interest in an office property that measures 850 square metres (NIA) and is occupied by a tenant on a fifteen-year lease that commenced two years ago. The current rent payable is £255,000 per annum on full repairing and insuring terms. Rent reviews are every five years to full market rent.

The freehold in a very similar nearby building has just been sold on the open market for £6 million. It currently provides a net rental income of £200,000 per annum. The lease on this property is also on FRI terms. The original length of this lease was twenty years, but there are only twelve years left to run on it. Rent reviews are again on a five-yearly basis to open market rent. The net internal floor area is 783 square metres.

Research into local market conditions leads to the conclusion that market rents for this type of building are currently £400 per square metre per annum when let on ten to fifteen year FRI leases and subject to five-year rent reviews.

Valuation

The market rent for each property can be estimated as:

Subject property = 850 sq.m. @ £400 per sq.m. = £340,000 per annum.
Comparable property = 783 sq.m. @ £400 per sq.m. = £313,200 per annum.

Analysis of sale of comparable property

To find the equivalent yield at which the sale took place, an estimate needs to be made to provide the first trial rate to use. Taking the market rent as a percentage of the sale price (the reversionary yield) gives:

£313,200 × 100/£6,000,000 = 5.22%

Say use 5.25 per cent as it can be looked up in *Parry's Tables*. The first trial valuation is:

Term

Net income =	£	200,000 p.a.	
× YP for 2yrs @ 5.25%		1.853	
		= £	370,569

Reversion to market rent

Net income =	£	313,200 p.a.	
× YP perp def 2yrs @ 5.25%		17.195	
		= £ 5,385,403	
	Market value =	£ 5,755,972	

This is reasonably close to the actual sale price, but not accurate enough so a second trial rate needs to be used. As the value is not high enough, a *lower* yield is necessary, but not too much lower otherwise the value will be far too high, so try 0.5 per cent lower, i.e. 4.75 per cent.

Term

Net income =	£	200,000 p.a.	
× YP for 2yrs @ 4.75%		1.866	
		= £	373,204

Reversion to market rent

Net income =	£	313,200 p.a.	
× YP perp def 2yrs @ 4.75%		19.187	
		= £ 6,009,247	
	Market value =	£ 6,382,451	

Now interpolate between the results to estimate the nominal equivalent yield:

The trial rate of 5.25 per cent produces a value of £5,755,972 and the trial rate of 4.75 per cent a value of £6,382,451. The difference between the values = £626,479 and the difference between the trial yields = 0.5 per cent. The difference between the value at the lower trial rate and the actual sale price is £382,451. Thus to find the rate at which the value = the sale price:

4.75% + (0.5% × (£382,451/£626,479)) = 5.05%

There is some inconsequential error in this figure in that the connection between the values at the two different rates is not a perfectly straight-line relationship as is implied by the above calculation. However, for most practical purposes, the mathematical error is too minor to be of any significance, providing the two trial rates are very close together. In this instance, using a computer spreadsheet proves that the actual nominal yield producing an exact value of £6 million is 5.043189 per cent.

Using the estimated nominal equivalent yield of 5.05 per cent results in a valuation of:

Term

Net income =	£	200,000 p.a.	
× YP for 2yrs @ 5.05%		1.858	
		= £	371,619

Reversion to market rent

Net income =	£	313,200 p.a.
× YP perp def 2yrs @ 5.05%		17.944
	=	£ 5,620,025
	Market value =	£ 5,991,644
	SAY =	£6,000,000 (the sale price)

Alternatively, the analysis could have been undertaken on a true equivalent yield (quarterly in advance) basis rather than the nominal equivalent yield (annually in arrear) one. Referring to the look-up chart in Table 11.3, an average of the figures at 5.0 and 5.1 per cent can be used to approximate the 5.05 per cent nominal figure. This leads to the conclusion that the true equivalent yield is 5.21 per cent (if the exact nominal yield of 5.043189 per cent is used and the quarterly in advance formula applied to it then the precise true yield will be found as 5.20625 per cent).

The valuation can be analysed using quarterly in advance formulae or tables at the TEY of 5.21 per cent as follows:

Term

Net income =	£	200,000 p.a.
× YP for 2 yrs @ 5.21%		1.914
	=	£ 382,775

Reversion to market rent

Net income =	£	313,200 p.a.
× YP perp def 2 yrs @ 5.21%		17.901
	=	£ 5,606,598
	Market value =	£ 5,989,373
	SAY =	£ 6,000,000 (the sale price)

Freehold valuation of subject office building

Having established the current market equivalent yield from comparison, the subject property can now be valued, on an annually in arrear basis.

Term

Net income =	£	255,000 p.a.
× YP for 3yrs @ 5.05%		2.721
	=	£ 693,778

Reversion to market rent

Net income =	£	340,000 p.a.
× YP perp def 3yrs @ 5.05%		17.081
	=	£ 5,807,636
	Market value =	£ 6,501,414
	SAY =	£ 6,500,000

Alternatively, on a quarterly in advance basis:

Term

Net income =	£	255,000 p.a.
× YP for 3 yrs @ 5.21%		2.800
	= £	714,085

Reversion to market rent

Net income =	£	340,000 p.a.
× YP perp def 3 yrs @ 5.21%		17.015
	=	£ 5,784,949

Market value = £ 6,499,034

SAY = **£ 6,500,000**

11.12 'Hardcore' or 'layer' approach

Basis of method and why used

The basic principle of the hardcore approach is that the lowest level (usually the initial) rental is considered the 'hardcore' or 'bottom layer' and is capitalised in perpetuity for a freehold or the full length of the unexpired term for a leasehold. This is the most secure portion of the income as it is below the rack rental, and a relatively low yield for the property type is used on its capitalisation to reflect this. On reversion, the difference between this lower rental and the rack rent is known as the 'top slice' or 'top layer'. This is capitalised separately from the hardcore at a higher yield rate to reflect the greater risk and uncertainty associated with the collection of this additional rental sum. The sum of the capitalised hardcore and top slice incomes provides the assessment of market value.

Where the rent may be increased on one or more occasions to a figure still less than the rack rent, and before the final reversion to rack rental, each slice or tranche of income would be valued separately, with appropriate yields applied to reflect the relative uncertainty and risk associated with each slice or layer.

Example

Hardcore or layer approach example 1

A freehold property is let at £100,000 per annum for the next three years, after which the rent is to be increased under a stepped rental agreement to £125,000 per annum for the following four years. After seven years there will be a rent review at which time the market rent, which is at present estimated at £150,000 per annum, will become payable. The hardcore rent is adjudged to represent a 6 per cent risk rate, with each rental increase being considered an additional 1 per cent risk.

Valuation

Hardcore rent =	£	100,000 p.a.
× YP perp @ 6%		16.667
		= £ 1,666,700
Middle slice income =	£	25,000 p.a.
× YP perp def 3 yrs @ 7%		11.661
		= £ 291,525
Top slice income =	£	25,000 p.a.
× YP perp def 7 yrs @ 8%		7.294
		= £ 182,350
	Capital value =	£ 2,140,575
	SAY =	£ 2,140,000

Advantages and disadvantages of method and comparison with term and reversion calculations

The hardcore method is in some ways arithmetically simpler to calculate and is favoured by many practitioners over the traditional term and reversion method as being more 'modern'. However, the following are some of the issues and problems that must be considered in using either method of valuation:

- What yield rate should be used in each and should this be on an equivalent yield basis or should differential yields be used for the term and hardcore parts of the respective calculations?
- Yields derived from traditional term and reversion analysis of comparables cannot be used in hardcore valuations as this will then not be comparing like with like.
- Analysis and valuation must be undertaken using the same method for consistency and because figures derived in one way do not directly apply when employed in a different approach.
- There is only one 'correct' value for every investment (its achieved sale price in the open market) and so, whichever method is used, the result should be the same!
- Discounted cash flow should be used to support values derived from term and reversion or hardcore valuations and to show the potential problems and differences in investments.

These points are considered in more detail based on the following scenario.

Industrial premises were let three years ago on a fifteen-year full repairing and insuring lease. The rent passing is £220,000 per annum and rent reviews are five-yearly. The lettable area is 3,500 sq.m. A directly comparable unit on the same estate has recently been let on the same terms as the subject property at a rent of £260,000 per annum. It has a floor area of 3,250 sq.m. The freehold all risks yield at market rent is estimated at 6 per cent.

Value the freehold interest in the subject property using:

- the conventional term and reversion approach; and
- the hardcore or layer approach.

Analysis of comparable:

3,250 sq.m. recently let at £260,000 p.a. = rental of £80.00 psm p.a.
Therefore estimated market rent of subject property is 3,500 sq.m. × £80.00 psm
= £280,000 per annum

Assuming the yields on term and reversion and the hardcore and top slice incomes are the same (in other words the equivalent yield approach), the valuations will be as shown in Figure 11.5.

Thus it can be seen that when the same yields are used for both the term and reversion and for the hardcore and top slice incomes, the same result is obtained from both methods. This will not be so if the term and reversion yields differ, to reflect the additional security offered by the term income only being 78.6 per cent of the reversionary income, unless the incremental rent in the hardcore method is valued at the appropriate marginal yield rate. However, deciding what rates are to be used in the layer approach is not as straightforward as it may appear.

Using the same data as above, but assuming 5 per cent is adopted as the yield on the term, being a conventional 1 per cent below the reversion all risks yield to reflect the more secure nature of

Valuation by conventional approach		
Term		
Net income =	£220,000 p.a.	
× YP for 2yrs @ 6%	1.833	
	=	£403,346
Reversion to market rent		
Net income =	£280,000 p.a.	
× YP perp def 2yrs @ 6%	14.833	
	=	£4,153,317
	Market value =	£4,556,663
	SAY =	**£4,557,000**
Valuation by hardcore approach		
Hardcore		
Net income =	£220,000 p.a.	
× YP perp @ 6%	16.667	
	=	£3,666,667
Top slice		
Net income =	£60,000 p.a.	
× YP perp def 2yrs @ 6%	14.833	
	=	£889,996
	Market value=	£4,556,663
	SAY=	**£4,557,000**

Figure 11.5   Valuation of a property using the equivalent yield in the conventional approach and hardcore method

Term

Net income =	£220,000 p.a.	
× YP for 2yrs @ 5%	1.859	
	=	£408,980

Reversion to market rent

Net income =	£280,000 p.a.	
× YP perp def 2yrs @ 6%	14.833	
	=	£4,153,240
	Market value=	£4,562,220
	SAY-=	**£4,560,000**

Figure 11.6   Valuation by conventional approach using differential yields

the term income, the valuation is as shown in Figure 11.6. Logically, if an income of £220,000 per annum is considered a 5 per cent risk under this approach, it would follow that the same yield is used on the hardcore income, as it is the same amount. Unfortunately, this conclusion would be incorrect. It will be observed that the use of a lower yield in the term has only made a negligible difference to the final value.

Assuming the valuation produced by the conventional approach provides the 'correct' value, how could the same sum be arrived at using the layer method? To do this, the marginal rate of interest is calculated to apply to the top slice income. The yield on the incremental rent will equal the incremental income divided by the capital value of the incremental income:

Incremental income = £280,000 – £220,000 = £60,000 p.a. (market rent less term rent).

Capital value of incremental income = Capital value of rack rent in perpetuity minus capital value of hardcore rent in perpetuity (see Figure 11.7).

This is within £15,000 (or 0.3 per cent) of the value produced by the term and reversion approach, so is virtually the same, but is it correct or logical? How can the receipt of an additional £60,000 per annum income be considered such a risky investment that the yield needs to be increased by 17.5 per cent from that used to value the hardcore rent? It is illogical and yet if the yield is not adjusted to 22.5 per cent, but 6 per cent used, the hardcore valuation would be as shown in Figure 11.8.

And yet: Market rent × YP perp @ 6% = £280,000 × 16.667 = £4,666,760!

In other words, a property let at less than market rent has been valued substantially higher than if it were let at that full rent! This is obviously grossly incorrect.

An alternative approach that may seem logical is to adopt the all risks yield of 6 per cent on the hardcore income and increase it by 1 per cent for the top slice to reflect the additional risk represented by the receipt of the extra £60,000 per annum (see Figure 11.9).

This is clearly better, but it is still £140,000 (3 per cent) adrift of the figure produced by the term and reversion value, although this time an under rather than overvaluation. So what would be the correct yields to adopt if it is known that the property has actually sold for £4,560,000 (the term and reversion valuation using differential yields) and the sale is being analysed for comparable purposes? Figure 11.10 shows the corrected hardcore valuation, found by trial and error.

Rack rent =	£280,000 p.a.	
× YP in perp @ 6%	16.667	
	=	£4,666,760
LESS Hardcore rent	£220,000 p.a.	
× YP in perp @ 5%	20	
	=	£4,400,000
Capital value of incremental rent	=	£266,760

Yield on incremental rent = £ 60,000/£266,760 × 100%
= 22.49%; say = **22.5%**

So adopting this yield on the top slice income produces:

Hard core rent =	£220,000	p.a.
× YP perp @ 5%	20	
	=	£4,400,000
Incremental rent =	£60,000	p.a.
× YP perp def 2 yrs @ 22.5%	2.962	
	=	£177,720
	Capital value =	£4,577,720
	SAY =	**£4,575,000**

Figure 11.7   Valuation by hardcore approach

Hardcore		
Net income =	£220,000 p.a.	
× YP perp @ 5%	20	
	=	£4,400,000
Top Slice		
Net income =	£60,000 p.a.	
× YP perp def 2yrs @ 6%	14.833	
	=	£889,996
	Market value =	£5,289,996
	SAY =	**£5,300,000**

Figure 11.8   Unadjusted hardcore valuation

This assumes the yield on the top slice will be just 1 per cent higher than that on the hardcore, to bring it more into line with the more usual practice of reflecting the relative security of the income flows. It is found that the hardcore market yield is just over 5.8 per cent, with around 6.8 per cent used for the top slice portion of the calculation. This is clearly more logical than using 5 and 22.5 per cent! However, it would not be easy to estimate that these would be the required yields just from the knowledge that the all risks yield is 6 per cent.

Hardcore		
Net income =	£220,000 p.a.	
× YP perp @ 6%	16.667	
	=	£3,666,667
Top Slice		
Net income =	£60,000 p.a.	
× YP perp def 2yrs @ 7%	12.478	
	=	£748,662
	Market value =	£4,415,328
	SAY =	**£4,420,000**

Figure 11.9   Alternative hardcore valuation

Hardcore		
Net income =	£220,000 p.a.	
× YP perp @ 5.809%	17.215	
	=	£3,787,227
Top Slice		
Net income =	£60,000 p.a.	
× YP perp def 2yrs @ 6.809%	12.874	
	=	£772,418
	Market value =	£4,559,644
	SAY =	**£4,560,000**

Figure 11.10   Corrected hardcore valuation

Conclusions

- Subjective rate adjustments, when solely based on term and reversion analysis, are always likely to produce an inaccurate value in a hardcore calculation.
- Making subjective variations to the rates to reflect security can be unsafe, especially if the reversion is more than five years away.
- Valuation of two investments, both with identical term and reversion incomes, but where the term years differ considerably and using the traditional 1 per cent differential adjustment to the term and reversion yields, can produce erroneous results.
- Using the equivalent yield in either the term and reversion or hardcore/layer approaches is preferable as it is simpler and provides more reliable and accurate results than subjective yields.
- The use of a discounted cash flow (DCF) analysis is recommended as a check on any hardcore valuations to ensure no gross inconsistencies occur.

Example

Hardcore or layer approach example 2

The freehold term and reversion Example 1, given in Section 11.10 above, valued a present rental income of £100,000 per annum that was receivable for the next two years followed by an increase to the market rent of £150,000 per annum thereafter. The valuation was made using an all risks yield of 8 per cent on the reversionary income, when market rent was being received, and produced a capital value of £1,788,000. How could a similar figure have been arrived at using the hardcore or layer approach as illustrated in Figure 11.11?

First, as has just been proven, the yields to be adopted would need to be derived from analysis of market comparables utilising the same layer valuation approach. To derive market information on a term and reversion basis and then use it with a different analysis approach will produce incorrect results. Assuming the receipt of the top slice income represented an additional 2 per cent risk (being a 50 per cent increase from £100,000 to £150,000 per annum), the following valuation could be produced:

Hardcore

	£	
Net income =	100,000 p.a.	
× YP perp @ 7.43%	13.459	
	= £ 1,345,895	

Top Slice

	£	
Net income =	50,000 p.a.	
× YP perp def 2yrs @ 9.43%	8.856	
	= £ 442,777	
Market value =	£ 1,788,672	
SAY =	£ 1,788,000	

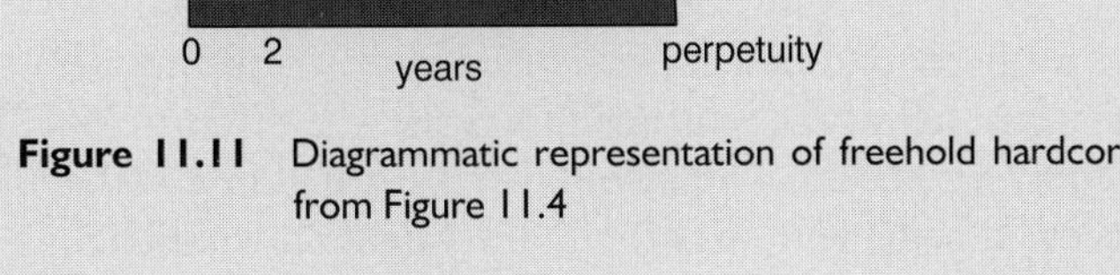

Figure 11.11   Diagrammatic representation of freehold hardcore or layer method applied to example from Figure 11.4

11.13 Gross versus net of tax calculations and effects of tax on leasehold valuations

In 'traditional' methods, allowance is made in leasehold valuations for the effects of tax on sinking fund instalments and interest earned on them by the use of the years purchase dual rate tax adjusted formula and tables. However, no deduction for tax is made from the profit rent. Why is this and what about freehold valuations? Clearly rental income from property is liable to be taxed, so should a deduction for this tax be made? When tax is deducted, it is known as using a 'net of tax' approach. Does such an approach need to be used though to arrive at market value? Sometimes, but mostly it is unnecessary, as it usually does not affect the capital valuation.

The conventional method of valuation is to capitalise the net income, or net profit rent, gross of Income Tax or Corporation Tax, by a years purchase representing the gross of tax all risks yield required. A similar practice is adopted on the Stock Market. Thus if an investor is considering investing in gilts, say 2.5 per cent stock currently standing at £25 in the market and with a face value of £100, he or she is thinking in terms of a yield of 10 per cent *gross of tax*. In other words, the 10 per cent means that the gross income is 10 per cent of the purchase price:

- £100 face value with annual interest paid at 2.5 per cent of this figure = £100 × 2.5% = £2.50
- Annual income of £2.50 compared to current market value of £25 = £2.50 × 100/£25 = 10 per cent

The annual income of £2.50 per stock held will be taxable in the UK as 'unearned income' from investment, but no deduction is made for this tax to calculate the annual percentage return.

Similarly if an investor investing in shops requires an all risks yield (ARY) of 5 per cent, it means that the gross of tax income is 5 per cent of the purchase price. When the income is £200,000 per annum and the price paid is £4,000,000, the gross of tax income is 5 per cent of the capital value. The valuation would therefore be in full:

Gross of tax income =	£	200,000 p.a.
Gross of tax yield required is 5%		
= YP perp. @ 5% =		20
	Market value =	£4,000,000

This valuation would be the basis adopted by a prospective purchaser, whatever his or her tax position. It is possible to value using a true net income approach, but generally there is little to be gained in making the adjustments required by this method, as the same market value will be arrived at.

A net of tax valuation of the shop property outlined above carried out for a person who pays tax at 40 per cent, would show a 3 per cent net of tax return (5% less tax = 5 × 0.6 = 3%). This equates to a years purchase in perpetuity of 33.333. The income after tax will be £120,000 per annum (£80,000 of the £200,000 received being paid in tax) and the valuation would be:

Net of tax income =   £   120,000 p.a.
Net of tax yield required is 3%
= YP perp. @ 3% =   33.333
                                            Market value =   £4,000,000

It will be seen that the final value is the same as when the gross of tax approach was used.

Even if the investor pays tax at 20 per cent, it will make no difference to the end result. This is because he or she will think in terms of a 4 per cent return after tax (5% × 0.8), or a YP of 25 (100/4). This will be applied to a net income after tax of £160,000 per annum (£200,000 less £40,000 tax) and the result is still £4,000,000:

Net of tax income =   £   160,000 p.a.
Net of tax yield required is 4%
= YP perp. @ 4% =   25
                                            Market value =   £4,000,000

Thus in these circumstances, there is no advantage in using a true net basis, even though this does indeed show the investor's actual net retained income and net achieved yield, it does not alter the assessment of capital value. Using this basis actually makes it much harder to compare investments one with another, as individual investors would be considering the same investment in completely different terms. It is more straightforward to make comparisons on the same gross basis.

What about deducting tax from the profit rent in a leasehold valuation? Suppose the leasehold interest in a property sublet at the full market rent of £100,000 per annum was to be valued. The lease has seven years unexpired without review at a rent of £65,000 per annum and the lessee pays tax at 30 per cent. A valuation of the interest on a gross of tax basis at a gross all risks leasehold market rent yield of 10 per cent and a net sinking fund rate of 3 per cent would be:

Rent receivable (market rent) =   £   100,000 p.a.
Less rent payable =   £   65,000 p.a.
Net profit rent =   £   35,000 p.a.
× YP 7 yrs @ 10% & 3% (tax 30%)   3.4912
                                            Market value =   £122,192
                                                     SAY =   **£122,200**

A true net valuation of the leasehold would be:

Rent receivable (market rent) =   £   100,000 p.a.
Less rent payable =   £   65,000 p.a.
Profit rent =   £   35,000 p.a.
Less tax @ 30% =   £   10,500 p.a.
Net profit rent =   £   24,500 p.a.
× YP 7 yrs @ 7% & 3% net *   4.9874
                                            Market value =   £122,191
                                                     SAY =   **£122,200**

*Yield adjusted by (100 − t); annual sinking fund already on net basis and thus tax-adjusted tables not used.

Again the net of tax valuation produces the same result as the gross of tax calculation.

Outcome

Use gross of tax income and gross of tax remunerative yields when valuing freehold and leasehold interests at market rents. However, there are times when valuing on a true net basis may be worthwhile. This is when the valuation is of a freehold with a reversionary rent (Millington 2000: 167–9)

Example

Consider the valuation, on a gross and net of tax basis, of a freehold interest in a property subject to an unexpired lease of seven years, where the rent currently paid by the tenant is £200,000 per annum and the market rent is estimated at £350,000 per annum. Allowing for an all risks rack rent yield of 10 per cent and a tax rate of 40 per cent the valuations are:

Gross valuation:

Term

Income =   £   200,000 p.a.
× YP 7 yrs @ 9%   5.033
                                    = £1,006,600

Reversion to market rent

Income =   £   350,000 p.a.
× YP perp def 7 yrs @ 10%   5.13
                                    = £1,795,500
                      Capital value = £2,802,100
                              Say = £2,800,000

True net valuation:

Term

Income =   £   200,000 p.a.
Less tax @ 40% =   £   80,000 p.a.
                Net of tax income = £   120,000 p.a.
                × YP 7 yrs @ 5.4%   5.703
                                    = £   684,360

Reversion to market rent

Income =   £   350,000 p.a.
Less tax @ 40% =   £   140,000 p.a.
                Net of tax income = £   210,000 p.a.
                × YP perp def 7 yrs @ 6%   11.085
                                    = £2,327,850
                      Capital value = £3,012,210
                              Say = £3,000,000

- Conventional term and reversion approach used with differential yields to reflect relative security of term income compared to market rent.
- Net of tax yields taken, using gross yields of 9 and 10 per cent for term and reversion, as in gross calculation, less tax at 40 per cent.
- Allowing for tax has produced a different final value, and it will be deduced that the higher the tax rate, the larger the difference between the gross and net valuations becomes.

Conclusions

- Most standard capital valuations are carried out on a 'gross' basis using a 'gross' all risks yield, which is taken to reflect all the inherent advantages and disadvantages, including tax, of the property.
- In a term and reversion freehold valuation it may be worth considering assessing the value on a true net of tax basis if the worth to an individual investor is being appraised.
- Investment decisions should be considered from both a gross and net of tax position, particularly when an investor has a choice of opportunities, some of which offer tax allowances or benefits that may affect the investment returns.
- The tax position of prospective purchasers can affect the likely sale price achievable for a property.
- Investors may be subject to different tax rates on income and capital and this will influence the assessment of net worth of an investment to them.
- Net of tax valuations using a discounted cash flow approach are useful to compare between investors with dissimilar tax liabilities and to find the true return on an investment.
- Even where gross of tax valuations have been used, valuers should state that no explicit allowance for tax has been made and the clients should consider their potential tax liabilities on an investment in consultation with their accountants (and legal advisers).

11.14 Different types of lease repairing and insuring terms and effects of property outgoings on net rental income

Property outgoings are the periodic expenditures incurred in the course of occupation, management or ownership of real property.

The main types of lease covenants concerning outgoings are:

- FRI – full repairing and insuring (the most common form of UK business lease);
- IRI – internal repairing and insuring only; or
- IRT – internal repairing terms only.

Outgoings need to be deducted from gross rental income received to arrive at a net figure, before the freehold or leasehold interest in a property can be valued. In theory, full repairing and insuring leases do not require any deduction for outgoings since the tenant is responsible for all the relevant expenditures. However, such properties still need to be managed on behalf of the freeholder and the fees paid for this service should be deductible, although this is not always done as explained below in the section on 'Management fees and charges'.

Rents paid by UK tenants may be described as 'exclusive' or 'inclusive'. 'Exclusive' rents may be shown by the suffix p.a.x. and indicates that the tenant is entirely responsible for payment of the business rates, completely separate from and in addition to the payment of the rent. This is the usual and normally implied basis for UK business leases. 'Inclusive' is shown by a p.a.i. suffix and indicates that the business rates are included in the amount of rent to be paid under the lease. This is very uncommon. Rates are only deducted as an outgoing if the rent is inclusive, since the landlord then is responsible for paying the rates to the local authority.

Business rates

In England and Wales, non-domestic occupiers pay rates to the local authority based on the rateable value (RV) of their property. It is a form of local taxation. Each property's RV is broadly similar to the net open market rental figure on FRI yearly lease terms as at a specific date. This is the Antecedent Valuation Date or AVD and pre-dates the effective date of the list by two years. Unless the building, its facilities or surroundings have changed since the last rating assessment, the RV remains unchanged from year to year. Rateable values for all non-domestic properties are proposed to be reassessed every five years. They are shown on the Valuation List, which can be viewed on the VOA website (Valuation Office Agency 2008a). The list operative on 1 April 2005 and the RVs on it have an AVD of 1 April 2003.

The rates paid each year are calculated by multiplying the national non-domestic rate (NNDR), which is commonly referred to as the uniform business rate (UBR), by the RV for the property. For example, if the rate is 45p and the property has a RV of £100,000, the rates paid to the local authority that financial year will be £45,000 (0.45 × £100,000).

The statutory definition of rateable value (RV) is: 'The rent at which it is estimated the hereditament might reasonably be expected to let from year to year, if the tenant undertook to pay all usual tenant's rates and taxes and to bear the cost of the repairs and insurance and other expenses, if any, necessary to maintain the hereditament in a state to command that rent' (Local Government Finance Act 1988 Section 56).

Hereditament is defined in Section 115(1) of the General Rate Act 1967 as: 'property which is or may become liable to a rate, being a unit of such property which is, or would fall to be, shown as a separate item in the valuation list'. *From year to year* is not necessarily the yearly rent paid under a lease (although this may be acceptable), but the annual rent that a tenant *would* pay for a yearly tenancy *with the prospect of continuance.*

The hereditament is assumed to be vacant and to be let, even if actually occupied. The rental offer of a hypothetical tenant on the statutory terms, taking account of market conditions, is then assessed. The value arrived at should represent the result of bargaining between the landlord and prospective tenant, reflecting supply and demand (the 'higgling of the market').

Rating assessments are on the basis of *rebus sic stantibus,* meaning the hereditament is valued in its existing physical state. Minor changes of a non-structural nature may be assumed, but the possibility of change of use is precluded. Every intrinsic quality and circumstance that may affect the value up or down must be taken into account: *Williams (VO)* v. *Scottish & Newcastle Retail Ltd and Another (2001)*.

The *Tone of the List* principle is that the valuation is made at the date of the proposal but taking account of the prevailing values on the current Valuation List and thus what the value would have been on the AVD.

Council Tax

Domestic property occupiers in England pay Council Tax under the Local Government Finance and Valuation Act 1991. The tax is assessed on the estimated capital value of the property as at 1 April 1991. Basis of valuation is open market value of the unencumbered freehold interest (or 99-year leasehold for flats) with vacant possession. Each property is allocated a band letter and the tax payable each year is directly proportional to the band value and varies between local authorities.

Valuation band	Range of values
A	up to and including £40,000
B	£40,001 – £52,000
C	£52,001 – £68,000
D	£68,001 – £88,000
E	£88,001 – £120,000
F	£120,001 – £160,000
G	£160,001 – £320,000
H	more than £320,000

Repairs, maintenance and decorations

By means of express lease covenants, a tenant can be made liable for internal, and/or external, and/or structural repairs in any combination, except where statute specifically provides otherwise. 'Repair' in a lease generally includes 'maintain' and 'decorate'. When let on FRI terms (full repairing and insuring), the tenant is responsible for all three categories and the landlord's outgoings in this respect should be zero.

There are three methods for assessing the costs of repairs:

1. actual costs
2. spot figure
3. a percentage of market rent.

Using actual costs is the best method where full records for the property are available. The valuer can assess the appropriate average annual costs, taken from the records for at least the last three to five years, allowing a weighting for the effects of inflation and any 'exceptional' figures.

Value Added Tax (VAT)

VAT is a complex subject. For UK matters, valuers are advised to consult the various publications and information made available by HM Revenue and Customs (2008) and to consider liaising with a client's accountants and legal advisers on the implications and application of the tax.

The existence of VAT increases the cost of the majority of outgoings. Most building work, including repairs, maintenance, alterations and improvements plus management and other professional fees are all taxable. Insurance premiums, however, are subject to a special (lower) tax rate. Some buildings will have VAT added to the rent where the owner has exercised the option and 'elected' to tax it.

Firms registered for VAT must charge the tax and pay it to HM Revenue & Customs but generally may recover the VAT they pay on purchase of all their equipment and supplies. Effectively, it means that their expenses are net of VAT. Thus for standard valuation purposes no allowances for VAT are customarily made as it affects all parties equally. Again, it is the gross or net of tax principle and whether it alters the final value.

Where parties are not equally affected (for example, one is VAT registered and another is not) then adjustments *will* need to be made to valuation calculations. The Valuation Office Agency (2006a: part 3, VAT, 3.6) confirms this point:

> However, for tenants who cannot recover all or any of the VAT on their expenditure, occupation costs will rise. The types of tenants likely to be most affected include banks, building societies, insurance companies, bookmakers, charities and small companies having a turnover below the VAT registration threshold. Such tenants are frequently referred to as VAT-averse. If a tenant sub-lets all or part of a building or land the VAT incurred on the rent paid to its landlord will become irrecoverable unless it also elects to tax the sub-letting.

Other outgoings

Water and sewerage rates and charges

Paid to local water authority, these can be calculated as pence in the £ of rateable value as for business rates, set sums charged according to type of occupier or, in areas where water meters have been installed, based on actual consumption. Large users (such as industrial) are usually given a reduction in the rate per litre/gallon.

Square or garden rate

This is not a common outgoing. Where it exists it is a special rate levied for the use of an adjoining garden or square, where only those with keys pay the rate.

Rent charge

Also called 'fee farm rent' or 'chief rent'. Very seldom encountered in practice nowadays. It is a charge against the freeholder's interest payable by the current owners of the freehold to the holder of the rent charge. They are usually charges in perpetuity that run with the land and the liability passes on transfer of ownership to the new owner. There may be a right by the owner of the charge to repossess in cases of non-payment. Under the Rent Charge Act 1977, if they do not terminate before, all rent charges will end in 2037.

Example

Example layout for valuations including outgoings

For a freehold valuation where the landlord is responsible for all outgoings (very unusual in practice and non-applicable items should be deleted from the list below):

Gross income =   £   p.a.

Less outgoings:

Rates RV × UBR =	£	p.a.
Repairs @ say % MR =	£	p.a.
Insurance @ say % MR=	£	p.a.
Management @ say % rent=	£	p.a.

$$= \text{£} \quad \text{p.a.}$$
$$\text{Net income} = \text{£} \quad \text{p.a.}$$
$$\times \text{YP single rate} \underline{\qquad}$$
$$\text{Capital value} = \text{£}$$

For a leasehold valuation where the landlord is responsible for all outgoings:

Rent receivable =   £   p.a.

Less rent payable =   £   p.a.

Gross profit rent =   £   p.a.

Less outgoings:

Rates RV × UBR =	£	p.a.
Repairs @ say % MR =	£	p.a.
Insurance @ say % MR =	£	p.a.
Management @ say % rent =	£	p.a.

$$= \text{£} \quad \text{p.a.}$$
$$\text{Net profit rent} = \text{£} \quad \text{p.a.}$$
$$\times \text{YP single rate (tax adjusted)} \underline{\qquad}$$
$$\text{Capital value} = \text{£}$$

Example

Valuation with outgoings example 1

Smith owns office premises currently let to Brown at £150,000 per annum on internal repairing terms. The lease expires in three years' time. Value the freehold (on both term and reversion, and hardcore approaches) and leasehold interests assuming:

- Market rent on internal repairing terms (from comparables) = £250,000 p.a.
- Freehold rack rent all risks yield = 10 per cent.
- Lessee pays tax at 28 per cent and can obtain 3 per cent net annual sinking fund rate.

Valuation of (Brown's) leasehold interest:

Rent receivable IRT=	£	250,000 p.a.
Less rent payable IRT =	£	150,000 p.a.
Net profit rent =	£	100,000 p.a.
× YP 3yrs @ 12% + 3% (tax 28%)		1.7564
	Capital value =	£175,640
	SAY =	£176,000

Note: leasehold yield assumed as 2 per cent higher than freehold yield.

Valuation of Smith's freehold interest:

Term

Gross income=		£ 150,000 p.a.
Less outgoings:		
Ext/struct repairs @ say 5% MR =	£	12,500 p.a.
Insurance @ say 3.5% MR =	£	8,750 p.a.
Management @ say 4% rent =	£	6,000 p.a.
	= £	27,250 p.a.
	Net income = £	122,750 p.a.
× YP for 3yrs @ 9%		2.531
		= £310,716

Reversion

Gross income =		£ 250,000 p.a.
Less outgoings:		
Ext/struct repairs @ say 5% MR=	£ 12,500	p.a.
Insurance @ say 3.5% MR =	£ 8,750	p.a.
Management @ say 4% rent =	£ 10,000	p.a.
	= £ 31,250	p.a.
	Net income = £218,750	p.a.
× YP perp def 3yrs @ 10%	7.513	
	=£1,643,469	
	Capital value =£1,954,185	
	SAY =£1,950,000	

Note: term yield reduced by 1 per cent as income considered 'more secure' being considerably less than market rent.

Example

Valuation with outgoings example 2

A twenty-year lease of an industrial unit is to be assigned. The lease provided for only one rent review, after ten years, and it has an unexpired term of four years. The tenant is responsible for internal repairs and insurance only. Current rent payable is £65,000 p.a.x., and from comparables you estimate the market rent to be £100,000 per annum, let on FRI terms. Nearby properties let at market rent show a leasehold all risks yield of 12 per cent. What price will be paid on assignment of the leasehold interest assuming a prospective assignee pays tax at 28 per cent and can obtain 3.5 per cent net on a sinking fund?

Valuation

Market rent value on FRI terms=   £   100,000 p.a.
Adjust to gross terms as present lease:
Add external/structural repairs
@ 5% market rent=   £   5,000 p.a.
Add management @ 3% of rent =   £   1,950 p.a.
          Market rent (rent receivable) on internal
                    repairing and insuring basis (IRI) = £   106,950 p.a.
                              Less Rent payable = £   65,000 p.a.
                                        Net profit rent = £ 41,950 p.a.
                    × YP 4 yrs @ 12% + 3.5% (tax 28%)   2.2246
                                        Capital value = £ 93,322
          Therefore SAY capital sum paid on assignment = £ 93,500

- This value excludes any potential additions for fixtures and fittings, goodwill and/or 'key money'.
- 'Outgoings' are added back to bring comparable value in line with internal repairing and insuring terms as under existing lease.
- Prospective tenants would be prepared to pay more than the market rent calculated on FRI terms since they are not responsible for some of the matters a tenant on FRI would be. In other words, they are 'saving' money on not having to pay out for those outgoings and this would be reflected in them being prepared to pay more rent in the first place.
- This can be equated to the market rent on FRI terms *plus* the value of the outgoings they are *not* responsible for.
- Even better than making these adjustments is to find comparables recently let on IRI basis so that a straightforward comparison can be made with market rent on these terms.

Progress check questions

- What is the fundamental basis of the investment method of valuation?
- What types of market evidence do you need to undertake a property valuation?
- What is a leasehold 'profit rent'?
- When and why would a capital payment on assignment of a leasehold interest be calculated?
- What is an annual sinking fund?
- How does taxation affect annual sinking fund calculations?
- What relevance has annual sinking funds to leasehold valuations?
- Why are they called all risks yields?
- Why will the market ARY on a leasehold interest usually be higher than the corresponding freehold interest in the same property?
- How do you know what ARY to use in a valuation?
- Why are 'equated yields' different to 'all risks yields' and in what way?
- What is a 'FRI' lease?
- What is the basis of the 'term and reversion' approach to valuing reversionary freehold interests?
- How does the 'hardcore' or 'layer' approach to valuing reversionary interests differ from the 'term and reversion' one?
- What is an 'equivalent yield' and why may it be preferable to use this when valuing a reversionary interest?
- Why are conventional valuations undertaken on a 'gross of tax' basis?
- What types of annual expenses are incurred in the ownership and occupation of land and buildings and how may these be estimated when deducting 'outgoings' from a rental income?
- Why is the figure at which a property is insured likely to differ from its market value?

Chapter summary

The investment method of valuation is the next-best 'conventional' approach after the comparison method. It still relies on obtaining some data from comparables but is used when the required value cannot be found simply through comparison. The basic premise is that capital value is found by capitalising the income flows derived from the property, using the years purchase multiplier. For freeholds, a single rate approach is used and traditionally for leaseholds a dual rate tax adjusted one adopted, with the all risks yield providing the return rate on investment. Other yields are also used to measure performance, such as initial, reversionary, equivalent and equated.

When the rental on a property is not the market rent, in present value terms, there will be a reversionary interest. This can be valued using a term and reversion 'block' income or a 'hardcore' or 'layer' approach. The choice of yields for each tranche of income is critical to the final value. For consistency and to minimise subjectivity, the equivalent yield is recommended as well as use of discounted cash flow analysis to confirm the found capital values.

Tax affects all property incomes and receipts and the valuer should have an understanding of the range and nature of taxes involved. In many cases no explicit adjustment needs to be made to valuations to allow for taxation as gross of tax rents and yields can be used. Nevertheless, clients must be made aware that tax will affect their investment.

Buildings require regular repair, maintenance, insurance and management and the costs of these outgoings may need to be estimated by the valuer and suitable adjustments incorporated into the valuations to allow for the related expenditures.

Further reading

Armatys, J. (2008) 'Mainly for students: How to avoid zero points', *Estates Gazette*, 0827 (12 Jul.): 112–13.

Askham, P. (1990) 'Mainly for students: conventional and contemporary methods of investment valuation', *Estates Gazette*, 9011 (17 Mar.): 84-6; reprinted in P. Askham and L. Blake (eds), *The Best of Mainly for Students* (London: Estates Gazette, 1993), 346–53.

—— (1990) 'Mainly for students: conventional and contemporary methods of investment valuation (2)', *Estates Gazette*, 9023 (9 Jun.): 76-8.

—— (1991) 'Mainly for students: service charges', *Estates Gazette*, 9127 (13 Jul.): 131-2; repr. in P. Askham and L. Blake (eds), *The Best of Mainly for Students* (London: Estates Gazette , 1993), 101–7.

Enever, N. and Isaac, D. (2002) *The Valuation of Property Investments,* 6th edn, London: Estates Gazette.

Fraser, W.D. (1993) *Principles of Property Investment and Pricing,* 2nd edn, Basingstoke: Palgrave Macmillan.

Rich, J. (1999) 'How to calculate the true equivalent yield', *Estates Gazette*, 9949 (11 Dec.): 84–5.

Royal Institution of Chartered Surveyors (1997) *Commercial Investment Property: Valuation Methods,* Coventry: RICS Books.

Sayce, S., Smith, J., Cooper, M.R.R. and Venmore-Rowland, P. (2006) *Real Estate Appraisal: From Value to Worth,* Oxford: Blackwell.

Scarrett, D. and Smith, M. (2007) *Property Valuation: The Five Methods,* 2nd edn, London: Routledge.

Investment method
Discounted cash flow

In this chapter ...

- The valuation method based around an accountancy technique and introduced into UK practice since the 1970s, having been in standard use in the US beforehand.
- Using discounted cash flow for calculation of worth appraisals as well as objective valuations.
- The difference between the present value and the internal rate of return approaches to discounted cash flow appraisals.
- The role of computer spreadsheets and other software in both facilitating the use of the method and in becoming the standard tool for valuations that will influence adopted methodology.
- Equated yield calculations and the need for and difficulty in forecasting future market performance.
- Applying discounted cash flow to property investments and comparison with conventional valuation methods.
- Combining all the investment method approaches in one valuation.

12.1 Discounted cash flow methods

Discounted cash flow (DCF) is an accounting technique that can be used for appraising and comparing between investments. It has a long history of use by accountants and financial institutions and for real estate appraisals in the US. With the increasing overlapping of roles of the various professions in investment management and assessment, the method has gradually become established as a valid method of property valuation within UK practice since the mid-1970s.

There was some initial reluctance to use a different approach to the 'traditional' valuation methods, as these had been tried and tested over a very long period and been found reliable. In addition, DCF required a series of seemingly complex calculations to be undertaken, which would need to be reworked each time one of the variables was altered. The additional work and time involved was a disincentive to its use before personal computers and spreadsheet programs became available. With the use of these the multitude of calculations required is no longer a problem. There are a number of commercial software packages now available for DCF valuation purposes, or any standard spreadsheet program can be readily adapted.

As a result, DCF has become established not only as a standard additional valuation tool that is useful in its own right, but one that can help highlight factors in an appraisal which may otherwise not be fully appreciated. It is also a useful means of confirming traditional valuation figures, or of explicitly incorporating a number of factors into the calculation which in the past were dealt with on an implicit basis. In other words, the calculation reflected the presence of these various factors, but no specific allowance was made for them in the calculations.

The DCF method can include allowances for inflation, tax liability, irregular receipts or payments and for anticipated rental and capital growth. In 'traditional' valuations such factors are generally implied in the choice of all risks yield. The equated yields that can be used in DCF valuations reflect the true overall return from the property, after taking into account the effects of inflation. Since they are founded on different bases, it is important that each type of yield is only used within the context from which it has been derived. Thus equated yields are used in growth-explicit DCF valuations and all risks yields in 'traditional' growth-implicit investment method calculations. As they incorporate an allowance for growth, equated yields will always be higher than the corresponding all risks yield on any given investment. The two are different ways of assessing the investment and are thus mutually exclusive in that they should only be used within their relevant valuation scenario.

Equated yields may represent a more meaningful expression of the performance of an investment. It may be easier to comprehend that a commercial property let with five-yearly rent reviews is producing a true return, or equated yield, of 10 per cent, than to say that an investor can expect to receive an all risks yield of 7.5 per cent on the same investment. Both of these percentages are consistent, but the 10 per cent includes an explicit allowance for anticipated future rental growth whereas this is only implied by the all risks figure. The market is prepared to purchase based on a nominal 7.5 per cent return, knowing that the real return will be higher. Valuers are very aware that this is the basis for the selection of an all risks yield, but to non-property people it may seem strange that the yield achieved can appear to be less than that obtained on other investments. Once the true return is shown to be the equated yield, the apparent anomaly is made clear. It will also enable true comparisons to be made in the performance of property as an investment, compared with other types of investment.

A fundamental role of discounted cash flow in real estate appraisal is its ability to calculate value or worth. When all the factors used are directly obtained from market comparisons, the objective value should be calculable. Conversely, a subjective approach can be adopted where the factors relate to a specific client's situation and investment requirements. The calculation of worth to that particular investor can then more fully inform the decision of whether to purchase an investment at a given price.

There are two basic approaches to use of discounted cash flow:

- Present value (either gross present value or net present value) is when the final capital value needs to be found and a yield, either based on comparable market evidence or a 'target' figure sought by the investor, is used to calculate this value; or
- Internal rate of return (IRR) is used when the final capital value of the investment is known and the calculation finds at what rate of return (yield) this price will apply.

Present value approach

This calculates the total present value of all the discounted receipts and payments cash flows over the period of the investment at the required target interest rate, or *equated* yield.

Net present value (NPV) is found when the initial outlay, or purchase price, is included in the analysis. The NPV is then the surplus or deficit present value of the discounted cash flows. When the initial outlay is not included as a cash flow, the present value of the all the discounted cash flows provides the total worth of the investment at the specified rate of return and is known as the gross present value (GPV). Where this present value at the target yield rate is less than the purchase price of the investment, it means that the target yield cannot be achieved and thus the purchase price is too high for that particular investor.

Gross present value is found by taking the following steps:

- Estimate the size and timing of all future income sums and any costs that will be incurred in attaining that income (excluding the initial purchase price and costs).
- Allow for future 'growth', or increases in costs/values due to inflationary effects over time.
- Growth is calculated at compound interest (using Amount of £1 formula or table): thus market rent in three years = market rent now × A £1 for 3 yrears at estimated annual rental growth rate.
- Decide on the rate of return required by the investor, otherwise known as the 'target rate'. This percentage will need to be at least equal to the investor's opportunity cost of capital, that is, what could be obtained on the next-best form of investment. Where money is being borrowed it will need to be at least the rate of interest that is to be paid.
- Allowance would normally also be made in the target rate for likely effects of inflation and taking account of returns available on alternative types of investment – wherever possible, use comparable evidence to inform this choice of rate.
- Discount (multiply by present value) all future incomes and costs at the target rate.
- Where possible, cumulative PV factors can be used to discount 'blocks' of fixed incomes or costs rather than calculating on a year-by-year basis.
- Add up all the discounted incomes and costs.
- This leaves the GPV at the target rate.
- When the target rate is the market equated yield found from direct comparison with open market sales of similar property types, the GPV will represent the estimated capital value or market value for the property.
- When the target rate is a subjectively selected figure specific to the circumstances and requirements of a particular investor, the GPV is the calculated *worth* to that investor.

The net present value will be found when the initial purchase price and incidental costs of acquisition are deducted from the discounted cash flows. When the NPV is zero, it indicates that

Table 12.1   Basic layout of yearly discounted cash flow calculations

End of year	Cash outflow	Cash inflow	Net cash flow (inflow– outflow)	PV £1 @ x%	Discounted cash flow (net cash plow × PV)
0	–£	+£	+/–£	0.xxx	+/–£
I	–£	+£	+/–£	0.xxx	+/–£
2	–£	+£	+/–£	0.xxx	+/–£
3	–£	+£	+/–£	0.xxx	+/–£
4	–£	+£	+/–£	0.xxx	+/–£
5	–£	+£	+/–£	0.xxx	+/–£
etc.					
Gross present value (GPV)				= £	
Less initial purchase price and incidental costs of acquisition			= – £		
Net present value (NPV)				= £	

the target rate will be met. A positive NPV shows that the investment will perform better than the target rate, by which it could be interpreted that the investor can afford to pay a higher purchase price to acquire the investment. Alternatively, if the purchase is made at the stated price, the return on investment will be higher than the target rate; the exact return being found by an IRR calculation. When the NPV is a negative sum, it indicates that the target rate cannot be met and the investor should not purchase at the currently stated price.

The basic layout of the calculations is as shown in Table 12.1.

This example shows cash flows on a 'traditional' yearly in arrear basis. However, there is no reason why annually in advance, or half yearly, or quarterly in advance, or arrear, cash flows could not be used instead providing the appropriate present value is also used. All payments and receipts, allowing for growth where appropriate, are entered into the cash outflow and inflow columns. Growth rates for the future need to be estimated and future sums calculated by use of the compound interest formula or table. Varying growth rates could be used to reflect the expected changes in the market in the future, but for simplicity, the *average* increase per annum that may be expected over the time period involved is adopted (see Section 12.3 for further explanation of this).

The net cash flow column is the sum of the difference between the cash outflow and inflow columns in that row of the table. The PV of £1 at the target equated yield rate is inserted in the allocated column, with the figure in each row being different according to the elapsed time period indicated at the start of the row. The discounted cash flow column is the result of the sum of multiplying the net cash flow by the PV in each row. The present value is the sum of all the figures in the discounted cash flow column.

Individual years are shown but when 'blocks' of years are used, they will be multiplied by cumulative PV factors (otherwise known as *Years Purchase*) to discount rather than single PV figures. For example, net cash flow of £5,000 per annum received in years 3, 4 and 5 can be calculated as: £5,000 × YP 3 years × PV 2 years.

Internal rate of return approach

The object here is to find at what rate of interest the present value of the 'outs' equals the present value of the 'ins', that is, the net present value (NPV) = 0.

The 'rate of return' on an investment indicates how much annual income is generated per £100 of capital invested. The rate returned is determined entirely by the cash flows themselves with no reinvestment of any part of the income in a project other than the subject investment. The basic rule would be to accept an investment as a worthwhile purchase if its IRR is greater than or equal to its 'target' rate, but reject it if it is less than the 'target' rate.

Normally the IRR is calculated by using a computer spreadsheet. Should a manual approach be used, the appropriate rate of interest is obtained by trial and error and interpolating between the trial rates. The basis of this calculation was considered in an example in Section 11.11 above. It is necessary to carry out the calculation at least twice, until one negative NPV and one positive NPV have been obtained (preferably both as close to 0 as possible). The IRR can then be found either through plotting the results graphically or through mathematical interpolation. It assumes a straight-line relationship between the two points, whereas in reality it is a slight curve when plotted graphically. Thus the closer the two initial trial figures are to producing a zero NPV, the less the error in the final figure.

Example

The trial rate of 10 per cent produces a NPV of +£10,000 and the trial rate of 15 per cent a NPV of –£20,000.

The difference between the NPVs = £30,000 and the difference between the trial yields = 5 per cent. The difference between the NPV at the lower trial rate and an NPV of 0 is £10,000. Thus to find the rate at which NPV is 0:

10% + (5% × (£10,000/£30,000)) = 11.67%

Conclusion is that the Internal Rate of Return (IRR) is 11.67%

The gap between the trial rates of 5 per cent is quite large. Ideally, to reduce the mathematical error from it not being a perfectly straight-line relationship between the values at the two different rates, both trial NPVs need to be closer to zero.

A computer spreadsheet's solver function can produce an accurate answer to any given number of decimal places, although in practice it is seldom necessary to calculate a yield to more than two places.

12.2 Use of computer spreadsheets and other software

All spreadsheets include functions for the calculation of discounted cash flow appraisals. Examples of the code inputs that could be used in Microsoft Excel to carry out basic valuation calculations including DCF are provided in Table 12.2 (on pp. 246–7) on an annually in arrear basis. Table 12.3 (on pp. 248–9)provides similar examples on quarterly in advance basis for conventional freehold and leasehold valuations only. Commercial software, such as available from Argus (Argus Software

2008b) and KEL (KEL 2008), provides more sophisticated DCF and conventional investment method property valuations, including the ability to value multiple income streams using the techniques.

Dr Tim Havard's view is that, as 'we are moving towards a more global valuation profession, this will lead to a converging of valuation techniques and the software platforms that we use may well lead rather than accommodate this move'. He further explains that the world's two appraisal 'superpowers' are the UK and US. Valuation practice in the UK uses a capitalisation approach that is market driven and reliant and which implies future performance from purchaser behaviour rather than requiring forecasting. Methods in the US are heavily reliant on cash flow forecasting and the appraiser's opinion where forecasting future rents, capital values and costs is important. Computer software enables both approaches to be combined more easily than in the 'pen and paper manual era'. This leads him to conclude that 'software has moved from being a mere tool – a calculator or a slide rule substitute – to being THE tool in each market that is the natural thing used to conduct valuations/appraisals' and it is 'the technology that determines how something is done rather than being led by the profession/users' (Havard 2007).

12.3 Future property rental and capital growth

Over the long term it is expected that inflation will cause prices of goods and services to rise. Generally, this will equally be true of property rents and capital values, although in the short or medium terms these values can fall. Any valuation that ignores the effect of these expected long-term increases would clearly be wrong. However, with the conventional methods of property valuation there is no need to make any explicit adjustments as the expectation of future changes in value are implied in the all risks yield that investors accept in purchasing or selling a property investment. The yield reflects the market's attitude to the particular investment and inflation. Thus first class prime shop yields may be low because the market expects nearly full protection against inflation whereas secondary industrial yields are high because there is a much more limited 'hedge against inflation'.

However, when using equated yields and the discounted cash flow approach there is a need to make an explicit allowance for inflationary increases on the rents, costs and values likely to be obtained in the future. This means anticipating and forecasting market growth, which in turn creates a difficulty when using the equated yield DCF approach – the need to forecast the future! By definition, this is an inexact art; otherwise everybody would make perfect decisions all the time. The best that can be done is to make informed 'guesstimates' of what will happen in the future. The usual method of doing this is to look at historical performance data and try to identify trends. Sophisticated statistical techniques, such as regression analysis, may assist in this. Historical evidence of average changes in value over time is provided by property research organisations, such as IPD, and by research departments of property consultants. Ultimately though it will not be known whether the forecast was correct until after the event. Accordingly, forecasts should be continually revisited and reconsidered as events unfold, to check their accuracy against actual performance and to make adjustments to subsequent forecasts.

For simplicity, the growth rate adopted in most discounted cash flow valuations is assumed to be an average annual figure that will apply throughout the period assessed. In reality rental levels seldom increase uniformly each and every year. During some years, rents may remain static or even fall, whereas in other years they may increase by modest or large percentages. Should such detail need to be incorporated into the DCF model used it can be, with varying rates of growth included from year to year. This will extend the calculations required and given the difficulty of

Table 12.2   Annually in arrear basic valuation code inputs in Microsoft Excel spreadsheet

	A	B	C	D	E	F	G
1	VALUATION ANALYSIS - Annually in Arrear Basis						
2	Only enter data in shaded boxes!						
3	Freehold Data:						
4	Term Income						
5	Reversion Income				Leasehold Data:		
6	Term Yield			%	LH ARY		%
7	Reversion ARY			%	asf rate gross		%
8	Term Years				tax		%
9							
10	Freehold Valuation						
11	Term						
12	Net income			=		p.a.	
13	x YP for		yrs @				
14							
15	Reversion to Market Rent						
16	Net income			=		p.a.	
17	x YP perp def		yrs @				
18							
19					Market Value =		
20					SAY =		
21							
22	Equivalent Yield approach: Freehold Valuation						
23	Term						
24	Net income			=		p.a.	
25	x YP for		yrs @				
26							
27	Reversion to Market Rent						
28	Net income			=		p.a.	
29	x YP perp def		yrs @				
30							
31					Market Value =		
32					SAY =		
33							
34	Nominal Equivalent Yield =				% (annually in arrear)		
35	True Equivalent Yield =				% (quarterly in advance)		
36							
37	Equated Yield Approach: Freehold Valuation						
38	Assumed annual rental growth rate =					%	
39	Equated Yield (annually in arrear) =					%	
40	True Equated Yield (quarterly in advance) =					%	
41							
42		Years		Income p.a.	PV factor	DCF	
43		1 to					
44		to					
45		to					
46		to					
47		to					
48		to	perp *				
49					GPV =		
50							
51							
52	* = income capitalised @ ARY to reflect continued future rental growth but then discounted at equated yield						
53							
54	Hardcore or layer approach: Freehold Valuation						
55	Hardcore						
56	Net income			=		p.a.	
57	x YP	perp	@				
58					=		
59	Top Slice						
60	Net income			=		p.a.	
61	x YP perp def		yrs @				
62					=		
63					Market Value =		
64					SAY =		
65							
66	Leasehold Valuation						
67							
68	Rent receivable (Market Rent)			=		p.a.	
69	Less rent payable			=		p.a.	
70	Net Profit Rent			=		p.a.	
71	x YP for		yrs @				
72	+		& tax @				
73					Market Value (assignment price) =		
74					SAY =		

Formulae for Table 12.2

Cell	Code
A44	=C43+1
A45	=C44+1
A46	=C45+1
A47	=C46+1
A48	=C47+1
B13	=C8
B17	=B13
B25	=C8
B29	=B25
B61	=C8
B71	=C8
B72	=INT((F7*(100-F8)/100)/0.5)*0.005
C6	=IF(C4<0.81*C5,C7-1,C7)
C43	=C8
C44	=C43+5
C45	=C44+5
C46	=C45+5
C47	=C46+5
D13	=0.01*C6
D17	=C7*0.01
D25	=D34*0.01
D29	=D34*0.01
D35	=(1/(1-D34*0.01/4)^4-1)*100
D43	=C4
D44	=C5*((1+E38/100)^C43)
D45	=D44*((1+$E$38/100)^(C45-C44))
D46	=D45*((1+$E$38/100)^(C46-C45))
D47	=D46*((1+$E$38/100)^(C47-C46))
D48	=C5*((1+$E$38/100)^C47)
D61	=IF(E60>0.2*C5,D57+0.02,D57)
D71	=F6*0.01
D72	=F8*0.01
E12	=C4
E13	=((1+C6/100)^C8-1)/ (((1+C6/100)^C8)*C6/100)
E16	=C5
E17	=1/(((1+C7/100)^C8)*C7/100)
E24	=E12
E25	=((1+D34/100)^C8-1)/ (((1+D34/100)^C8)*D34/100)
E28	=E16
E29	=1/(((1+D34/100)^C8)*D34/100)
E39	=E38+C7-0.5
E40	=(1/(1-E39*0.01/4)^4-1)*100
E43	=((1+$E$39/100)^C43-1)/ (((1+$E$39/100)^C43)*$E$39/100)
E44	=(((1+$E$39/100)^(C44-C43)-1)/ (((1+$E$39/100)^(C44-C43))*$E$39/100))/ ((1+$E$39/100)^(A44-A43))

Cell	Code
E45	=(((1+$E$39/100)^(C45-C44)-1)/ (((1+$E$39/100)^(C45-C44))*$E$39/100))/ ((1+$E$39/100)^C44)
E46	=(((1+$E$39/100)^(C46-C45)-1)/ (((1+$E$39/100)^(C46-C45))*$E$39/100))/ ((1+$E$39/100)^C45)
E47	=(((1+$E$39/100)^(C47-C46)-1)/ (((1+$E$39/100)^(C47-C46))*$E$39/100))/ ((1+$E$39/100)^C46)
E48	=1/(((1+$E$39/100)^C47)*C7/100)
E50	=E20
E56	=C4
E57	=1/D57
E60	=C5-E56
E61	=1/(((1+D61)^B61)*D61)
E68	=C5
E69	=C4
E70	=E68-E69
E72	=1/(D71+((B72/((1+B72)^B71-1))*100/ (100-F8)))
F6	=(INT(C7*4.8))*0.25
F14	=E13*E12
F18	=E17*E16
F19	=F18+F14
F20	=IF(MOD(F19,1000)<500,INT(F19/1000)*1000, INT((F19/1000)+1)*1000)
F26	=E25*E24
F30	=E29*E28
F31	=F30+F26
F32	=F20
F43	=E43*D43
F44	=E44*D44
F45	=E45*D45
F46	=E46*D46
F47	=E47*D47
F48	=E48*D48
F49	=SUM(F43:F48)
F50	=F20
F58	=E57*E56
F62	=E61*E60
F63	=F62+F58
F64	=F20
F73	=E72*E70
F74	=IF(MOD(F73,1000)<500,INT(F73/1000)*1000, INT((F73/1000)+1)*1000)

Table 12.3   Quarterly in advance basic valuation calculations code inputs in Microsoft Excel spreadsheet

	A	B	C	D	E	F	G
1	VALUATION ANALYSIS - Quarterly in Advance basis						
2	Only enter data in shaded boxes!						
3	Freehold Data:						
4	Term Income						
5	Reversion Income				Leasehold Data:		
6	Term Yield			%	LH ARY		
7	Reversion ARY			%	asf rate gross		%
8	Term Years				tax		%
9							
10	Freehold Valuation: quarterly in advance basis						
11	Term						
12	Net income			=		p.a.	
13	x YP for		yrs @				
14							
15	Reversion to Market Rent						
16	Net income			=		p.a.	
17	x YP perp def		yrs @				
18							
19					Market Value =		
20					SAY =		
21							
22	Equivalent Yield approach QIA: Freehold Valuation						
23	Term						
24	Net income			=		p.a.	
25	x YP for		yrs @				
26							
27	Reversion to Market Rent						
28	Net income			=		p.a.	
29	x YP perp def		yrs @				
30							
31					Market Value =		
32					SAY =		
33							
34		True Equivalent Yield =			% (quarterly in advance)		
35							
36	Leasehold Valuation: quarterly in advance basis						
37							
38	Rent receivable (Market Rent)			=		p.a.	
39	Less rent payable			=		p.a.	
40	Net Profit Rent			=		p.a.	
41	x YP for		yrs @				
42	+		& tax @				
43			Market Value (assignment price) =				
44			SAY =				

Formulae for Table 12.3

Cell	Code
B13	=C8
B17	=B13
B25	=C8
B29	=B25
B41	=C8
B42	=INT((F7*(100-F8)/100)/0.5)*0.005
C6	=IF(C4<0.81*C5,C7-1,C7)
D13	=0.01*C6
D17	=C7*0.01
D25	=D34*0.01
D29	=D34*0.01
D41	=F6/100
D42	=F8/100
E12	=C4
E13	=(((1+C6/100) ^ C8-1)/(((1+C6/100) ^ C8)*C6/100))*(C6*0.01/(4*(1-(1/(1+C6/100) ^ 0.25))))
E16	=C5
E17	=(1/(((1+C7/100) ^ C8)*C7/100))*(C7*0.01/(4*(1-(1/(1+C7/100) ^ 0.25))))
E24	=E12
E25	=(((1+D34/100) ^ C8-1)/(((1+D34/100) ^ C8)*D34/100))*(D34*0.01/(4*(1-(1/(1+D34/100) ^ 0.25))))
E28	=E16
E29	=(1/(((1+D34/100) ^ C8)*D34/100))*(D34*0.01/(4*(1-(1/(1+D34/100) ^ 0.25))))
E38	=C5
E39	=C4
E40	=E38-E39
E42	=1/(4*(1-(1/(1+F6/100) ^ 0.25))+((4*(1-(1/((1+B42) ^ 0.25)))/(((1+B42) ^ C8)-1))*100/(100-F8)))
F6	=(INT(C7*4.8))*0.25
F14	=E13*E12
F18	=E17*E16
F19	=F18+F14
F20	=IF(MOD(F19,1000)<500,INT(F19/1000)*1000,INT((F19/1000)+1)*1000)
F26	=E25*E24
F30	=E29*E28
F31	=F30+F26
F32	=F20
F43	=E42*E40
F44	=IF(MOD(F43,1000)<500,INT(F43/1000)*1000,INT((F43/1000)+1)*1000)

predicting future growth may be considered unnecessarily complex for most standard valuation purposes.

12.4 Equated yield and implied rental growth formulae

The equated yield can be considered to comprise three elements:

- compensation for the effects of inflation;
- a risk-free rate of return for giving up the use of the invested capital now; and
- a payment for the risks and uncertainties associated with the specific investment involved.

Accurate calculation of an equated yield is achieved by finding the internal rate of return (IRR) from DCF analysis. A *very* approximate method of estimating freehold equated yield as an IRR trial rate is:

- All risks yield (ARY) *plus* forecasted annual rental growth rate *minus* half a percent.

For example, if all risks yield is 8 per cent and rental growth is forecast at 3 per cent per annum, then equated percentage yield *approximately* will be: 8 + 3 − 0.5 = 10.5.

This can be used as one initial trial rate to find the IRR.

To find the all risks yield when the equated yield and annual growth rate are known and the market rent is receivable, the formula devised by Marshall (1988) can be used:

$$Y = E - E\left(\frac{(1+G)^n - 1}{(1+E)^n - 1}\right)$$

To find the implied annual percentage rental growth rate when the all risks yield and equated yield are known and the market rent is paid, Marshall (1988) provided another formula:

$$G = \left(1 + \frac{E - Y}{asf}\right)^{1/n} - 1$$

where:
Y   = all risks yield (/100)
E   = equated yield (/100)
G   = annual growth rate(/100)
n   = rent review period in years
asf = annual sinking fund for *n* years @ *E* per cent

Table 12.4 (on p. 252) provides an example of how both these formulae can be entered on a spreadsheet to ease their calculation.

Example

A client wants to purchase a freehold shop investment let at a rack rent of £100,000 per annum and subject to five-yearly upwards-only rent reviews. Her target equated yield is 13 per cent. The probable average future growth rate in rental income has been assessed to be 5 per cent per annum compound.

Using the formula shown in Section 12.4:

$Y = 0.13 - 0.13((1.05^5 - 1)/(1.13^5 - 1))$
$= 0.13 - 0.13 \times 0.3279 = 0.13 - 0.042627 = 0.087373$
Therefore 8.7373%, or say 8.74%, is the all risks yield (ARY)

'Traditional' valuation

Net income=	£	100,000 p.a.
× YP perp @ 8.74%		11.442
	Market value=	£1,144,200
	SAY=	£1,144,000

DCF valuation

Years	Income	PV @ 13%	DCF
1 to 5	£100,000	3.517	£351,723
6 to 10	£127,628	1.909	£243,644
11 to 15	£162,889	1.036	£168,776
16 to 20	£207,893	0.562	£116,913
21 to 25	£265,330	0.305	£80,988
26 to perp	£338,635	0.538	£182,290
		GPV =	£1,144,333
		SAY =	£1,144,000

Conclusion: Market value is SAY £1,144,000

12.5 Application of the method to property investments, including holding period and exit yield

In property valuation analysis, when no explicit allowances are included for future rental or capital growth, then the yield (IRR) used will be the equivalent yield. When growth is expressly calculated, then the yield will be the equated yield. Both of these will be different to the 'standard' all risks yield as they are derived from a different form of analysis.

When valuing freehold interests, the main 'positive' cash flow will be provided by the rental income, either actual or deemed. As rent is generally fixed for set time periods in between rent

Table 12.4 Spreadsheet calculator for equated yield and implied rental growth formulae

	A	B	C
1	ARY from Equated Yield Calculator		
2			
3	Equated Yield =	9.2	%
4	Annual Rental Growth rate =	2.5	% p.a.
5	Rent Review Period in Years =	5	years
6	ARY to reflect above figures =	7.01	%
7			
8			
9	Implied Annual Rental Growth Calculator		
10	Equated Yield =	10	%
11	All-Risks Yield =	8	%
12	Rent Review Period in Years =	5	years
13	Implied Annual Rental Growth =	2.33	% p.a.

Code for tinted boxes

Box 1: $= (B3/100-(B3/100)*(((1+B4/100)\wedge B5)-1)/(((1+B3/100)\wedge B5)-1))*100$

Box 2: $=((1+(((B10-B11)/100)/(B10/100/(((1+B10/100)\wedge B12)-1))))\wedge(1/B12)-1)*100$

reviews it simplifies calculations if the years where the same rental would be receivable are 'blocked' together. This reduces the number of rows needed in the DCF appraisal.

Although with a freehold interest the valuation is assessed over perpetuity, clearly an indefinite number of rows of years cannot be included in the discounted cash flow. Mathematically, it will be found that the net cash flow receivable on a freehold investment after 20 to 25 years from the present can be aggregated and calculated as receivable in perpetuity thereafter. To continue to add more year rows of calculations does not significantly affect the GPV.

Thus if market rent from year 22 to perpetuity was to be capitalised, growth is added to current market rent (MR × A £1 22yrs) and YP in perpetuity deferred 22 years would be the discount multiplier in the PV column. To reflect the implicit continued rental growth into perpetuity, the all risks yield (ARY) is used for this years purchase in perpetuity applied to the 'terminal rental'. The present value of this capital sum is then found by discounting at the equated yield (IRR), as are all other cash flows in the appraisal, in order to find its value today.

Example

Value a freehold interest that produces a current net rental income of £100,000 per annum but which is subject to a rent review to market rent in three years' time and five-yearly thereafter. Estimated full rental value now is £150,000 per annum and average rental growth per annum is anticipated at 3 per cent per annum. Current market all risks yield is 8 per cent and equated yield is 10.5 per cent.

The annual net cash flow will be the rental income. For the next three years it will be £100,000 per annum, then £150,000 per annum in present value terms for the next five years, then increased to market rent every five years afterwards, in perpetuity. To estimate what the actual expected rent will be in three years', eight years', thirteen years' time, and so on, requires use of the compound interest formula, otherwise known as Amount of £1.

Therefore, the expected market rent in three years' time will be:

£150,000 × Amount of £1 in 3 years @ 3% = £163,909 p.a.

and in eight years time it will be:

£150,000 × Amount of £1 in 8 years @ 3% = £190,016 p.a.

and so on.

The expected cash flows can be set out on a year-by-year basis:

Years	Net Cash Flow	PV	DCF
1	£100,000		
2	£100,000		
3	£100,000		
4	£163,909		
5	£163,909		
6	£163,909		
7	£163,909		
8	£163,909		
9	£190,016		
etc			

However, this would require many rows of calculations. Much simpler is to block together those rows where the same cash flow is received and use the years purchase (Present Value of £1 per annum) factor as well as the single present value one to calculate the discounted cash flow. For the required valuation, the layout is as in the table.

Years	Net Cash Flow	PV factor	DCF
1 to 3		£YP 3yrs @ E%	
4 to 8		£YP 5yrs × PV 3yrs @ E%	
9 to 13		£YP 5yrs × PV 8yrs @ E%	
14 to 18		£YP 5yrs × PV 13yrs @ E%	
19 to 23		£YP 5yrs × PV 18yrs @ E%	
24 to perp		£YP perp @ ARY% × PV 23yrs @ E%	

E% = equated yield; ARY% = all risks yield

The full calculation then becomes:

Years	Income p.a.	PV factor	DCF
1 to 3	£100,000	2.465	£246,512
4 to 8	£163,909	2.774	£454,694
9 to 13	£190,016	1.684	£319,959
14 to 18	£220,280	1.022	£225,148
19 to 23	£255,365	0.620	£158,432
24 to perp	£296,038	1.258	£372,327
		GPV =	£1,777,074
		SAY =	£1,777,000

Proof that aggregating all years from around 20 to 25 from today through to perpetuity does not significantly affect the mathematical outcome can be shown through analysis of another simple example. Suppose a freehold property with a current market rent value of £100,000 per annum has just been sold with vacant possession for £1.25 million. Conventional freehold valuation analysis would show:

Net income =   £   100,000 p.a.
× YP perp @ 8%   12.5
                              Capital value = £1,250,000

Therefore, the sale was at an all risks yield of 8 per cent.

The equated yield approach to analysing this sale, on the assumptions that average annual rental growth rate is estimated at 3 per cent and similar properties let on leases would have rent reviews every five years, will be as shown in Table 12.5. By trial and error, the internal rate of return or equated yield that produces this valuation was found to be 10.58 per cent.

Now instead of aggregating all years to perpetuity after year 25, suppose the income rows were continued through to year 200. This time span could certainly be considered perpetuity. Does it affect the gross present value?

DCF analysis shown in Tables 12.5 and 12.6 assumptions:

Equated Yield	10.58%
Rent Review Period	5 years
Annual Rental Growth Rate	3%
Estimated market rent	£100,000 p.a.
Current Rent Payable	£100,000 p.a.

The conclusion that can be reached is that this longer calculation does not produce a different value. The same GPV is found even when rental growth is explicitly calculated all the way through to year 200 – see Table 12.6. Therefore, the shorter version is preferred as it reduces the amount of calculation required without affecting the accuracy of the outcome.

Table 12.5   The equated yield approach to analysing a sale

Years	Income per annum	PV factor @ 10.58%	DCF
1 to 5	£100,000	3.735	£373,528
6 to 10	£115,927	2.259	£261,894
11 to 15	£134,392	1.366	£183,624
16 to 20	£155,797	0.826	£128,746
21 to 25	£180,611	0.500	£90,268
26	£209,378	1.012	£211,800
		GPV =	£1,249,861
		SAY =	**£1,250,000**

* = income capitalised @ ARY to reflect continued future rental growth but then discounted at equated yield.

Table 12.6   Discounted cash flow analysis of sale of freehold property using explicit rental growth through to year 200

Years	Rental per annum	YP×PV @ 10.58%	DCF
1 to 5	£100,000	3.735280093	£373,528.01
6 to 10	£115,927	2.259124915	£261,894.49
11 to 15	£134,392	1.366335390	£183,624.05
16 to 20	£155,797	0.826369708	£128,745.71
21 to 25	£180,611	0.499794487	£90,268.44
26 to 30	£209,378	0.302279387	£63,290.59
31 to 35	£242,726	0.182820800	£44,375.41
36 to 40	£281,386	0.110571367	£31,113.26
41 to 45	£326,204	0.066874377	£21,814.67
46 to 50	£378,160	0.040446116	£15,295.09
51 to 55	£438,391	0.024462109	£10,723.96
56 to 60	£508,215	0.014794864	£7,518.97
61 to 65	£589,160	0.008948042	£5,271.83
66 to 70	£682,998	0.005411842	£3,696.28
71 to 75	£791,782	0.003273122	£2,591.60
76 to 80	£917,893	0.001979608	£1,817.07
81 to 85	£1,064,089	0.001197282	£1,274.01
86 to 90	£1,233,571	0.000724125	£893.26
91 to 95	£1,430,047	0.000437956	£626.30
96 to 100	£1,657,816	0.000264879	£439.12
101 to 105	£1,921,863	0.000160201	£307.88
106 to 110	£2,227,966	0.000096891	£215.87
111 to 115	£2,582,823	0.000058600	£151.35
116 to 120	£2,994,200	0.000035442	£106.12
121 to 125	£3,471,099	0.000021435	£74.40
126 to 130	£4,023,955	0.000012964	£52.17
131 to 135	£4,664,866	0.000007841	£36.58
136 to 140	£5,407,859	0.000004742	£25.65
141 to 145	£6,269,190	0.000002868	£17.98
146 to 150	£7,267,710	0.000001735	£12.61
151 to 155	£8,425,268	0.000001049	£8.84
156 to 160	£9,767,194	0.000000635	£6.20
161 to 165	£11,322,855	0.000000384	£4.35
166 to 170	£13,126,292	0.000000232	£3.05
171 to 175	£15,216,971	0.000000140	£2.14
176 to 180	£17,640,639	0.000000085	£1.50
181 to 185	£20,450,336	0.000000051	£1.05
186 to 190	£23,707,544	0.000000031	£0.74
191 to 195	£27,483,541	0.000000019	£0.52
196 to 200	£31,860,957	0.000000011	£0.36
		GPV =	£1,249,831
		SAY =	**£1,250,000**

With freehold valuations a choice will also need to be made on whether to assess the investment in perpetuity or only over a specific 'holding period'. Institutional funds and similar investors tend to adopt a policy on acquisition of a freehold property that they will retain ownership of it for a specific period and then look to sell it and acquire a newer investment. In this way, they dispose of a building before depreciation has significantly eroded its value and before the need for redevelopment arises. This length of time during which they own the property is the holding period. Typically, this is in the range of ten to fifteen years. It is a time span that mirrors medium-term gilt investments.

The question then remains of at which particular ARY percentage rate is the capitalisation undertaken? Usually this will be the ARY which would apply to the market rent income for a property of this type as at today. However, some investors prefer to use a slightly higher 'exit yield', say plus 1 per cent. Why do they do this? One reason is that it reflects the increased uncertainty of predicting long-term future market conditions and the possible difficulties of finding a buyer at the optimum sale price. Another is that it implicitly allows for the building depreciating over time due to physical deterioration and/or functional or economic obsolescence. As a result, its market value on resale will be adversely affected compared to the building as it now stands – the longer the 'holding period' (the time period from now to the chosen sale date) the higher the probability that depreciation of the building will occur. Askham (1989c: 52) explains that there will be 'a progressive diminution in rental value and thus a fall in capital value, aggravated by an increase in yield which results from the increased risk resulting from the anticipation of a future fall in rental growth'.

The effect of using a higher 'exit yield' will, of course, be that the present market value is reduced or the overall return from the investment, as reflected by IRR (equated yield) from a discounted cash flow appraisal, will be lower if the market value or price remains fixed.

Example

A pension fund is considering purchasing the freehold interest in an office building that is currently let at a rent of £500,000 per annum, although the market rent is estimated at £600,000 per annum. The lease is on full repairing and insuring terms and has eighteen years left to run. There are rent reviews due in three, eight and thirteen years' time. Market all risks yields for similar investments are currently 5 per cent. Average annual rental growth is forecast at 2.5 per cent. The fund are looking to achieve a target equated yield of 7 per cent (annually in arrear) and anticipate holding the investment for the next thirteen years before selling it at the end of that year at an 'exit yield' 1 per cent higher than the present all risks rate. What is the worth of the investment to the fund?

Valuation

Assumed annual rental growth rate =	2.50 %
Equated Yield (annually in arrear) =	7.00 %
True Equated Yield (quarterly in advance) =	7.32 %

Years	Income	PV factor	DCF
1 to 3	£500,000	2.624	£1,312,158
4 to 8	£646,134	3.347	£2,162,600
9 to 13	£731,042	2.386	£1,744,523
13 sale*	£13,785,116	0.415	£5,720,332
		GPV =	£10,939,613
		SAY =	£10,940,000

* At end of year 13 property to be sold at a price estimated as the then market rent × YP in perpetuity @ 6% (current ARY + 1%) = £827,107 × 16.6667

An alternative view suggested by Crosby (2006) is that the annual growth rate adopted should be related to the building in question rather than the location and could explicitly be net of depreciation.

12.6 DCF method compared with 'traditional' property valuation methods

The traditional methods of valuation are not always the best or the most accurate. It is therefore often useful to carry out a discounted cash flow analysis of an investment valuation to check on its accuracy and reliability.

Suppose a valuation of the freehold interest in each of two properties, A and B, is required. They are both worth £105,000 per annum market rent on FRI terms. A is let at £100,000 per annum for the next five years and B at £100,000 per annum for the next fifteen years, without rent review. They are to be valued using the traditional term and reversion method and a rack rent all risks yield of 6 per cent, then again using the DCF method and an equated yield of 10.5 per cent. Assuming average rental growth is expected at 5 per cent per annum compound, and at the expiry of the existing leases A and B will both be relet on new leases at market rent with five-yearly rent reviews, what conclusions can be drawn from the two sets of valuations?

Using the traditional term and reversion method the valuations are as shown in Figure 12.1. The equivalent yield approach has been adopted because the term and reversion rentals are almost the same and thus the term rent is no 'more secure' than the reversionary one. Conclusion: using the traditional method of valuation, the investment that has a fixed income for the longer term possesses only a very marginally lower market value, which does not seem logical.

Using the DCF method of valuation, the valuations are as shown in Table 12.7. The GPV of property A from DCF analysis is virtually the same as the capital value produced by the term and reversion method, but the GPV of B is considerably lower than that found originally. Given that the income on one property is fixed for a period of ten years longer than on the other, the gross present values found from the discounted cash flow analysis look consistent. It is reasonable to expect that Property A would have a considerably higher market value than property B. Without undertaking the DCF approach though, this error may have gone unnoticed.

This shows that DCF offers advantages over the traditional method of valuation in certain instances and is a useful check on the accuracy of the figures produced by a conventional valuation. To produce an accurate conventional valuation of property B will require a higher ARY to be used

Valuation of property A:

Term

Net income	£100,000 p.a.	
× YP for 5 yrs @ 6%	4.212	
	=	£421,236

Reversion to market rent

Net income	£105,000 p.a.	
× YP perp def 5 yrs @ 6%	12.454	
	=	£1,307,702
	Market value =	£1,728,938
	SAY =	**£1,729,000**

Valuation of property B:

Term

Net income	£100,000 p.a.	
× YP for 15 yrs @ 6%	9.712	
	=	£971,225

Reversion to market rent

Net income	£105,000 p.a.	
× YP perp def 15 yrs @ 6%	6.954	
	=	£730,214
	Market value =	£1,701,439
	SAY	**£1,701,500**

Figure 12.1   Valuations of properties A and B using the traditional term and reversion method

Table 12.7   Discounted cash flow valuations of properties A and B

Years	Income A	Income B	YP×PV@ 10.5%	DCF for A	DCF for B
1 to 5	£100,000	£100,000	3.743	£374,300	£374,300
6 to 10	£134,009	£100,000	2.272	£304,468	£227,200
11 to 15	£171,030	£100,000	1.379	£235,850	£137,900
16 to 20	£218,290	£218,290	0.837	£182,709	£182,709
21 to 25	£278,600	£278,600	0.508	£141,530	£141,530
26 to perp	£355,570	£355,570	1.373 [*]	£488,198	£488,198
			GPV =	£ 1,727,055	£1,551,837
			SAY =	**£1,729,000**	**£1,552,000**

[*] YP perp @ ARY of 6% def 25 yrs @ 10.5%

on that property to reflect the additional risk and uncertainty caused by having a fixed income for an extra 10 years. At its 'true' market value of £1,552,000, property B would analyse at an equivalent yield of around 6.63 per cent compared to the 6 per cent used on property A. This difference in yields reflects the additional risk and uncertainty associated with the longer fixed income on B which makes it a less attractive investment; and follows the basic premise that the worse an investment is, the higher its ARY will be.

12.7 'Short-cut' freehold DCF valuations

This is a quick approach to producing a discounted cash flow property valuation. It is not as accurate as a full appraisal, but allows for an approximate estimation of gross present value for comparison with figures produced by 'traditional' methods.

The calculations required are:

Current rent

    × YP for term years @ equated yield
    plus

Market rent

    × rental growth to next rent revision *
    (rent review or end of lease whichever sooner)
    × YP perp @ reversionary ARY
    × PV for term years @ equated yield

    * = Amount of £1 @ estimated annual rental growth rate for number of years between now and next rent revision.

Example

In Section 12.4 above, a freehold shop investment let at its market rent of £100,000 per annum was valued at an equated yield of 13 per cent, where rental growth was assumed at 5 per cent. Using the relevant formula it was found that the all risks yield equated to 8.74 per cent in that specific case. The market value was calculated at £1.144 million. This is how the same property could be assessed using the short-cut DCF approach:

	£	
Current rent=	100,000 p.a.	
× YP for 5 years @ 13%	3.5172	
		= £ 351,720
plus		
Market rent=	£ 100,000 p.a.	
× Amount of £1 for 5yrs @ 5%	1.2763	
	= £ 127,630 p.a.	
× YP perp @ 8.74%	11.4416	
× PV for 5 years @ 13%	0.5428	
	= £ 792,646	
	Market value= £1,144,366	
	SAY= **£1,144,000**	

12.8 Leasehold DCF valuations

Leasehold valuations using a discounted cash flow approach are potentially more complex than freeholds due to making an allowance for sinking fund instalments. This will also significantly affect the equated yield obtained from such investments.

Example

Leasehold DCF example 1

A leasehold interest is to be valued. The market rent for the property is estimated at £20,000 per annum but the tenant only pays a rent of £15,000 per annum at present. The lease has five years left to run without rent review. All risks leasehold yields for similar investments are 9.5 per cent and rental growth rate is expected to average 3 per cent per annum. The valuation is to assume that the lessee pays tax at 30 per cent and can obtain 4.3 per cent gross on a sinking fund.

Net annual sinking fund interest rate = 4.3% × (100 − 30)/100 = 3.01% = SAY 3% net

The conventional valuation will be:

Rent receivable (market rent) =	£	20,000 p.a.
Less rent payable =	£	15,000 p.a.
	Net Profit Rent = £	5,000 p.a.
	× YP for 5 yrs @ 9.5%	
	+ 3.0% & tax @ 30%	2.747
	Market value (assignment price)=	£13,733
	SAY=	**£13,750**

The discounted cash flow valuation is as shown in the table.

Years	Rent receivable	Rent payable	ASF instalments	Net Cash Flow	PV @ 17.10% (gross of tax)	DCF
1	£20,000	£15,000	£3,693	£1,307	0.854	£1,116
2	£20,600	£15,000	£3,693	£1,907	0.729	£1,390
3	£21,218	£15,000	£3,693	£2,525	0.623	£1,572
4	£21,855	£15,000	£3,693	£3,161	0.532	£1,681
5	£22,510	£15,000	£3,693	£3,817	0.454	£1,734
5	Capital replaced by asf at end of lease =	£13,728			0.454	£6,235
					GPV =	£13,728
					SAY =	£13,750

From trial and error or using a spreadsheet solver function, the internal rate of return (equated yield) is found to be 17.1 per cent.

Growth each year has been calculated on the rent receivable as the tenant will occupy premises that will be worth that figure but still be paying £15,000 per annum. The annual sinking fund instalments need to be grossed up to allow for income before tax required for the instalments:

£13,728 × asf for 5 years @ 3% = £2,585.73 annual instalments needed.

Before tax the amount that needs to be earned to provide £2,585.73 will be:

£2,585.73 × 100/(100 – 30) = £3,693.90.

With long leaseholds in particular it is arguable whether an allowance for annual sinking fund should be incorporated and a single rate approach adopted instead. This is considered in more detail in Section 16.11.

Example

Leasehold DCF example 2

A client is interested in purchasing a ground leasehold investment. The ground lease has an unexpired term of 22 years at a fixed ground rent of £50,000 per annum. A sub-tenant on a lease occupies the premises for a term of 25 years, granted three years ago, on full repairing and insuring terms and with five-yearly rent reviews. The current rent payable under the lease is £220,000 per annum and the present market rent for the property is estimated at £280,000 per annum. All risks yields for similar leasehold property investments are 11 per cent on a 'single-rate' basis (in other words a slightly higher yield being adopted to offset no sinking fund with tax deductions being used). What price should the client expect to pay to purchase this ground lease?

Conventional valuation:

Term

Rent receivable	£	220,000 p.a.		
Less rent payable	£	50,000 p.a.		
	Net profit rent=	£	170,000 p.a.	
	× YP for 2 yrs @ 11%		1.7125	
			=	£ 291,125

Reversion

Rent receivable	£	280,000 p.a.		
Less rent payable	£	50,000 p.a.		
	Net profit rent=	£	230,000 p.a.	
	× YP for 20 yrs @ 11%		7.9633	
	× PV for 2 yrs @ 11%		0.8116	
			=	£ 1,486,493
	Market value (assignment price)=			£ 1,777,618
			SAY=	£ 1,780,000

Analysing the rental growth since the commencement of the lease and assuming this rate of growth is maintained on average over the remaining unexpired term of the ground lease, what equated yield would be achieved on this investment if the lease is purchased now at the price suggested by the valuation above?

Analysis: Current rent receivable from sublessee is £220,000 per annum, agreed when the lease commenced three years ago. Current estimated market rent is £280,000 per annum. Thus, over three years the increase has been £280,000/£220,000 = 1.27273. The Amount of £1 for three years would represent this compounded growth and so $(1+i)^3 = 1.27273$.

$$1+i = \text{inv log} (\log 1.27273/3) = 1.0837$$

Therefore $i = 0.0837$ and rental growth rate = 8.37 per cent per annum. This is obviously a high rate and it may be questionable whether it could be maintained over next 22 years, but assuming it is then the discounted cash flow valuation at equated yield will be as shown in the table.

DCF Valuation

Year	Market rent	Rent paid	Profit rent	PV @ 18.90%	DCF
1	£220,000	£50,000	£170,000	0.84104	£142,977
2	£328,834	£50,000	£278,834	0.70735	£197,234
3	£328,834	£50,000	£278,834	0.59491	£165,882
4	£328,834	£50,000	£278,834	0.50035	£139,514
5	£328,834	£50,000	£278,834	0.42081	£117,337
6	£328,834	£50,000	£278,834	0.35392	£98,686
7	£328,834	£50,000	£278,834	0.29766	£82,999
8	£491,498	£50,000	£441,498	0.25035	£110,528
9	£491,498	£50,000	£441,498	0.21055	£92,959
10	£491,498	£50,000	£441,498	0.17708	£78,183
11	£491,498	£50,000	£441,498	0.14894	£65,755
12	£491,498	£50,000	£441,498	0.12526	£55,303
13	£734,627	£50,000	£684,627	0.10535	£72,126
14	£734,627	£50,000	£684,627	0.08860	£60,661
15	£734,627	£50,000	£684,627	0.07452	£51,018
16	£734,627	£50,000	£684,627	0.06267	£42,909
17	£734,627	£50,000	£684,627	0.05271	£36,088
18	£1,098,025	£50,000	£1,048,025	0.04433	£46,462
19	£1,098,025	£50,000	£1,048,025	0.03729	£39,077
20	£1,098,025	£50,000	£1,048,025	0.03136	£32,865
21	£1,098,025	£50,000	£1,048,025	0.02637	£27,641
22	£1,098,025	£50,000	£1,048,025	0.02218	£23,247
				GPV =	£1,779,451
				SAY =	£1,780,000

Conclusion is that equated yield return on investment over next 22 years, assuming purchase price is £1.78 million and annual rental growth averages 8.37 per cent per annum, will be 18.9 per cent per annum

12.9 Combining the methods in a single valuation appraisal: block income, layer, equivalent yield and equated yield/DCF approaches

It is always advisable to calculate the value of a property using at least two different methods. This acts as a means of checking the reliability of the final recommended value and can offer an alternative view of how that value can be justified.

Example

An investor wishes to purchase the freehold interest in an office building that currently produces a rental income of £450,000 per annum. The property is let on full repairing and insuring terms on a 20-year lease granted two years ago and is subject to five-yearly upwards or downwards rent reviews. The present market rent value of the property is estimated at £525,000 per annum. Market evidence indicates that the following can be assumed for this property:

- all risks yield of 8 per cent; and
- equivalent yield of 7.97 per cent; and
- equated yield of 11 per cent; and
- rental growth of 3.5 per cent per annum.

Using this information undertake a market value assessment of the freehold interest in the property using:

- 'traditional/conventional' methods on both term and reversion and equivalent yield approaches; and
- 'hardcore' or 'layer' method; and
- discounted cash flow analysis.

Freehold term and reversion valuation

Term

Net income =	£	450,000	
× YP for 3 yrs @ 7%		2.624	
		=	£ 1,180,942

Reversion

Net income =	£	525,000	
× YP perp def 3 yrs @ 8%		9.922	
		=	£ 5,209,524
		Capital value =	£ 6,390,466
		SAY =	£ 6,400,000

Note: term yield decreased by 1% as term income 'more secure' than reversionary income.

Equivalent yield approach

Term

Net income	£	450,000
× YP for 3 yrs @ 7.97%		2.578
	=	£1,160,322

Reversion

Net income	£	525,000
× YP perp def 3 yrs @ 7.97%		9.968
	=	£5,233,493

Capital value = £6,393,815
SAY = £6,400,000

Hardcore or layer approach using equivalent yield

Hardcore

Net income	£	450,000
× YP perp @ 7.97%		12.547
	=	£5,646,173

Top Slice

Net income	£	75,000
× YP perp def 3 yrs @ 7.97%		9.969
	=	£ 747,642

Market value = £6,393,815
SAY = £6,400,000

Hardcore or layer approach using differential yields

Hardcore

Net income	£	450,000
× YP perp @ 7.83%		12.77
	=	£5,747,126

Top Slice

Net income	£	75,000
× YP perp def 3 yrs @ 9.83%		8.579
=	=	£ 643,486

Capital value = £6,390,613
SAY = £6,400,000

Using 2 per cent differential in the yields, to reflect the additional risk associated with the top slice income of £75,000 per annum, the hardcore yield is found to be 7.83 per cent from trial and error. This is 'back fitting' the yields to match the capital value found from other approaches. It is useful to show the achieved return at the suggested value, but also shows the difficulty of yield selection when using this method.

Equated yield approach (given annual rental growth rate = 3.5 %)

Years	Income	YP×PV factor	DCF @ 11%
1 to 3	£450,000	2.444	£1,099,672
4 to 8	£582,077	2.702	£1,573,009
9 to 13	£691,325	1.604	£1,108,710
14 to 18	£821,077	0.952	£781,457
19 to 23	£975,182	0.565	£550,797
24 to perp	£1,158,210	1.134	£1,313,012
		GPV =	£6,426,658
		SAY =	£6,400,000

Conclusions

- It is unwise to rely on only one method of valuation – using two or more methods and comparing the results is likely to give a more accurate and balanced view.
- Equivalent yield approach is useful to provide overall return on investment, which many investors would like to know and avoids having to decide whether term income is 'more secure' and by how much the ARY needs adjustment.
- It is difficult to decide on top slice yields to be used in layer approach as these have to be derived from analysis of comparables using the same analysis method.
- Discounted cash flow enables valuers to communicate with banks, accountants and other financial advisors in a format they understand and use themselves and helps identify particular drawbacks in specific cash flows.
- The conventional term and reversion method has worked well for many years, but current practice recommendations (including in RICS *Red Book*) also demands use of DCF.
- Deciding on rental growth rate to adopt is difficult as it involves forecasting the future and assuming the same average rate will apply each and every year, whereas in practice it is likely to fluctuate.
- Conventional methods rely on observation of price and present value in the current market, which are more certain and easier to analyse.
- But the equated yield derived from DCF analysis gives the total overall return from the investment, which enables a truer comparison to be made with other non-property investments.

Progress check questions

- How does gross present value differ from net present value?
- What does the internal rate of return in a discounted cash flow appraisal measure and how is it found?
- What is the difference between an equivalent and an equated yield?
- What is meant by the yield being 'implicit' in traditional valuations and 'explicit' in discounted cash flow valuations?
- Why will a fixed income for a long time tend to be overvalued when using traditional term and reversion valuation?
- What is a holding period and exit yield in relation to financial fund property investments?
- How is the continued growth of rental income in perpetuity allowed for in DCF calculations when a 'terminal income' is only calculated up until around years 20 to 25?
- Why may using more than one valuation approach be useful when analysing a property investment?
- Using the DCF approach requires some forecasting of the future – how can this be done?
- Why do banks and accountants find the DCF approach more similar to their assessment methods?
- What is the difference between the historical valuation methodology used in the UK compared to the US and how are the two increasingly being used together?

Chapter summary

Discounted cash flow is an accounting technique that has long been used for property appraisals in the US. It was introduced to the UK in the 1970s but initially was slow to be adopted by the property profession. With the availability of computer spreadsheets and other specialist software, it is now used extensively as an additional procedure to supplement the conventional methods of valuation. Two approaches can be used with DCF: present value or internal rate of return. In the first a value at a target yield rate is found. In the second, a known value is analysed to discover the overall rate of return provided to the owner of that investment.

DCF can calculate the objective value or the subjective worth of a property, depending on whether the yield used is market-derived or a client's target rate of return. The equated yield approach effectively requires a forecast of future values to be made since all rents and other figures need to take account of expected market growth from the date of valuation until receipt or payment. This explicit adjustment to take account of inflationary increases contrasts with the implicit approach adopted in conventional valuation methods where the all risks yield reflects the prospects of growth. Various formulae are available to help deduce the yield or growth rate implied by known market data.

The basis of the method is to set out the yearly cash flows attributable to an investment and then discount at equated yield each flow to arrive at its present value. The sum of these discounted cash flows provides the present value of the investment. The method allows for the exact timing of all cash inflows and outflows to be taken into account and provides a more overt summary of expected flows than is provided by traditional investment method calculations. When used together with conventional valuation methods it provides a checking and balancing system that can highlight errors or deficiencies in either and help to ensure the calculated values are more robust and reliable.

Further reading

Baum, A. and Crosby, N. (2007) *Property Investment Appraisal,* 3rd edn, London: Wiley Blackwell.

Baum, A., Nunnington, N. and Mackmin, D. (2006) *The Income Approach to Property Valuation,* 5th edn, London: Estates Gazette.

Bornand, D. (1988) 'Mainly for students: internal rate of return and external rates', *Estates Gazette,* 843 (29 Oct.): 88; repr. in P. Askham and L. Blake (eds), *The Best of Mainly for Students* (London: Estates Gazette, 1993), 111–15.

Bowcock, P. and Bayfield, N. (2000) *Excel for Surveyors,* London: Estates Gazette.

—— (2004) *Advanced Excel for Surveyors,* London: Estates Gazette.

Bowcock, P. and Rose, J.J. (1979) *Valuing with the Pocket Calculator,* Oxford: Technical Press.

Crosby, N. (1991) 'Valuing commercial investment property', *CSM* (Oct.): 9–12.

Curwin, J. and Slater, R. (1996) *Quantitative Methods for Business Decisions,* 4th edn, London: International Thomson Business Press.

Estates Gazette (1976) 'Mainly for students: discounted cash flow techniques and model building – Part I', 238 (26 Jun.): 984–5.

—— (1995) 'Mainly for students: spreadsheets and valuation', 9503 (21 Jan.): 116–19.

Fawcett, S. (2003) *Designing Flexible Cashflows,* London: Estates Gazette.

Fraser, W.D. (2004) *Cash-flow Appraisal for Property Investment,* Basingstoke: Palgrave Macmillan.

Gilbertson, B., Preston, D. and Howarth, A. (2006) *A Vision for Valuation* (RICS Leading Edge Series), London: RICS. Online: <http://www.rics.org/NR/rdonlyres/BBEBD43B-11CA-4A2E-BDAC-B37BB2505CA1/0/vision_for_valuation.pdf>

Havard, T. (2002) *Investment Property Valuation Today,* London: Estates Gazette.

Hutchison, N. and Nanthakumaran, N. (1996) *The Cutting Edge: Estimation of the Investment Worth of Commercial Property,* London: University of Aberdeen and RICS Research.

Ifediora, B.U. (2005) *Valuation Mathematics for Valuers and Other Financial and Investment Analysts,* Enugu: Immaculate Publications.

Imber, A. (1997) 'Mainly for students: value of theory and practice', *Estates Gazette,* 9744 (1 Nov.): 195–8; repr. in P. Askham and L. Blake (eds), *The Best of Mainly for Students* (London: Estates Gazette, 1999), vol. 2, pp. 358–68.

Isaac, D. and Steley, T. (1999) *Property Valuation Techniques,* 2nd edn, Basingstoke: Palgrave Macmillan.

James, D. and Imber, A. (2002) 'Mainly for students: how much will I make?', *Estates Gazette,* 0232 (10 Aug.: 80; repr. in L. Blake and A. Imber (eds), *The Best of Mainly for Students* (London: Estates Gazette, 2004), vol. 3, pp. 196–202.

Lumby, S. (1994) *Investment Appraisal and Financing Decisions: A First Course in Financial Management,* 5th edn, London: Chapman & Hall.

Morley, S. (1995) 'The future for rental growth', *Estates Gazette*, 9503 (21 Jan.): 110–12.

Mott, G. (1997) *Investment Appraisal*, 3rd edn, London: Pitman.

Parsons, G. (ed.) (2004) *The Glossary of Property Terms*, 2nd edn, London: Jones Lang LaSalle and Estates Gazette.

Rose, J. (1994) 'A valuation revolution? Arithmetic v. algebra', *Estates Gazette*, 9445 (12 Nov.): 141–2.

Royal Institution of Chartered Surveyors (1997) *Commercial Investment Property: Valuation Methods*, Coventry: RICS Books.

Royal Institution of Chartered Surveyors and Investment Property Forum (1997) *Calculation of Worth: An Information Paper*, London: RICS. Online: <http://www.rics.org/NR/rdonlyres/B59119AF-A281-4E8A-A079-1FD98394CD09/0/Calculation_of_Worth_1997.pdf> (accessed 5 March 2009).

Whipple, R.T.M. (1995) *Property Valuation and Analysis*, Sydney: Law Book Co of Australasia.

Residual method

In this chapter ...

- The method of valuation used to assess the value of development, redevelopment or refurbishment properties.
- How to calculate the value of the completed development and estimate the costs of the required construction work and associated expenditure.
- Why the residual value obtained should also be checked against available comparables or other market evidence.
- The simplified approach to this method of valuation and how more complex approaches can and should be used.
- Using probability, more sophisticated analytical techniques and computer software to improve reliability and accuracy of development appraisal valuations.

13.1 Basis of method and when used

'A method of determining the value of a property which has potential for development, redevelopment or refurbishment' (Parsons 2004: 220). It is used to calculate the site value where direct market comparable evidence is not available and also by developers to ascertain the feasibility of a proposed scheme. That is, to establish what they can *afford* to pay for the land, which may produce a different figure to a market value on a comparable basis, or what potential *profit* may be achieved if the land is purchased at a specified price.

It combines a comparison or investment method with a costs approach. The basis of valuation is as follows:

- Suppose a site is developed or redeveloped, then its value will be £x when the work is completed.
- If the inclusive costs of developing it, with a profit margin included, are £y, the developer can afford to pay £x – £y for the site as it stands.
- The sum of £x – £y is the residue; thus this approach is called the 'residual method'.

The basic calculation is:

Gross development value (GDV)
Less all the costs of development, including:
- compensation to existing occupiers;
- planning and other statutory consents charges;
- demolition, site clearance and site preparation;
- promotion and marketing costs;
- construction costs;
- architect, quantity surveyor and other construction professional fees;
- legal and estate agent fees on site purchase and final letting and/or sale;
- miscellaneous or contingency costs;
- interest charges on development finance; and
- developer's risk and profit.
= Balance
Less costs of site acquisition and interest payments on site funding
= Residual site value

The following summaries consider each of the above factors in more detail.

Gross development value

The first step in a residual valuation is to calculate the gross development value (GDV) of the proposed project. This is the capital value of the proposed project when it is *completed*. It is found from direct comparison with buildings of the same type to be built, located in the same area. Alternatively, the investment method is used so that the estimated market rent and all risks yield are found from comparables and the capital value calculated using years purchase (that is, MR × YP perp @ ARY per cent = GDV).

It should be noted that capital and rental values are often assessed on a different floor area measurement to construction costs and care is needed to use the appropriate basis. For instance, rents on offices are normally found from the net internal floor area (NIA), whereas construction costs are based on the gross internal area (GIA). The rent per square metre and all risks yield may be overall figures for the whole development, or separate calculations for each part.

The purchaser of a completed commercial development (assuming the investment will be sold on after being fully let) will incur 'fund costs'. These will include agent's fees, valuation fees and solicitor's fees and in the UK Stamp Duty Land Tax. They will total around 3 to 7 per cent of the GDV. An allowance for these factors can be made to calculate the net investment yield from the gross investment yield. For example, purchaser pays £100 million to buy a completed development producing a net rental income of £5 million per annum and incurs costs on purchase of 6 per cent. The purchase is made on a 5 per cent all risks yield basis, but total outlay has been £10.6 million. Therefore net yield is £5m × 100/£10.6m = 4.72 per cent.

Costs: compensation

To obtain vacant possession of the site, compensation payments may need to be made to existing tenants and occupiers. Statutory compensation payments are appropriate in some cases, such as disturbance payments under the UK's Landlord and Tenant Act 1954 Part II, otherwise amounts payable are by voluntary private agreement. The actual agreed figures should be used if possible, or if not an estimate.

Costs: consents

Amount to be paid in planning, building regulation and similar consent application fees.

Costs: demolition and site clearance

Any costs to demolish existing buildings and to clear the site ready for new construction work to commence. This may involve ground remediation work where contamination exists. Actual figures are again best; otherwise, an estimate must be used.

Costs: promotion costs

Expenditures incurred in advertising and marketing the scheme to find potential tenants and/or purchasers.

Costs: construction costs

Floor area to be constructed is multiplied by the relevant cost per square metre. Residential property is normally assessed on GEA (gross external floor area) and non-residential on GIA (gross internal floor area) basis. However, check which method of measuring area is used by the costing source. Different buildings within the same development will have separate cost rates, for example, retail space is assessed at a different figure to offices. An allowance may also need to be made for any site works such as new roadways, extra drainage, landscaping, environmental impact reduction works, etc. Actual figures supplied by a contractor are the most reliable, but when not available, an estimate can be made using BCIS data (Building Cost Information Service 2008b) or similar sources, such as *Wessex* (BCIS Wessex 2008) or *Spon's* (Davis Langdon 2008) as explained in Section 11.14.

Costs: construction professionals' fees

Payable for services provided by architects, quantity surveyors, project managers, civil, structural and mechanical and electrical engineers, landscape architects, traffic engineers, planning consultants, environmental consultants and others to design, cost and manage the development project. Commonly amount to 10–13 per cent of the construction costs although sometimes can be lower due to competitive fee tendering.

Costs: fees on letting and/or sale

Legal and estate agents fees paid when completed buildings are let or sold. For letting commercial and industrial property, the charge is usually around 10 per cent of the first year's rent. For the sale of either residential property for owner-occupation or business premises as a fully let investment or for owner-occupation, the fees are normally in the range of 1 to 3 per cent of the GDV.

Costs: miscellaneous or contingency

An allowance for additional items not already included in the other categories or for payments arising from unforeseen circumstances over the whole development period.

Costs: finance

Interest charges to be paid as the cost of borrowing money to finance the project. The interest payments will increase over the development period as more money is borrowed. For example, the initial sum borrowed will be to cover the cost of land acquisition, clearing it and obtaining consents. The next sum to be borrowed could be the first 'stage' payment due to the contractor. In large building contracts, it is common practice for the contractor to be entitled to several of these 'stage' payments at specified intervals during the contract, as certain stages of the work are completed. It would obviously be inequitable for a developer to expect the contractor to fund all the costs of the building work and only receive a single payment at the end. The contractor needs to purchase materials, hire machinery and pay the workforce and so on at regular intervals. A contractor would find it difficult to finance the whole project from start to finish. The stage payment system overcomes this problem, as the total sum owing to the contractor at any one time will only be a percentage of the total contract figure.

In a housing development, typical 'stages' at which payments become due are on completion of:

1. foundations
2. first floor level
3. roof
4. final completion.

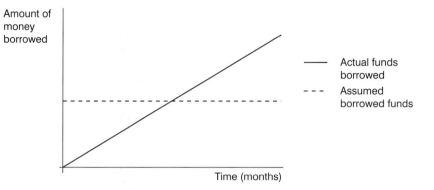

Figure 13.1   Basis of estimated sum borrowed for development costs excluding site purchase

As it is not necessary for the developer to borrow the total development cost sum at the outset of the project, it would be incorrect to calculate interest on this sum over the whole development period. To calculate accurately the actual interest charges that will be payable, it is necessary to carry out a month-by-month analysis of the project, identifying when money is borrowed and adding interest charges accordingly. In a discounted cash flow analysis, this approach would be adopted. However, for the 'simplified' residual valuation approach, this calculation is overly complex.

Traditionally it has therefore become common practice to *take the full interest charges due on half the development costs over the full development period* as an approximation of the effects of borrowing the money gradually over the period. This is a balance between less than half the money being borrowed in the first half of the development project period and the rest in the second half, all at the same rate of interest, and is illustrated in Figure 13.1. With this approach there is an overcalculation of interest charges in the first half of the development period that is approximately counterbalanced by an undercalculation in the second half.

The full development period covers:

- the pre-build void period between acquiring the site and starting construction work on site;
- the building period, taken from the start on site to practical completion of the construction; and
- the letting or sales void, which is the period once the buildings are completed before they are fully let and/or the finished buildings are sold on to owner-occupiers or to an investor.

The pre-build void occurs after purchase of the site as it can take time for the construction contract to be agreed and for the final legal and administrative documentation to be completed before the contractor can move on to the site.

To reduce the effects of the lettings or sales void, most developers will complete larger developments in phased batches of units that can be marketed, let or sold and occupied whilst construction work continues on the remainder of the site. Thus on completion of the development, there will (hopefully) only be a few units remaining to be let or sold. However, in a slack market, this is not always achievable and it can be many months, or in extreme cases, much longer, before disposal of the final unit. The inherent risks to the developer's cash flow from the potential difficulties posed by such a market need to be adequately reflected in the risk and profit element of the calculation.

Interest rate per cent charged by the lending financial institution will vary according to the type of development, the identity and perceived security of the developer, other prevailing rates and the state of the economy. Generally, it will be around 1 to 5 per cent above base lending rate.

The calculation of the finance element is usually taken on the yearly rate of interest payable. Where it is assessed on a different basis, such as monthly, the equivalent annual percentage rate is quoted. To calculate the sum of money paid as interest, the conventional annually in arrear formula can be used, or any other basis such as quarterly in advance that reflects the way the interest will actually be calculated.

Costs: developer's risk and profit

This is the developer's reward for carrying out the project together with an allowance for the necessary risks accepted in undertaking the work. Financial losses can be incurred from a number of unforeseen circumstances, such as the costs of materials or labour rising, delays due to strikes, inclement weather or accidents or changes in economic and market conditions. The effects of some of these eventualities can be mitigated through insurance, but this will never adequately cover all the risks.

The sum for risk and profit can be calculated in two different ways. The method that will tend to be used by an investor will be to express it as a percentage of the total investment value of the project (the GDV). In other words, how much of the investment value they are willing to surrender to the developer. Contractors, on the other hand, will more likely express it as a proportion of the total costs of the construction contract, since they will look for a return for the risk incurred on the total outlay.

Both approaches should produce a similar sum, and for a balanced opinion, it may be worth the valuer adopting a compromise figure between the two resultant amounts. In the end, it depends on what negotiated sum the parties are willing to agree. The usually sought ranges of percentages used in the two methods are around: 10–20 per cent of GDV or 15–25 per cent of costs. The precise figures adopted, which could be greater or smaller than these 'normal' ranges, will depend on the degree of risk and uncertainty connected with the project.

The GDV and the costs can be considered the 'inputs' to the residual valuation calculation. These basic factors make up the total project. Once these have been ascertained, the results (or 'outputs') can be calculated:

1. balance
2. site value.

Balance

This is the difference between the GDV and the total costs, and is the sum available for the site purchase and the costs associated with it. Obviously, if the costs are larger than the GDV the project is not feasible. Specifically the balance contains:

- The site purchase money, or 'site value': this is the reason for carrying out the residual valuation and shows what the land is worth if it is purchased for the proposed development purpose.
- Legal fees, surveyors fees, Stamp Duty Land Tax and any other costs associated with the site purchase. As a guide these commonly amount to 3–7 per cent of the site value.
- Interest payable on borrowing money for the above two items. These sums need to be borrowed at the outset of the project to acquire the land before construction can commence and therefore interest will be payable for the total cost of land plus acquisition fees over the entire development period. Again, they can be calculated annually in arrear or using a formula reflecting actual way in which interest is calculated by lender.

Site value

Having found the balance, the site value needs to be extracted from it. To achieve this, consider the site value as 'x'. Then the fees on site purchase will be say $0.03x$ to $0.07x$. That is 3 to 7 per cent of 'x'. The third element in the balance is the interest. To remove this from the total, the balance needs to be 'discounted', which is achieved by multiplying the balance by the Present Value of £1 for the development period at the appropriate rate of interest, as used in calculating other borrowed finance.

Thus: balance × PV of £1 = site value + fees and other costs of acquisition

So if site value = x and costs of acquisition = 0.03 to $0.07x$

Then: present value of balance = 1.03 to $1.07x$

Or in other words site value = PV of balance/1.03 to 1.07

Example 1: A residential development site valuation

Value the freehold interest in a site with a gross area of 0.33 ha, which is being sold with the benefit of full planning permission for residential development at a density of 250 habitable rooms per hectare. The local planning authority is prepared to allow a development of one and two bedroomed flats. Similar flats in the area are currently selling for £175,000 for one-bedroom properties and £235,000 for the two bedroom units. Construction costs are estimated at £1,000 per sq.m.

The client has decided to provide an equal number (as far as possible) of one and two bedroomed units. The floor area is to be 60 sq.m. GEA for the smaller flats and 80 sq.m. for the larger units. Total development period is expected to be 22 months and finance has been obtained at 8 per cent per annum.

Other costs are estimated as follows:

- initial site clearance and preparation: £50,000;
- other initial costs: £100,000;
- architect, QS and other construction professional fees: 13 per cent of construction costs;

- agents and legal fees on sale: 3 per cent of GDV;
- developer's risk and profit: an average between 25 per cent of costs or 12.5 per cent of GDV;
- site purchase costs: 6 per cent.

Calculation of the GDV:

Gross site area = 0.33ha @ 250 habitable rooms per ha
= 83 habitable rooms
@ 2 hr per 1 bed flat × 16 flats = 32 hr
and @ 3 hr per 2 bed flat × 17 flats= 51 hr
Total = 83 habitable rooms

Project total development period (years) = 22/12 = 1.833 years

Gross development value:
One bedroom flats
 number of units
 = 16 @ sale price per unit of   £   175,000
          Market value =   £ 2,800,000

Two bedroom flats
 number of units
 = 17 @ sale price per unit of   £   235,000
          Market value =   £ 3,995,000
           Total GDV =   £ 6,795,000
             SAY =   **£ 6,790,000**

Less costs of development:
Site clearance/remediation/preparation £   50,000
Other initial/miscellaneous/
contingency costs =   £   100,000
Construction costs:
One bedroom flats
 constructable floor area per unit in sq.m.
 = 60m²@ construction cost per sq.m.
 = £1,000.00/m²
Costs =   £   960,000
Two bedroom flats
 constructable floor area per unit in sq.m.
 = 80m² @ construction cost per sq.m.
 = £1,000.00/m²
Costs =   £   1,360,000
Arch./QS/other prof. fees
 @ 13% of const. costs   £   301,600

Agents/legals on sale
  @ 3% of GDV   £   203,700
Finance charges on half costs over
  full development period @ 8%
  = $((1.08)^{1.833} - 1) \times £2,975,300/2$ £   225,386
Developer's risk and profit:
as 25% of costs =   £800,171
as 12.5% of GDV =   £848,750
                    SAY = £   820,000
                       Total Costs = £ 4,020,686
                         Balance = £ 2,769,314

Less site purchase finance over
    development period @ 8%
= Balance – (PV of £1 for 1.8333 years
      @ 8% × Balance) =   £   364,362
  =   £ 2,404,952
Less site purchase costs @ 6%
  = £2,404,952 – £2,404,952/1.06 =   £   136,129
                             Site Value = £   2,268,823
                                   SAY = £   2,270,000
                        = per hectare: £   6,848,485

The found value per hectare should be compared to that achieved on other similar sites. The calculated value for the site represents 33 per cent of the GDV (£2,270,000 × 100/£6,790,000). This is relatively high and suggests the land is situated within an inner-city area. Again, comparisons should be sought to ascertain whether this percentage is in line with other plots.

Example 2: Offices development valuation

A property developer is interested in acquiring the freehold interest in a site measuring 0.2 hectare that is for sale at a price of £3.25 million. Town planning permission has been granted for the erection of 5,000 sq.m. gross external area (GEA) of offices on the land to a maximum height of six storeys. This will provide a gross internal area (GIA) of 4,750 sq.m. and a net internal area (NIA) of 3,750 sq.m. The local planning authority has imposed a requirement that no more than 40 car parking spaces are provided. The overall size of these car spaces, allowing for turning room, is to be 9 metres by 3 metres, and is to be provided on the surface of the site, with any remaining land area being landscaped.

Site clearance and preparation works at a cost of £100,000 are required before construction works can commence. Construction costs are anticipated to be £2,000 per sq.m. (GIA) for the building itself, £150 per sq.m. for the car park area and £100 per sq.m for the remainder of the site for landscaping and ancillary works. The development project is expected to take 21 months from start to completion, full letting of the finished building and investment sale of the freehold. Promotion costs and a contingency sum will amount to £250,000. Market rental for the completed building is estimated at £300 per sq.m. (net) per annum. Freehold all risks yields for investments of this type are currently 6 per cent. The developer can borrow funds at 8.5 per cent.

Other costs are estimated as follows:

- architect, QS and other construction professional fees: 12 per cent of construction costs;
- agents and legals on letting: 10 per cent of market rent;
- agents and legals on investment sale: 3 per cent of GDV;
- site purchase costs: 6 per cent.

What will be the developer's risk and profit margin if the land is purchased at the asking price?

Valuation
Assessment of site layout:
5,000 sq.m. /6 storeys = 833.3 sq.m. site coverage
Site area = 0.2 ha = 2,000 sq.m.
Leaves 1,166.7 sq.m. for car parking and landscaping
Car parking = 40 × 9 × 3 = 1,080 sq.m.
Thus landscaping = 86.7 sq.m.
Market rent = £300 × 3,750 sq.m. = £1,125,000 per annum

Gross development value:

Market rent =	£	1,125,000	p.a.
× YP perp @ 6%		16.667	
	=	£	18,750,000
	SAY=	£	18,750,000

Less costs of development:

Site clearance/remediation/preparation	£	100,000
Other initial/miscellaneous/ contingency costs =	£	250,000
Construction costs:		
Offices		
constructable floor area in sq.m. =4750m²		
@ construction cost per sq.m. =£2,000/m²		
Costs =	£	9,500,000
Surface car parking		
constructable floor area in sq.m. =1080m²		
@ construction cost per sq.m. =£150/m²		
Costs =	£	162,000

Landscaping
   constructable floor area in sq.m. $=86.7m^2$
   @ construction cost per sq.m. $=£100/m^2$
   Costs =   £   8,670
Arch., QS and other prof. fees
   @ 12% of construction costs
   $= 0.12 \times £9,670,670$ £ 1,160,480
Agents/Legals on letting
   @ 10% of MR   £   112,500
Agents/Legals on investment sale
   @ 3% of GDV   £   562,500
Finance charges on half costs over
   development period @ 8.5%
   $= ((1.085)^{1.75} – 1) \times £10,696,830/2$ £   820,761
   Developer's risk and profit =   £_____x
                              Total Costs= $£\ \underline{11,517,591+x}$
                                  Balance=   £   7,232,409 – x

Less Site Purchase finance over development period @ 8.5%
= Balance – (PV for 1.75 yrs @ 8.5% × Balance)
= (£7,232,409 – x) – (PV for 1.75 yrs @ 8.5% x (£7,232,409 – x))
= (£7,232,409 – x) – (0.86696 × (£7,232,409 – x))
= (£7,232,409 – x) – £6,270,193 – 0.86696x $=\ \underline{-(£\ \ 962,216 – 1.86696x)}$
                                                        = £ 6,270,193 – 0.86696x

Less site purchase costs @ 6%:
= (£6,270,193 – £6,270,193/1.06) – (0.86696 – 0.86696x/1.06)
                                            = $£\ \underline{354,917 – 0.04907x}$
                          Site Value =   £ 5,915,276 – 0.81789x
                                            =   £ 3,250,000 (sale price)
                          = per hectare: £ 16,300,000

If sale price = calculated site value then:
£3,250,000 = £5,915,276 – 0.81789x
0.81789x = £5,915,276 – £3,250,000
x = £2,665,276/0.81789 = £3,258,722
SAY Developer's risk and profit margin will be £3.25 million

- As a double-check as to whether the asking price for the land is reasonable, comparable site values per hectare are needed.
- The calculated site value assessed as a percentage of GDV in this case is 17 per cent, which is not an excessive figure subject to confirmation by direct comparables.
- The developer's risk and profit if the land is purchased at the asking price will equate to 28.2 per cent of costs or 17.3 per cent of GDV, which appears high, but the developer will need to judge whether it is reasonable for this project and compared to other similar developments.

13.2 Simple and more complex methods

The simplest residual method calculation is undertaken using pen, paper and a calculator. However, apart from being limited in its scope and level of detail, it also lacks flexibility to test out 'what if' scenarios. For example, what if the interest rate paid on borrowing increased or decreased by as much as 1.5 per cent from the figure used in the initial calculation? Even more interesting, what if it is considered how each 0.25 per cent change in rate, up or down, until it reached 1.5 per cent either way, would affect the findings? This would require twelve new sets of calculations. Undertaken manually this would be slow and laborious.

To improve on the simple manual residual valuation, the data can be input into a computer spreadsheet program. This will enable all the basic calculations to be instantly performed whenever one variable is changed. It also permits valuation formulae to be input so that tables are not needed to calculate compound interest and present values. Examples of simplified residual valuations using the Microsoft Excel program and the cell coding employed are shown in Table 13.1 (residential property) and Table 13.2 (commercial property).

Although better than 'back of an envelope' calculations, nevertheless the spreadsheets illustrated are still quite elementary. They may provide a quick check on the basic feasibility of a project and some rough estimates of the effects of potential changes in the factors involved, but they are not that sophisticated. Developers have increasingly looked for comprehensive methods of appraisal. With any development project involving high financial risks, the participants are naturally keen to explore every aspect of the proposal before committing themselves. There are other texts and sources that cover all these more complex approaches to development valuations, but a brief consideration of the main aspects that can be incorporated into an appraisal are considered below.

Architects, engineers and quantity surveyors have for some time used 'network analyses' for their projects. These show in a graphical form the various activities involved in a project, their inter-relationship and the anticipated time for completion of each. This will help to highlight those factors which will most affect the flow of the development and its timing, and from this can be derived a 'critical path analysis' which shows the critical stages in the project. Once a project has been broken down in network analysis fashion, it is possible to identify on a month-by-month basis the activities that are expected to take place. Supplied with this information, it is possible for a valuer to provide a more detailed and accurate appraisal of the residual value using a discounted

Table 13.1 Simplified development appraisal valuation spreadsheet for residential property

	A	B	C	D	E
1	SIMPLIFIED DEVELOPMENT APPRAISAL VALUATION OF FREEHOLD SITE FOR RESIDENTIAL PROPERTY				
2					
3	Enter Data in ALL SHADED BOXES only!				
4	Project total development period (years) =	0	yrs		
5	Cost of borrowing (% per annum)	0	%		
6	Developer's Risk & Profit as % of costs =	0	%		
7	Developer's Risk & Profit as % of GDV =	0	%		
8	Site area in hectares =		ha		
9					
10					
11	Insert description of property to be valued				
12					
13	Gross Development Value:				
14	description of building type				
15	number of units =	0			
16	@ sale price per unit of	£0			
17	Market Value =			=B15*B16	
18	description of building type				
19	number of units =	0			
20	@ sale price per unit of	£0			
21	Market Value =			=B19*B20	
22	description of building type				
23	number of units =	0			
24	@ sale price per unit of	£0			
25	Market Value =			=B23*B24	
26	Total GDV =			=SUM(D17:D25)	
27				SAY =	=INT(D26/10000)*10000
28	Less Costs of Development:				
29	Compensation to existing site occupiers =			£ -	
30	Demolition of existing buildings =			£ -	
31	Site clearance/remediation/preparation =			£ -	
32	Other initial/misc/contingency costs =			£ -	
33	Construction costs:				
34	=A14				
35	constructable floor area per unit in sq.m. =	0	m²		
36	@ construction cost per sq.m. =	£0.00	/m²		
37	Costs =			=B36*B35*B15	
38	=A18				
39	constructable floor area per unit in sq.m. =	0	m²		
40	@ construction cost per sq.m. =	£0.00	/m²		
41	Costs =			=B40*B39*B19	
42	=A22				
43	constructable floor area per unit in sq.m. =	0	m²		
44	@ construction cost per sq.m. =	£0.00	/m²		
45	Costs =			=B44*B43*B23	
46	Arch/QS/other prof fees as % of const. Costs	0	%	=SUM(D37:D45)*B46/100	
47	Agents & Legals on sale as % of GDV	0	%	=B47*E27/100	
48	Finance charges on costs over devt. period @	=B5	%	=SUM(D29:D47)*0.5*(((1+B5/100)^B4)-1)	
49	Developer's Risk & Profit:				
50	as % of costs	=SUM(D29:D48)*B6/100			
51	as % of GDV	=E27*B7/100			
52	SAY =			=INT(((B50+B51)/2)/10000)*10000	
53				Total Costs =	=SUM(D29:D52)
54				Balance =	=E27-E53
55	Less Site Purchase finance over devt. period @	=B5	%	=	=E54*(1-(1/((1+B55/100)^B4)))
56				=	=E54-E55
57	Less site purchase costs @	0	%	=	=B57*E56/((1+B57/100)*100)
58				Site Value =	=E56-E57
59				SAY =	=INT(E58/10000)*10000
60					
61				= per hectare:	=E59/B8

cash flow approach. All the anticipated costs and returns can be shown on a monthly basis from project commencement to completion.

Sensitivity analysis is a technique that tests the effects on the 'outputs' of anticipated or accepted variations in the 'inputs' to the residual valuation. It indicates the effect on the overall feasibility of a scheme resulting from these variations. For instance, it may reasonably be anticipated that the rental value of the completed development could be higher or lower than the figure used in the residual valuation. Similarly there could be changes in the interest rate payable, and so on.

Table 13.2 Simplified development appraisal valuation spreadsheet for commercial property

	A	B	C	D	E
1	SIMPLIFIED DEVELOPMENT APPRAISAL VALUATION OF FREEHOLD SITE FOR COMMERCIAL PROPERTY				
2					
3	Enter Data in ALL SHADED BOXES only!				
4	Market Rent per sq.m. =	£0	/m²		
5	Rentable floor area sq.m. =	0	m²		
6	Freehold All-Risks Yield % =	0	%		
7	Project total development period (years) =	0	yrs		
8	Cost of borrowing (% per annum) =	0	%		
9	Developer's Risk & Profit as % of costs =	0	%		
10	Developer's Risk & Profit as % of GDV =	0	%		
11	Site area in hectares =		ha		
12					
13	Insert description of property to be valued				
14					
15	Gross Development Value:				
16	Market Rent =	=B5*B4			
17	x YP perp	=100/B6			
18			=	=B16*B17	
19				SAY =	=INT(D18/10000)*10000
20					
21	Less Costs of Development:				
22	Compensation to site's existing occupiers =		£	-	
23	Demolition of existing buildings =		£	-	
24	Site clearance/remediation/preparation =		£	-	
25	Other initial/miscellaneous/contingency costs =		£	-	
26	Construction costs:				
27	description of building type				
28	constructable floor area in sq.m. =	0	m²		
29	@ construction cost per sq.m. =	£0.00	/m²		
30	Costs =		=B29*B28		
31	description of building type				
32	constructable floor area in sq.m. =	0	m²		
33	@ construction cost per sq.m. =	£0.00	/m²		
34	Costs =		=B33*B32		
35	description of building type				
36	constructable floor area in sq.m. =	0	m²		
37	@ construction cost per sq.m. =	£0.00	/m²		
38	Costs =		=B37*B36		
39	Arch, QS & other prof. fees as % of const. costs	0	%	=SUM(D30:D38)*B39/100	
40	Agents & Legals on letting as % of MR	0	%	=B16*B40/100	
41	Agents/Legals on investment sale as % of GDV	0	%	=E19*B41/100	
42	Finance charges on costs over devt. period @	=B8	%	=SUM(D22:D40)*0.5*(((1+B8/100)^B7)-1)	
43	Developer's Risk & Profit:				
44	as % of costs	=SUM(D22:D42)*B9/100			
45	as % of GDV	=E19*B10/100			
46	SAY =		=INT(((B44+B45)/2)/10000)*10000		
47				Total Costs =	=SUM(D22:D46)
48				Balance =	=E19-E47
49	Less Site Purchase finance over devt. period @	=B8	%	=	=E48*(1-(1/((1+B49/100)^B7)))
50				=	=E48-E49
51	Less site purchase costs @	0	%	=	=B51*E50/((1+B51/100)*100)
52				Site Value =	=E50-E51
53				SAY =	=INT(E52/10000)*10000
54					
55				= per hectare:	=E53/B11

A sensitivity analysis is used to try to highlight those input factors which are the most sensitive, in other words those which if varied will have the greatest effect on the outputs of site value or profit. In its basic form, the analysis will require repetitive recalculation of the residual valuation, each one incorporating a change in the relevant variables. However, the more variables amended at the same time, the more numerous the outcomes that need to be calculated. For example, a matrix comprising changes in rent over a range of £10 per square metre in steps of £1 together with changes in interest rates over a 2 per cent range in 0.25 per cent steps produces ten times eight, or eighty separate outcomes. Add in a third variable tested over a range of ten steps and eight hundred are now needed.

On completion, the sensitivity analysis will provide a good guide to the most 'sensitive' variables and show the expected outcome from each envisaged scenario. With this knowledge, developers are more fully aware of the risks being undertaken and can endeavour to find ways to reduce those risks.

A further refinement is to use sensitivity analysis, together with probability measures, to analyse more effectively the risks and uncertainties. This involves several cash flows being constructed based on different assumptions, each of which is assigned a probability factor according to the likelihood of it occurring.

The basic principles of probability and its application to investment appraisal were considered in Section 3.8 above. Property investments are usually only made infrequently under relatively limited conditions and therefore a 'long-run' outlook of probability is rather limited in relevance. However, by taking historical evidence from different types of investments, the probability of various returns can be assessed and incorporated into a DCF appraisal for comparative purposes.

Probability (P) is calculated as the number of ways success can occur divided by the number of possible outcomes. Thus if a six-sided die was thrown, the probability of each of the numbers (1, 2, 3, 4, 5 or 6) occurring would be 1 in 6 (0.16667 probability) on each throw. Instead of an objective approach such as this, a subjective assessment may be used, where the likelihood of various outcomes is estimated based on knowledge, experience and informed opinion rather than mathematical certainties. This approach frequently has to be adopted in investment analysis.

It is necessary to establish which elements in an appraisal are uncertain and to what extent. Then a series of likely or possible alternatives is considered around what is regarded as the 'best', 'worst' and 'most likely' estimates of value for each particular variable, and probability factors

Example 3: Applied probability

A residual valuation has been carried out using an all risks yield of 10 per cent to calculate the GDV. Having carefully analysed the market, the valuer has concluded that between now and the completion of the project this yield may vary by +1.5 per cent to –1 per cent. The 'worst' scenario would be with a yield at 11.5 per cent and the 'best' with it at 9 per cent. Using steps of 0.25 per cent the range of options would thus be:

9%, 9.25%, 9.5%, 9.75%, 10%, 10.25%, 10.5%, 10.75%, 11%, 11.25% and 11.5%.

For each of these yields, the valuer must then decide on the probability of each occurring, and weighting each yield accordingly. For ease of calculation, it is usual to allocate these 'weights' out of a total score of 100 (that is 100 per cent) and convert into a probability factor. Supposing that the valuer decides the yield of 10 per cent chosen for the initial valuation is the most likely to occur and feels it has a 50 per cent chance of being correct. The yield of 10 per cent is thus weighted by 50 per cent. Multiplying the yield by the probability factor, expressed as a decimal, does this. A 50 per cent chance out of total 100 per cent certainty = 50/100 = 0.5 is used as the multiplier. The valuer then decides to allocate these estimated probabilities to each of the other yields in the chosen range, ensuring summated total probability is 1.00 (or 100 per cent):

As the range of probability is 1 per cent to 100 per cent, random integer (whole) numbers within this range now need to be generated by a computer. As each number is generated, it is then compared to the probability of occurrence in the table, and the appropriate yield selected. Thus if the number 32 is produced, this would indicate a yield of 10 per cent would occur; the number 5 indicates a yield of 9.5 per cent, and so on.

Each time a random number is generated, the valuation is recalculated using the appropriate yield and the result noted in the computer database. The operation is performed many times (thousands or even millions) and the average valuation at the end will reflect the long-term probable result. Effectively, it has 'modelled' the future over vast numbers of periods. Additionally a frequency distribution can be produced to show the overall spread of the results from each run of the valuation.

This example has only considered the yield, but the simulation can be applied to any number of the input factors that may be deemed to be subject to variation. This will enable an investor to take a more considered and informed viewpoint of the investment and give a greater insight into the assumptions made in the valuation.

13.3 Reliability and limitations of method

The residual method has often been criticised because greatly different outcomes can be reached from relatively small variations in most of the factors included in the valuation. With the large number of variable inputs involved and the susceptibility of the resultant site value to alter in a major way from comparatively minor changes in any of these constituent figures, the Lands Tribunal has in the past looked on the method with disfavour.

As outlined above, several techniques that are more complex can be employed to improve the reliability and accuracy of a basic residual method valuation. Use of computer spreadsheets and devoted software programs enable these many calculations to be easily undertaken. With the high levels of risk and potential return involved in property development, it is essential appraisals in professional practice are as thorough as possible and explore all probable scenarios. Whatever additional mathematical approaches are used, it is also vital to base the figures used on sufficient evidence taken from the market. Comparable evidence of site values of similar development opportunities is clearly of particular importance in this respect. When considering ground leases in Section 10.2 above it was seen that site value can be expressed as a percentage of the capital or rental value of the completed development and this simple check should always be made to confirm the validity of a value found from a residual approach.

Two main examples of commercial programs used in practice are ARGUS and KEL delta. ARGUS Developer (previously known as CircleDeveloper) is a 'real estate development project system comprising appraisals, residuals, cash flows, sensitivity and scenario analysis, budgets & forecasting, debt finance, equity partnership management and reporting' (Argus Software 2008). KEL delta is 'a powerful development appraisal program which is equally capable of dealing with single unit developments or large multi-phased mixed-use projects' (KEL Computing Limited 2008).

Progress check questions

- What is the basis of the residual method of valuation and when is it used?
- Do you understand what gross development value (GDV) is and how to calculate it?
- Why may the floor areas used to calculate market value and construction costs be different?
- What are the typical costs involved in a property development?
- Why does there need to be a sum of money allowed within a development appraisal for developer's 'risk and profit'?
- How can you estimate the costs of paying interest on money borrowed to undertake a development scheme?
- What 'void' periods can occur during the lifespan of a development project and why?
- How do developers reduce the risks that would be incurred if they borrowed all the money needed for a project at once and built all parts of a development simultaneously?
- How reliable is the 'residual method' of valuation and what other approaches can be used to value a development site?
- How can valuers produce more accurate and reliable development valuations that take account of all likely changes in variables?
- What role can applied probability play in improving accuracy of the residual valuations?

Chapter summary

The residual method of valuation is used to calculate the value of development properties either where direct market comparable evidence is not available or to supplement a value found from comparison. It can also assist developers to ascertain the feasibility of a proposed scheme. It combines a comparison or investment method with a costs approach. The basic calculation takes the estimated value of the completed development and subtracts all the costs incurred to find the residue or residual value.

The full development period runs from acquisition of the site through to completion and full letting and/or sale of the buildings. Large financial obligations are incurred and even quite small changes in the critical factors over this extensive time span can have a very significant effect on the successful outcome of the development. Accordingly, developers seek an adequate profit margin to recompense this risk-taking and will look for advice that fully accounts for all likely outcomes. To do this the valuer will need to use more sophisticated techniques than just the basic residual valuation calculation, which has limited accuracy and reliability. These will include discounted cash flow and complex numerical techniques, which attempt to take into account the mathematical probability of future events. Computer software, from spreadsheets to devoted commercial programs, is used for this purpose.

Further reading

Askham, P. (1997) 'Mainly for students: examine your technique', *Estates Gazette*, 9720 (17 May): 139–42; repr. in P. Askham and L. Blake (eds), *The Best of Mainly for Students* (London: Estates Gazette, 1999), vol. 2, pp. 351–7.

Bowcock, P. and Bayfield, N. (2000) *Excel for Surveyors*, London: Estates Gazette.

—— (2004) *Advanced Excel for Surveyors*, London: Estates Gazette.

Boyd, H. and Walker, K. (2003) 'Mainly for students: to build or not to build?', *Estates Gazette*, 0344 (3 Nov.): 164–5.

—— (2004) 'Mainly for students: to develop or not to develop?', *Estates Gazette*, 0406 (7 Feb.): 140–1.

Cadman, D., Austin-Crowe. L., Topping, R. and Avis, M.R. (1995) *Property Development*, 4th edn, London: Spon Press.

Curwin, J. and Slater, R. (1996) *Quantitative Methods for Business Decisions*, 4th edn, London: International Thomson Business Press.

Estates Gazette (1976) 'Mainly for students: discounted cash flow techniques and model building – Part I', 238 (26 Jun.): 984–5.

—— 'Mainly for students: discounted cash flow techniques and model building – Part II', 239 (10 Jul.): 140–1.

—— (1987) 'Mainly for students: DCF appraisals', 284 (31 Oct.): 639–40.

—— (1995) 'Mainly for students: spreadsheets and valuation', 9503 (21 Jan.): 116–19.

Fawcett, S. (2003) *Designing Flexible Cashflows*, London: Estates Gazette.

Hayward, R. (ed.) (2008) *Valuation: Principles into Practice*, 6th edn, London: Estates Gazette.

Isaac, D. (1996) *Property Development: Appraisal and Finance*, Basingstoke: Palgrave Macmillan.

—— and Steley, T. (1999) *Property Valuation Techniques*, 2nd edn, Basingstoke: Palgrave Macmillan.

Johnson, T., Davies, K.I. and Shapiro, E. (2000) *Modem Methods of Valuation of Land, Houses and Buildings*, 9th edn, London: Estates Gazette.

Mott, G. (1997) *Investment Appraisal*, 3rd edn, London: Pitman.

Royal Institution of Chartered Surveyors (1997) *Commercial Investment Property: Valuation Methods*, Coventry: RICS Books.

—— (2007) *The Valuation of Development Land: Valuation Information Paper No. 12*, Coventry: RICS Books.

Scarrett, D. and Smith, M. (2007) *Property Valuation: The Five Methods*, 2nd edn, London: Routledge.

Wyatt. P. (2007) *Property Valuation in an Economic Context*, Oxford: Blackwells.

Profits method

In this chapter ...

- Valuations based on the receipts and expenditures accounts and profits of a business or enterprise occupying the property to be valued.
- Why 'specialised' properties in the leisure and associated sectors are often valued using this method.
- How the profit and loss accounts are assessed and adjusted to arrive at a rental value for the property.
- Calculating the 'share' of the receipts to be retained by the occupier as a reward for risk and entrepreneurship and its relationship to the value of any 'goodwill' inherent in the property.
- Turnover rents – an alternative approach to using the business accounts of the occupier to calculate rent to be paid.

14.1 Basis and variations of method and when used

With this method the capital or rental value of the property is related to the receipts and expenditure incurred and/or profit that can be made from occupying it for business purposes. It is used if there are no or insufficient direct comparables, and even then, the resultant value should be compared with other market information to see if it appears to be consistent. It is mostly used wherever there is an element of monopoly associated with the user of the premises, as there would be no true comparables for the use. Such a position is usually found with licensed and other 'specialised' properties in the leisure, entertainment, sports, tourism or hospitality industry sectors.

It is 'the preferred method of valuation in those cases where rental evidence is sparse or non-existent and the rent is likely to be dictated by the actual or anticipated profit of the business carried on at the hereditament'. However, its use need not be confined only to potentially profitable business operations properties. 'The aim to make a profit is not an absolute requirement' and even where a not-for-profit operation is involved, this does not mean the accounts may not be useful 'in deciding what the rent might be. In other classes where the occupier may have considerations in mind besides making a profit the accounts may still be useful' (VOA 2006b: 1.3).

The method can also be used for valuing the 'goodwill' element of a business separate to the values of the real property (land and buildings) it occupies and the other fixed assets (fixtures and fittings) used in the business.

Valuers will need to examine closely the accounts of the business, so a sound knowledge and understanding of these and of the type of business carried on is essential. Profit levels will vary according to differences in the type and style of business, and the valuer must have the requisite specialist knowledge and experience to understand the variations in the levels of specific items of income and expenditure. Full access to the 'books' of the business is desirable and it is usual to look at the last three to five years' accounts to obtain a balanced view of the average reasonably sustainable annual figures, calculated on a present value basis. Where no accounts are available, anticipated figures for a hypothetical operator can be estimated.

The basic calculation undertaken for the profits method is:

> Gross receipts
> *Less* purchases
> = Gross profit
> *Less* adjusted working expenses (exclude rent and rates and certain other items)
> = Net profit or divisible balance
> *Less* tenant's or proprietor's 'share' (their income/profit for running the enterprise) including interest on (their) capital
> = Balance for rent + rates
> Less rates payable (RV × UBR)
> = Market rent (MR)

When a capital value is required, this rental value can then be capitalized (multiplied by a YP) at an appropriate all risks yield, derived from comparable evidence. Some types of property can also be assessed on a 'shortcut' profits method, for example, on gallonage or litreage throughput for petrol filling stations or public houses, a price per seat or percentage of gross receipts for cinemas. Effectively this is valuing by comparison, but centred on trading figures. Even then, the presentation of the full accounts on which such methods are based is to be preferred whenever possible, and in particular if evidence is to be presented to court or tribunal.

The RICS have a generic guidance note relevant to the method (RICS 2007d GN1: 103–9) plus three Valuation Information Papers on particular property types that may be valued using the method (RICS 2003a, 2003b and 2004).

The initial stages of the valuation are to examine critically and analyse the last three to five years of business accounts to establish:

- Do they reflect the performance of a 'reasonably efficient operator'? That is a trader with reasonable competence in the particular line of business. Inefficient operators should not benefit from their incompetence by paying a lower rental. The actual performance of the current occupier is not adopted, but rather how a hypothetical reasonably efficient operator would perform and whether the actual accounts conform to such a hypothesis. 'The starting points for the valuation are the receipts and expenditure of the actual occupier. It is, however, the profit potential of the hereditament that has to be established and not the profit achieved by the actual occupier. It is therefore essential to guard against the inherent danger of valuing the actual occupier's business rather than the hereditament itself'(VOA 2006b: 2.1).

- During each year, have there been any unusual circumstances or occurrences that could have affected the profit and loss accounts?
- Have all proper expenses and allowances been included in the accounts?
- Are the sums entered in the accounts reasonable under each heading and if not how should they be adjusted?

As the value of money decreases with time due to inflation, it is recommended historic figures be brought in line with current sums by adjusting for the general increase in prices over the time period involved by use of a suitable index that measures inflation, such as the UK's Retail Price Index (RPI). An estimate of each item of receipts and expenditure in present value terms can then be made based on the average of the past expenditures and by looking at any trends in the figures and allowing for any unusual or occasional sums. These figures should also be checked against the returns obtained on similar businesses in the area where possible, to see if the subject accounts are consistent with accepted business practice and performance. Once this analysis is completed, the 'fair maintainable trade' (FMT) figures can be estimated for use in the remainder of the profits method assessment.

Gross receipts and purchases

Ascertaining what are the gross receipts and purchases for the business should be straightforward to arrive at the gross profit. No deduction is made for liability to Income Tax or Corporation Tax: R v. *Southampton Dock Co.* (1851) 17 QB 83. However, tax liability can be considered when estimating the tenant's or proprietor's 'share'. Whilst the receipts and expenditure of the actual occupier are examined, it is the profit potential of the property itself that has to be established and not the profit achieved by the actual occupier. The category of 'purchases' covers the 'raw materials' used in the business – for example, for a restaurant it will be the food ingredients and drink purchased.

Working expenses

Unlike gross receipts, working expenses provide scope for argument over what items of expenditure should and should not be included. Accordingly, the actual net profit as shown in the accounts usually needs adjusting.

The expenses will include such items as wages and salaries, advertising, electricity, gas, water and telephone charges as well as repairs and renewals both to the tenant's or occupier's chattels and to the property itself. Distinction must however be made between repair (which is carried out to maintain the property in its present state) and improvement, which will tend to increase the value of the property and is not allowable as an annual expense being a capital expenditure. Repairs and insurance cover should be carefully checked to see if they are adequate and reasonable. Neglecting repairs could enhance short-term profits, but the accumulative dilapidations will have an adverse effect on value in the longer term. On the other hand, excessive repair costs could indicate the work is more in the nature of renewal and improvement.

Where paid, the actual rent is not deducted as a working expense, even though it will appear as a legitimate business expense in the occupier's accounts, since the purpose of the valuation is to find the true market rent and the actual amount paid may be too little or too much. Similarly, in rating valuation, the rates payable cannot be found until the RV (rateable value) is known. As this is based on a rental figure, these cannot be calculated until the end and to deduct them earlier would be misleading, as the amount paid may not be correct once the true RV has been found. Business rates are a proper deduction though so, when not valuing for rating purposes and 'rent and rates' are shown as a single deduction in the accounts, only the rental element needs to be added back.

It was established in *Welwyn Garden City Electric Co. v. Barnet AC* (1938) 29 R & IT 88 that directors' fees are properly deductible as an expense, even if they do not appear in the actual accounts. However, expenses that are specific to the actual occupier, such as interest on overdrafts and hire purchase agreements, are normally 'added back'. They are not deducted from gross profit unless they are considered applicable to the 'hypothetical tenant' who would occupy the premises. Interest paid on plant, equipment and working capital is the subject of a separate calculation. Interest on a mortgage is a return on land and buildings, which is included in the final rental figure.

An allowance for the operator's remuneration is made at the end of the assessment, so any deductions in the accounts for this should be disregarded, as the figure paid may be too low or excessive. Any casual work undertaken by members of the operator's family in the business though must be properly costed.

Depreciation is 'the diminution in value of a fixed asset due to wear and tear or obsolescence over an accounting period' (Law 2006: 163). 'Fixed assets' are real estate and fixtures, fittings, plant, machinery and equipment installed therein. They are not things the company trades in, but enable business to be conducted. They are thus not bought and sold for profit and an allowance for their depreciation is normally included as an expense in the business accounts. However, is it an allowable working expense since it is a reduction of the capital invested in the business, rather than an income item? Certainly, renewals of plant and equipment are capital, not income expenditures, and should therefore be excluded from the profit and loss account. These expenses will instead be reflected in the value of the fixed assets that comprise the occupier's capital, on which interest is allowed in arriving at the divisible balance. On the basis that renewals are capital replacements and depreciation is loss of value of the capital, the balance between them will indicate whether there is a net capital addition or reduction each year.

There are various methods used to calculate depreciation. It is important that the assessment approach used is both reasonable and consistent over time. Three of the most popular alternative accepted methods of calculating depreciation are:

- Straight-line method of depreciation:
  - This is the simplest method although not entirely realistic in that it assumes the asset depreciates by the same sum each and every year – in practice, value reduces in a less regular fashion;

- It assumes the asset will be written-down to zero value over its life;
- The annual figure to be entered for each asset in the accounts = cost of the asset/ anticipated life of asset.
- Sum-of-the-digits method of depreciation:
  - This takes the anticipated life of the asset and adds the digits together that is anticipated life. For example, for life of 10 years, add 1, 2, 3, 4, 5, 6, 7, 8, 9, 10 together = 55;
  - Thus depreciation in year 1 = 10/55ths of cost, in year 2 = 9/55ths, etc.;
  - Again, asset is written-off to zero value over lifespan.
- Reducing balance method of depreciation:
  - A constant percentage of the asset balance is deducted, assessed as a reasonable rate for that asset;
  - Alternatively, it is found by: Rate % = $(1 - \sqrt[n]{[\text{scrap value/original cost}]}) \times 100$, where n = useful life of asset in years;
  - It is impossible to depreciate an asset completely using this method; there will always be some balance left.

Example

Calculate the annual depreciation of an asset with an expected useful life of ten years and an initial purchase cost of £50,000.

Straight line method

£50,000/10 = £5,000 per annum depreciation allowance

Sum-of-the-digits method

Sum of years 1 to 10 is 55
Year 1 depreciation = 10 × £50,000/55 = £9,090.91
Year 2 depreciation = 9 × £50,000/55 = £8,181.82
Year 3 depreciation = 8 × £50,000/55 = £7,272.73
etc.

Reducing balance method

Rate % = $(1 - \sqrt[10]{[£2,000/£50,000]}) \times 100$ = 27.52% (assuming a scrap value after ten years of £2,000)

Year 1 depreciation = 27.52% × £50,000 = £13,761.02
and reduced balance = £50,000 – £13,761.02 = £36,238.98
Year 2 depreciation = 27.52% × £36,238.98 = £9,972.97
and reduced balance = £36,238.98 – £9,972.97 = £26,266.01
Year 3 depreciation = 27.52% × £26,266.01 = £7,228.41
and reduced balance = £26,266.01 – £7,228.41 = £19,037.60
etc.

Tenant's or proprietor's 'share'

No operator will run a business without expecting some return or reward for his or her efforts. Three items must therefore be considered under this category. These are:

- A reasonable risk and remuneration return for the operator's services and enterprise. This will not necessarily be the same sum as entered in the accounts, since the sum actually drawn may be inadequate or excessive by 'normal' reasonable industry standards.
- An interest payment on the capital invested by the operator in the business. Were this capital not to be used in the business, the operator could have invested elsewhere and thus earned interest. He or she must therefore be adequately recompensed to be in no worse a position.
- Where the property concerned is a branch of a large company or organisation, a contribution to central office overheads will normally be allowed.

'The tenant's share is, broadly, the amount which the tenant will require out of the business to induce him or her to take the tenancy at the rent as evidenced by the valuation, bearing in mind the capital needed to operate the business and the risk to that capital' (VOA 2006b: 16.1).

The different ways in which the tenant's or proprietor's share can be assessed are:

- as a proportion of the divisible balance (50 per cent being a commonly adopted figure); or
- as a proportion of gross receipts; or
- an amount related to gross receipts and divisible balance plus a separate interest percentage return on the tenant's or proprietor's capital.

The choice of method is a matter of 'skill and judgement' and comparison with other similar businesses. Whatever method used, it is advisable to carry out a check using one of the alternatives. What is being estimated is the amount of remuneration for risk, profit and capital employed which would be sufficient to persuade the hypothetical tenant to occupy the property.

Tenant's or proprietor's capital comprises three categories:

1. fixed assets (fixtures and fittings including equipment, machinery and furniture)
2. stock
3. cash in hand and at bank for running business.

'Stock' will include:

- unused raw materials
- part-finished products (work in progress), and
- finished products ready for sale.

Stock figures need to be checked for consistency and reasonableness. It is normally valued as the *lower* of:

- cost of purchase, or
- net realisable value.

When the interest on invested tenant's or proprietor's capital is assessed as a separate item, it should not be deducted before the divisible balance. The interest rate applied should be the opportunity cost of putting the money in a secure investment. 'Bank base lending rates, interest rates payable on bank or building society deposits or yields on government bonds [at the relevant valuation date] may indicate the appropriate rate to apply' (VOA 2006b: 16.2).

Calculating rental value

After deducting the tenant's or proprietor's share from the divisible balance, the amount remaining is the payment for rent and rates. When the valuation is for rating assessment purposes, these figures are interdependent since rateable value (RV) is effectively a net rental value at a specific valuation date. Where this is the case, rental value is found from the formula:

Balance/(1 + UBR)

where UBR is national non-domestic rate multiplier in pence.

'Going concern' valuation

The required valuation may be only for the market rent value or the capital value of the building. For secured lending, it is usual for a 'bricks and mortar' only valuation to be sought as the lender is looking at the property's underlying asset value as security. Even then, the trading potential of the property will be important, as any loan would need to be repaid from out of the profits generated by the business occupying the premises.

This is confirmed in Appendix 4.2 of the International Valuation Application 2 (IVA 2) of the *Red Book*: 'the underlying principle of many valuations for financial reporting is the presumption that the entity will continue as a going concern. However, this would not usually be appropriate for valuations undertaken for lending purposes. Such a presumption has particular implications for specialised assets where the value and marketability of the secured property, separate from the business of which it forms part, may be limited' (RICS 2007d: 71). Also that 'A valuation may be required of a specialised property where the property is part of a going concern business. The lender should be alerted to the valuation being dependent on the continuing profitability (or otherwise) of the going concern. If the value on a vacant possession basis is potentially lower, this should be drawn to the attention of the lender' (RICS 2007d: 74).

Where the business is to be sold in its entirety what is assessed is often termed the 'going concern value', although the RICS recommends this term is only used within the context of business valuation rather than property (RICS 2007d: IVS GN 6). This entirety value can be deemed to comprise three components:

* land and buildings
* 'goodwill'
* fixtures, fittings, plant and equipment.

Each of these may be separately valued and totalled to arrive at the capital value for the property. However, this 'approach holds an implication that the fixed assets have a continuing value outside the business. This might be true in the case of a shop, for instance, whereas a swimming pool probably has a value only as part of the business complex containing it – there is no alternative market value' (Colborne 1989: 24). It also assumes goodwill is identifiable as a separate item to the value of the fixed assets. In reality, goodwill is a difficult to quantify marketable asset. It is particularly tricky to distinguish between personal goodwill associated with the present proprietor of the business and adherent or inherent goodwill of the business and property itself.

Adherent goodwill 'adheres as long as the reputation of the premises as a *business* exists, and will evaporate if that kind of business ceases to operate there' and is 'value attributable to attraction to customers arising from their knowledge that a business of the general relevant type is carried on there'. Inherent goodwill is the 'value attributable to the inherent advantages of the premises for the carrying on of such a business' and can be termed 'site advantages' (Marshall and Williamson 1997: 64).

Only those aspects of goodwill that have a saleable value, being based on the profits produced having a prospect of continuance, should attract a payment, as personal goodwill will move with the person or persons who attract it and be lost when they leave the premises. 'The valuer must exclude any element of trade attributable to personal goodwill, but include all or any additional potential that would be realised by an average competent operator' (Johnson *et al.* 2000: 581). Even then it is 'important to regard goodwill as a balancing figure after all other assets have been assessed' (Berney 1981a) and to be sure that allowance for it has not already been made within another aspect of the value. For instance, 'comparable open market lettings will inevitably include an element of payment for adherent goodwill unless the letting is of entirely new, untried, premises' (Marshall and Williamson 1997: 65).

The difficulty of separately identifying and valuing goodwill has led to the more prevalent approach of multiplying the estimated rental value, found from the adjusted net profit, by a years purchase number that is deemed to incorporate the value of all the fixed assets plus the goodwill. This capitalised adjusted net profit can be viewed as 'the lock-stock-and-barrel value of the business with all its working assets, including property. This is supported by the nature of current transactions, particularly of smaller businesses but also of hotels and leisure facilities on this basis' (Colborne 1989: 24).

Where it is deemed there is a goodwill element to be separately valued, it is traditionally calculated by multiplying the tenant's or proprietor's 'share' by a YP number. Note that although called a years purchase, no yield is quoted, just the YP numeral. The multiplier used depends on the purchaser's perception of how valuable this aspect of the property is. 'A high YP will be adopted if the profits show a rising trend. Conversely, falling profits justify a lower YP than would be normal' (Berney 1981b).

The Lands Tribunal decision of *Reynolds* v. *Manchester City Council* (1980) 257 EG 939 adopted a multiplier of 3.5. This was a high figure compared to other settlements in that area that had been 'narrowly within the range of 1.5 to 2.5 YP' and reflected the rising trend in profits in the case.

When a business is being totally extinguished, as may happen when compulsorily purchased, the multiplier normally will also be higher and then typically ranges between 2 and 5 YP (Johnson *et al.* 2000: 615). Half these figures, or even less, is likely to apply where the business is to continue, but no general rule can be applied and each case must be considered on its own individual merits.

The valuation of fixtures and fittings is a highly specialised activity undertaken by plant and machinery valuers. The balance sheet of the business will give the current figure entered against these assets, but often this is the depreciated historic cost rather than the current value if sold on the open market now. As with goodwill, sometimes the capitalised estimated rental value derived from the divisible balance is deemed to include the value of fixtures and fittings and no separate allowance or valuation for them is needed.

14.2 Reliability and limitations of method

As has already been indicated, the comparison method of valuation is always to be preferred as being the most dependable and the profits approach is thus not as reliable. Nevertheless, accurate results can be obtained providing the accounts are carefully and correctly analysed and adjusted. It is generally a more accurate method of valuation than the cost approach, since it is assessed on the profitability arising from potential occupation of that particular property which should at least indirectly reflect market supply and demand unlike using cost. Better quality premises in good locations should produce superior profits and have a higher rental value, which seems equitable.

An important point is that it is the property being assessed in terms of its annual value, not the profit of the business. In most types of property, the amount of profit made is dependent much more on the skill and business acumen of the tenant or proprietor (who in many cases could occupy another property just as profitably) rather than the characteristics of the property. It can thus be difficult to separate the personal compared to the intrinsic value of the business and premises.

Apart from the limitations imposed by the need to make a careful study and adjustment of the figures in any year's accounts, it is important to take into consideration differences in prevailing values when comparing these to previous years' figures. For example, if the adjusted profit two years ago was £125,000, and is now £150,000, it would appear that the business is growing and improving. Certainly, it is in nominal terms, but not necessarily in real terms. Suppose prices generally had risen by 20 per cent per over this two-year period, the increase would be £25,000 from the previous figure. The £150,000 now is just in line with other increases, so the business has remained level in real terms, not improved.

The Queens Moat Houses hotel group valuations in 1991–2 and the RICS *Mallinson Report* of 1994 (mentioned in Chapter 9) highlighted the advisability of not relying entirely on a profits method valuation, but seeking other ways to verify the figure found, including direct and indirect comparison, cost approach and especially discounted cash flow.

Example

A local entrepreneur owns the freehold interest in a sports club, comprising restaurant, bar, snooker room, two squash courts, swimming pool and gymnasium. It is situated on a site of 0.75 hectare. She is considering selling the property and the business as a going concern and has asked for an open market valuation for this purpose.

To assist in preparing the valuation, the client has provided income and expenditure accounts for the last three completed years of trading for the premises. In addition, from the latest balance sheet, it is noted that the current value attributed to fixtures and fittings is £200,000 and that on average there is £10,000 of stock and £20,000 in cash available for the day-to-day running of the business.

Accounts for years ended 30 September

Income	3 years ago	2 years ago	Last year
Gross receipts:			
Bar (drink and food)	£197,146	£178,380	£238,492
Restaurant	£300,749	£251,133	£284,358
Outside catering	£23,475	£26,353	£27,041
Members' subscriptions	£26,009	£25,390	£30,267
Room hire	£15,010	£13,895	£15,523
Other	£8,465	£8,126	£8,748
Total receipts:	£570,853	£503,276	£604,428
Purchases	£121,680	£133,549	£123,201
Gross Profit	£449,173	£369,727	£481,227
Expenditure:			
Wages and National Insurance	£219,686	£189,917	£243,223
Office expenses	£21,556	£20,624	£22,620
Proprietors' drawings	£53,901	£44,367	£57,747
Loan interest	£5,612	£6,078	£7,197
Bank charges	£728	£817	£877
Accountancy fees	£2,139	£2,352	£2,619
Telephone charges	£1,273	£1,363	£1,463
Electricity	£11,869	£12,308	£12,463
Business rates	£9,929	£10,451	£10,886
Environmental charge	£2,900	£3,206	£3,324
Repairs to land and buildings	£2,300	£7,166	£4,260
Vehicle servicing	£962	£1,006	£1,064
Petrol	£2,544	£2,678	£2,897

Licensing	£389	£454	£492
Insurance: premises	£3,242	£3,613	£3,720
Contents	£2,112	£2,420	£2,566
Other liability	£473	£523	£551
Bad debts	£1,738	£1,916	£1,882
Laundry	£2,117	£2,021	£2,106
Advertising	£3,078	£3,318	£4,080
New furnishings	£10,789	£0	£0
Total Expenditure	£359,337	£316,600	£386,036
Net Profit	£89,836	£53,127	£95,191

The information provided is minimal. In 'real life', the valuer would need to make further investigations to discover more, including examining the full accounts and notes to them, the cash flow forecasts and balance sheets as well as the profit and loss accounts. There would be a need to speak to the client, the client's solicitor and accountant or other professional persons, such as a plant and machinery valuer, and to base calculations around these facts, all of which would need to be fully stated and explained in a valuation report. However, the following provides the general basis of how a valuation could be undertaken.

'Fair maintainable annual figures' need to be compiled for all the items of income and expenditure. Adjusting historical prices to their current price equivalents can be a useful start point to judge whether an item is showing a rising or falling trend in real, not just nominal, terms. As each year's accounts cover a twelve-month period to end of September in each respective year, the latest retail prices index (RPI) available could be taken and divided by the September RPI figure for each year of the accounts to produce the multiplier to be applied to each individual figure.

Assume these are the relevant RPI numbers found:

Now	202.7
September last year	193.1
September 2 years ago	188.1
September 3 years ago	182.5

The straight average of the last three year's 'real' figures could then be taken, but there is a need to check whether there is a rising or falling trend in each case in deciding on fair maintainable figures. Table 14.1 shows a possible assessment. The average has been used as a reference, but the adopted figure chosen is considered to reflect the overall trend that appears to exist. Clearly, there is a good deal of subjective judgement involved here and each valuer may have different opinions on the fair maintainable figure to be adopted. This is one of the weaknesses of this method of valuation.

In practice, it is essential that the valuer asks more questions and carries out more research to find out why certain figures are what they have been in the past. Are there particular,

Table 14.1 Analysis of example accounts

Income	At present cost 3 yrs ago	2 yrs ago	Last year	Average	Estimated fair maintainable figure
Gross receipts:					
Bar (drink and food)	£218,967	£192,225	£250,348	£220,514	£220,000
Restaurant	£334,037	£270,625	£298,495	£301,052	£301,000
Outside catering	£26,073	£28,398	£28,385	£27,619	£27,000
Members' subs	£28,888	£27,360	£31,772	£29,340	£29,000
Room hire	£16,671	£14,973	£16,295	£15,980	£15,000
Other	£9,401	£8,757	£9,182	£9,114	£9,000
Total receipts:	£634,037	£542,338	£634,477	£603,619	£601,000
Purchases	£135,148	£143,915	£129,326	£136,130	£136,000
Gross profit	£498,889	£398,423	£505,151	£467,489	£465,000
Expenditure:					
Wages and National Insurance	£244,002	£204,658	£255,315	£234,658	£240,000
Office expenses	£23,942	£22,225	£23,745	£23,304	£24,000
Proprietors' drawings	£59,867	£47,811	£60,618	£56,099	£56,000
Loan interest	£6,233	£6,550	£7,555	£6,779	£6,000
Bank charges	£809	£880	£921	£870	£900
Accountancy fees	£2,376	£2,535	£2,749	£2,553	£3,000
Telephone charges	£1,414	£1,469	£1,536	£1,473	£1,700
Electricity	£13,183	£13,263	£13,083	£13,176	£13,500
Business rates	£11,028	£11,262	£11,427	£11,239	£12,500
Environmental charge	£3,221	£3,455	£3,489	£3,388	£3,800
Repairs to land and buildings	£2,555	£7,722	£4,472	£4,916	£5,000
Vehicle servicing	£1,068	£1,084	£1,117	£1,090	£1,200
Petrol	£2,826	£2,886	£3,041	£2,917	£3,000
Licensing	£432	£489	£516	£479	£500
Insurance: premises	£3,601	£3,893	£3,905	£3,800	£4,000
contents	£2,346	£2,608	£2,694	£2,549	£3,000
other liability	£525	£564	£578	£556	£600
Bad debts	£1,930	£2,065	£1,976	£1,990	£2,000
Laundry	£2,351	£2,178	£2,211	£2,247	£2,500
Advertising	£3,419	£3,576	£4,283	£3,759	£3,500
New furnishings	£11,983	£0	£0	£3,994	£4,000
Total expenditure	£399,111	£341,173	£405,231	£381,836	£390,700
Net profit	£99,778	£57,250	£99,920	£85,653	£74,300

unusual reasons for them? For example, notable political or economic events, both nationally and globally, could affect trade. So could trends and fashions in each year. Even the prevailing weather could be an influence. Are these likely to have repeatable effects on average figures in the coming years? What about the uneven expenditure on repairs? Have they been neglected some years (for example, three years ago) or have there been one-off jobs required which are unlikely to need to be repeated for many years? Why was no money spent on new furnishings in two out of the three years? Is it because that spent three years ago is sufficient for say a five-year period? In which case that expenditure could be averaged over five years and not three.

Once the fair expected net profit has been compiled, certain items need to be 'added back':

Personal drawings of proprietors	to avoid double-counting
Loan interest	as allowances for these
Bank charges	follow later in the
Annual rates paid	assessment

New furnishings and equipment	as this is adding to net value of fixed assets,
Less annual depreciation on	which are separately assessed both for sale
fixtures and fittings	and for annual interest on capital invested

Note that an amount for annual depreciation of fixed assets could be deducted, but some valuers dispute the validity of doing this. Consequently, this could be omitted from the deduction.

What remains is the 'divisible balance'. The proprietor's share needs to be deducted from this. As explained above, there is no single accepted way of doing this, although a 50 per cent figure is common, and deemed to include interest on capital invested.

Where a separate figure for this interest is required, the total value of capital needs to be estimated. In this example, figures have been provided for each relevant item and based on these the interest that could be earned can be calculated as follows (assuming a rate of 5 per cent is currently obtainable on a 'secure' investment):

Fixtures and fittings =	£	200,000
Stock SAY =	£	10,000
Cash SAY =	£	20,000
=	£	230,000
@ SAY 5% p.a. gross =	£	**11,500 per annum**

Value of stock usually kept on the premises (liquor, food with long shelf- or frozen-life, etc.) has been given here, but in reality the valuer would speak to the operator and have a look around to see what seems to be kept on average before deciding on a figure for this item when calculating interest on capital invested. Similarly, the amount of cash is given in the example, but further research would be needed to see what is 'normal' in this type of business and a check made with the client as to what amount is generally kept.

As it can be difficult accurately to assess the current value of capital invested, the 'all in' approach of taking a percentage share of the divisible balance to include interest on capital is more straightforward and preferred.

The balance left after the proprietor's share has been deducted is the rental value and the business rates of the land and buildings. Deduct the rates payable to find the estimated market rent. Multiply market rent by a years purchase in perpetuity at an appropriate all risks yield found from comparables to find the market value of the land and buildings.

The remaining 'going concern' value would comprise the fixed assets (plus possibly some of the stock) value and a possible payment for 'goodwill', although again this may be deemed to be included in the capitalised rent rather than separately assessed. The three sums provide the total value of the property *and business* as a 'going concern' and it assumes the purchaser would not only acquire the property but the whole enterprise and be able to continue running the business at a profit in line with the estimated fair maintainable figures.

Valuation

Estimated fair maintainable annual net profit = £ 74,300
Add back:

Proprietors' drawings	£ 56,000
Loan interest	£ 6,000
Bank charges	£ 900
Business rates	£ 12,500
New furnishings	£ 4,000
	= £ 79,400

Divisible balance = £153,700

Less Proprietors' share @ 50%
(including interest on capital invested) = (£76,850)

Rent + rates =	£ 76,850
Less business rates=	£ 12,500
Estimated rental value=	£ 64,350 per annum
Market rent SAY =	£ 64,000 **per annum**

A. Valuation of land and buildings ('bricks and mortar' valuation)

Market rent =	£64,000 p.a.
× YP perp @ 7.5%	13.333
	Market value = £853,312
	SAY = £850,000

Note: ARY obtained from current market evidence on similar leisure freehold properties.

B. Value of 'goodwill'

Proprietors' Share x SAY 1.25YP (found from comparables of similar properties and businesses)
= £76,850 x 1.25 SAY = **£96,000**

C. Value of fixtures, fittings and stock = **£210,000**

Total 'going concern' value (A+B+C) = £1,156,000

- Total value equates to 1.92 times estimated total receipts (£1,156,000/£601,000) and value of land and buildings only to a multiple of 1.41 (£850,000/£601,000).
- Both of these multipliers must be checked for consistency with other similar sales where price paid is on an all-inclusive basis for the three elements.

Bear in mind that this is just one possible method of valuing such a property as this. Obviously, it would be best to use direct comparison of value per square metre, but even when not possible, indirect comparables, such as the multiplier of receipts check, will help to confirm validity of estimated value. There is also the DCF approach, adopting the 'efficient operator's assessment' viewpoint to forecast future cash inflows and outflows to the business. Never rely on just one appraisal method. It is better to use alternative approaches to double- or even treble-check whether the final figure 'looks right'.

For this property, it may also be wise to consider an alternative use value. Maybe, subject to planning permission being obtainable, the site would have a greater value for redevelopment than in its existing use? Should this be the case, the client must be informed accordingly.

It will be observed that there are many decisions and variables involved in a profits method valuation, which makes it less reliable than straight comparison.

14.3 Turnover rents and leases

A variation on a profits method approach to assessing value is 'turnover rent'. Often it is used for non-standard retail locations such as units within airports and railway stations and increasingly for major new out-of-town or edge-of-town shopping centres. One reason for the selection of this method is that it is not dependent on comparable evidence, both to set the initial rental and for any subsequent rent reviews. In a new or untested market, or where there are no direct comparisons available, it enables a rental to be fixed against known criteria, which is the turnover of the business that occupies the premises.

Leases where the rent is partly or wholly calculated by reference to the annual turnover of the business are not profits-method valued as such. However, they are based on a similar principle, namely that the value of the property is related to the profit level of the business which operates from it.

This rental assessment method originated in the 1930s in the US during the great depression. It was introduced into the UK in the 1970s, having previously also been adopted in Australia. The main early users of the system in the UK were Capital and Counties for shopping centre schemes.

In principle, the rental will only increase if the turnover increases. The landlord thus has an incentive to encourage the tenant's business, and in return, the tenant is assured that the rent payable can be afforded, since any increase in rental is matched by an increase in trade. The disadvantage to the landlord from such a method of rental assessment is that if the entire rental was assessable on turnover and the retailer suffered a large drop in trade, the rental income could fall substantially. This insecurity of income would adversely affect the value of the investment.

To overcome this, turnover rental agreements in the UK are usually on the basis that a minimum rental is payable regardless of turnover volume, with an additional sum payable linked to the turnover once it exceeds a specified level. It has become customary for the base rent to be 75 to 80 per cent of the estimated market rent (Bennett and Rigby 1993; *Estates Gazette* 1998a: 128). The rental payable under the lease will then be an agreed percentage of the annual gross turnover, or the base rent, whichever is greater. This provides the landlord with a minimum guaranteed rental, plus the possibility of increased income when the retailer has good periods of trade.

The percentage of turnover payable as rent will depend on the nature and profit margins of the tenant's business. Different rates will apply to large volume businesses operating on low profit margins to those with small volumes but high mark-up rates. Typical percentages (VOA 2006b: Part 16, Appendix 1) used are shown in Table 14.2.

Table 14.2   Typical percentages of turnover used to assess rent

Trade	% of turnover
Baker	5
Booksellers	3–4
Butcher	4–5
Catering	7
Department stores	3–5
Food supermarkets	1–2
Furniture	2–3
Greengrocer	6
Hypermarkets	3–4
Jeweller	10–11
Ladies fashions	8–10
Men's fashions	8–9
Radio and electrical	7
Records	9
Shoes	8–10
Sports goods	9
Tailors	4–6
Toys	4–6
Variety shops	3–5

Progress check questions

- What types of properties are likely to be valued using the receipts and expenditure or profits method of valuation and why?
- What is the basic premise of the profits method?
- What checks can be made to ensure the valuation found from a profits method calculation is accurate?
- Why is it preferable to assess more than one-year's accounts when using the profits method?
- Why is it necessary to 'adjust' the accounts and 'add back' certain items shown as expenses in the business records?
- Why does the valuer need to understand the nature of the business conducted from the premises in order to analyse its accounts?
- What does the tenant's or proprietor's 'share' include?
- What is comprised in a 'going concern' valuation and why may this term be misleading?
- How and why are turnover rents and profits method valuations different?
- Why are turnover rents often used in new or non-standard retail locations?

Chapter summary

The receipts and expenditure or profits method is used when direct comparison cannot be made and is based on an analysis of the business accounts of the occupier. The types of property to which it is applied tend to be in the tourism, leisure, hospitality and entertainment sectors, or where there is an element of monopoly involved in the premises or its location, thus making comparison difficult. Even then, it is advisable to use other approaches, such as indirect comparison or discounted cash flow, to check the value found from the profits approach.

To obtain a balanced view of average receipts and expenditures, it is preferred to look at the last three to five years of accounts. The valuer needs to decide what are the fair maintainable trade figures for all categories of receipts and expenditure for a typical average year. It is the profit potential of the property itself to a prospective purchaser that has to be established and not the profit achieved by the actual occupier. Indeed, an assessment of the accounts can still be used as a basis for assessing property value even where the occupant is running a 'not-for-profit' operation. Accordingly, expenses not considered applicable to the 'hypothetical tenant' who would occupy the premises, and only relevant to the personal circumstances of the actual occupier, are not deducted from gross profit. Rent is also not deducted, when paid, since the objective of the valuation is to assess the current rental value, which may differ from the sum currently paid.

Once all allowable purchases and expenses are deducted from total receipts, the resulting divisible balance is divided between the tenant's or proprietor's 'share' and the rent plus rates. The 'share' is the occupier's remunerative reward for risk and enterprise. A capital value can be assessed through multiplying the estimated rental value by a years purchase calculated at an all risks yield found from comparable evidence. Where a separate sum is paid for 'goodwill', it will normally be based on a multiple of the 'share' figure. More frequently, a multiple of the total receipts or the estimated rent is used that is deemed to include the value of goodwill plus fixtures and fittings.

Turnover rents are an alternative approach to calculating rental using the accounts of the occupier. They are often used in retail locations where direct comparables are difficult to find, such as new out-of-town developments and in airports and railway stations. The rent paid is a percentage of the occupier's business turnover, often subject to a minimum 'base rent' if receipts fall below a pre-determined level.

Further reading

Askham, P. (1989) 'Mainly for students: depreciation in commercial buildings', *Estates Gazette*, 8933 (19 Aug.): 52–3; repr. in P. Askham and L. Blake (eds), *The Best of Mainly for Students* (London: Estates Gazette, 1993), 337–45.

—— (1990) 'Mainly for students: turnover rents', *Estates Gazette*, 9035 (1 Sep.): 77; repr. in P. Askham and L. Blake (eds), *The Best of Mainly for Students* (London: Estates Gazette, 1993), 94–100.

—— (1992) 'Mainly for students: public house valuation', *Estates Gazette*, 9217 (2 May): 99–100; repr. in P. Askham and L. Blake (eds), *The Best of Mainly for Students* (London: Estates Gazette, 1993), 326–32.

—— (2003) *Valuation: Special Properties and Purposes*, London: Estates Gazette.

Bond, P. and Brown, P. (2006) *Rating Valuation: Principles and Practice*, 2nd edn, London: Estates Gazette.

Brett, M. (1998) *Property and Money*, 2nd edn, London: Estates Gazette.

—— (2004) *Property under IFRS: A Guide to the Effects of the New International Financial Reporting Standards*, London: RICS.

Brown, P. (1992) 'Turnover rents', *IRRV Valuation Journal* (Jul.): 164–5.

Carr, W.D. (1994) 'What do machinery and equipment valuers do exactly?', *CSM* (Jun.): 42–3.

Estates Gazette (1992) 'Mainly for students: why study accounts?', 9237 (19 Sep.): 122–4.

—— (1992) 'Mainly for students: why study accounts?', 9241 (17 Oct.): 105–7.

—— (1999) 'Mainly for students: a price on the potential', 9925 (26 Jun.): 178–9; repr. in L. Blake and A. Imber (eds), *The Best of Mainly for Students* (London: Estates Gazette, 2004), vol. 3, pp. 450–5.

Harper, D. (2007) *Valuation of Hotels for Investors*, London: Estates Gazette.

Imber, A. (1998) 'Mainly for students: called to accounts', *Estates Gazette*, 9836 (5 Sep.): 142–4; repr. in L. Blake and A. Imber (eds), *The Best of Mainly for Students* (London: Estates Gazette, 2004), vol. 3, pp. 30–8.

Marshall, H. and Williamson, H. (1997) *Law and Valuation of Leisure Property*, 2nd edn, London: Estates Gazette.

May, R. (2006) *The Profits Method of Valuation: Does it Value the Property or the Business?* Online: <http://www.rics.org/NR/rdonlyres/E41B0818-AAE1-4BD2-9962-CB7FAEC488F6/0/RICSarticleJan06.pdf > (accessed 16 Jan. 2008)

May, R. (2008) 'A different approach', *RICS Commercial Property Journal* (Feb./Mar.): 16–17.

Royal Institution of Chartered Surveyors (1997) *Commercial Investment Property: Valuation*

Methods, Coventry: RICS Books.

Saunders, O. (2001) 'Earth, wind and water', *Estates Gazette,* 0106 (10 Feb.): 152–4.

Scarrett, D. and Smith, M. (2007) *Property Valuation: The Five Methods,* 2nd edn, London: Routledge.

Valuation Office Agency (2006) *Rating manual,* vol. 5, *Rating Valuation Practice, All Classes of Hereditament.* Online: <http://www.voa.gov.uk/instructions/chapters/rating_manual/vol5/frame. htm> (accessed 25 Jun. 2007).

Walter G. (1992) 'Turnover rents: prospects and implications', *CSM* (Jul.): 5–6.

Cost methods

In this chapter ...

- The cost approach to valuation.
- When and why it is used.
- The reliability of its basis.
- The contractor's basis and the depreciated replacement cost (DRC) method.
- How age and obsolescence of buildings can be assessed and suitable adjustments for it made in the valuation.
- The format and content of calculations using this valuation approach.
- Costs in use.

15.1 Basis of cost methods and when used

The cost approach to valuation is known under the alternative names of contractor's basis or depreciated replacement cost (DRC) method. Both of these adopt broadly similar methodologies. The main distinctions are that the contractor's basis has been developed in the context of rating valuations, and 'used to obtain a capital figure of which a percentage is taken to give a notional rental value'. Whereas the DRC method 'was and is conceptually different, in that it was designed to provide a capital figure equating to the "deprival value", that is to say, an estimate of the loss which would be suffered by an owner if deprived of the asset' (College of Estate Management 2004: 5).

Traditionally it was considered that the cost approach should only be used when value cannot be arrived at by any of the other valuation methods. As such, it was considered the least reliable and accurate. However, this is not necessarily true nowadays. In relation to rating valuation, it has been stated that:

Whilst in the past the Contractor's Basis has been criticised and referred to as a method of last resort, it has been refined over the years through a number of Lands Tribunal (and higher) appeals and is now an acceptable method to be employed in the valuation of certain classes of hereditament, in particular in the larger industrial field, for example, oil refineries and steelworks, and also public buildings, universities, etc.

(VOA 2006c: Section 7.2)

It is used on properties for which there is no market, or for which there is insufficient direct comparable market evidence or that will produce nil or insufficient profits for a prospective occupier, thus precluding the use of the comparison or profits methods of valuation. Such properties are often located in particular geographical areas for special reasons or are an unusual size, design or arrangement. It is primarily used in rating or asset valuation for the assessment of unusual or specialist properties, such as specialised industrial, operational local authority and public utility buildings. The contractor's basis, or sometimes known in the past as the contractor's test, is the approach used in rating valuation. In this respect, 'The usefulness of the contractor's basis in appropriate circumstances was approved by the Lands Tribunal in *Eton College* v. *Lane (VO) (1971) LT RA 186*' (VOA 2006c: Section 7.1.2). For asset valuation, the depreciated replacement cost is the method employed.

The methods equate cost with value, which is not always a reliable assumption. Capital value is found by calculating the current cost of constructing the property, less allowances for age and obsolescence and/or over-ornamentation or excess detailing, to allow for the fact it is not a new building, plus the site value. Assuming that the property is vacant and to let, the hypothetical rent at which it will let is found by taking a percentage of the capital value.

An argument in favour of the cost method is that tenants have an alternative to renting the property in that they could build an exactly similar building, borrowing the money to do so and paying interest on the amount borrowed. They would only do so if the rent chargeable on the property were at least marginally below the cost of interest. The method therefore is taken to indicate the maximum level of the rental value.

Costs of construction may be assessed on either the replacement or renewal approach. The bases of these were considered in Section 11.14 above, in relation to buildings insurance. The replacement cost approach assesses the costs, including fees, of reconstructing the existing building in exactly the same style and materials, as it currently exists; whereas the renewal approach is to construct a new building of the same size and to perform the same function as the present structure, but in modern materials and style.

Depreciated replacement cost (DRC) is the version of the cost valuation method defined in the RICS *Red Book* (RICS 2007d) and a separate Information Paper (RICS 2007e). DRC is 'The current cost of reproduction or replacement of an asset less deductions for physical deterioration and all relevant forms of obsolescence and optimisation' (RICS 2007d: 5) and is 'a method of valuing properties of unusual character or location for which evidence of comparable transactions does not exist' (Parsons 2004: 75). Where used for the valuation of properties for inclusion in a financial statement, the valuer must be satisfied that it is not practicable to prepare a valuation by any other more reliable method, such as comparison or even profits, before relying solely on DRC. Conversely, the method can also be useful to help support a valuation conclusion derived from another approach for a variety of purposes, although it is normally inappropriate to rely upon this approach if the valuation is required for secured lending.

'A valuation of a property in the private sector using a depreciated replacement cost method should be accompanied by a statement that it is subject to the adequate profitability of the business, paying due regard to the value of the total assets employed.' (RICS 2007d: PS6.5: 86). 'A valuation of a property in the public sector using a depreciated replacement cost method should be accompanied by a statement that it is subject to the prospect and viability of the continued occupation and use' (RICS 2007d: PS6.6: 87).

15.2 Age, obsolescence and depreciation

Allowances for the depreciation in the value of the buildings due to age and obsolescence must be incorporated in the valuation. There are two causes of this depreciation:

- physical deterioration ('age'): the wearing out of the fabric of the building due to age, condition and an increase in the likely costs of future maintenance;
- obsolescence: value decline that is not caused directly by use or passage of time (Baum 1991: 187).

The varieties of obsolescence that can be distinguished have been considered by many, including Askham (1989c), Baum (1991), Bond and Brown (2004), Parsons (2004), Weatherhead (1997), Whipple (1995) and most influentially by Britton *et al.* (1991) and the RICS (2007e).

- Physical obsolescence:
  - happens as 'a result of wear and tear over the years, which may be combined with a lack of maintenance';
  - the valuer should compare the decline of the value of buildings of a similar age and type for which there is a market with the value of the new buildings in that market (RICS 2007e: 13).
- Functional obsolescence:
  - 'problems which may be present in the design of the property which could be deficient by comparison with current requirements' (Bond and Brown 2006: 174);
  - caused by advances in technology that make it more efficient to deliver goods and services from a building of a different size or specification;
  - where the building is rendered fully or partially obsolete by modern production methods or 'changing work practices, new technologies and access to new utilities networks' (Parsons 2004: 182).
- Economic obsolescence:
  - arises from the impact of changing external macro and micro economic conditions on the property;
  - it 'is the loss in value due to factors outside the property itself' (Baum 1991: 57) or 'caused by influences external to the property' (Whipple 1995: 473);
  - also caused by 'inadequate repair and maintenance, poor design or construction or wear and tear' (Parsons 2004: 182).

Three other categories defined by Parsons (2004: 182) are:

- environmental obsolescence 'through flooding, health and safety regulations, pest and pollution';
- configurative obsolescence due to 'inadequate storage or parking space and poor road layout and congestion on site'; and
- strategic obsolescence caused by 'corporate policy, market change and financial constriction'.

Askham (1989c) and Weatherhead (1997, citing Salway 1986) also considered there is:

- aesthetic and social obsolescence: arising when 'occupiers begin to demand higher standards of visual appearance' that presents 'a modern image' (Askham 1989c: 53); or from 'incompatibility with current images of architecture, fashion or corporate aspirations' or through 'changing social patterns plus occupiers demanding a better quality environment (Weatherhead 1997: 131);
- legal obsolescence 'may result from the introduction of new legislation' (Askham 1989c: 53), such as 'changes in EU and national legislation and regulations plus local government decisions (Weatherhead 1997: 131).

Apart from the buildings, obsolescence will 'also apply to the plant and machinery, site works and all other constituent parts' and in certain instances 'may also apply to the land as well as the buildings' (Bond and Brown 2006: 174).

Askham (1989c), Weatherhead (1997) and Whipple (1995) suggest there are two classes of physical or functional obsolescence, namely curable and incurable. Curable is that which can be remedied 'at a cost less than the resulting increase in value' Whipple (1995: 471) or 'cured by means of maintenance, repair or piecemeal renewal' (Askham 1989c: 52). Incurable 'would entail an expenditure greater than the resulting increase in value' Whipple (1995: 471) or where it 'requires the introduction of new characteristics into a building which may not be compatible with the existing structure' (Askham 1989c: 52). Thus 'the test in both cases is the economic one' Whipple (1995: 471) in that 'curable elements of obsolescence can be resolved by means of capital expenditure but incurable elements cannot' (Askham 1989c: 53). However, Weatherhead (1997: 128) advises caution in believing expenditure on improvements necessarily cures the problem that caused the obsolescence rather than just extending the useful life of the building.

How the depreciation in the value of buildings should be calculated is the subject of some debate. Bond and Brown (2006: 175) state that it 'is one of the most subjective and difficult parts of the whole valuation process and the amount of adjustment is often down to the experience of the valuer'. Britton *et al.* (1991: 72) confirmed 'that there is no way of being precise in allowing for depreciation'.

It has been suggested that calculation can be undertaken in the same way as for other fixed assets, for example, using straight-line, sum-of-the-digits or reducing balance calculation methods as outlined in Chapter 14. Another alternative is the 'S curve' approach, which in some instances can present 'the most realistic representation of an asset's depreciation by assuming that in the early years depreciation is at a low rate, accelerates in the middle years and reduces again in the final years' (RICS 2007e: 9.26: 17). However, this approach also has its drawbacks.

Although it is normally accepted that the S-curve realistically represents the pattern of depreciation over the life of most assets, the percentage for any given year will depend on decisions made as to the rates of depreciation at different times and when these change. In the absence of empirical evidence in support of these inputs, the exact pattern of the curve may be dependent on subjective inputs, and be no more relevant than the other methods discussed.

(RICS 2007e: 9.27: 17)

A comparison-based method put forward by Whipple (1995: 474) is possible when sufficiently similar properties have been sold and land value can be reliably estimated. Taking the sale price minus the land value and comparing this with the current replacement cost can then measure the total depreciation suffered by each building. This can then be expressed per square metre and applied to the subject property. Unfortunately, this approach relies on enough evidence of bare sites being available, which as Baum and Crosby (2007) noted is 'rare in the UK'.

Britton *et al.* (1991: 73) observed that for rating valuations 'allowance for depreciation is made generally by percentage deduction for specific items' and 'there is no method of evaluating the accuracy of particular allowances' that are 'hallowed by legal precedent, custom or practice'.

Physical deterioration is continuous over the life of the asset and more readily fits these calculation methods. However, predicting obsolescence can be difficult due to its irregular nature. For instance, a building may be perfectly sound structurally but without value due to functional obsolescence. A question raised by Crosby (2006) is whether more efficient buildings hold their value through time better than less efficient ones, due to inefficient ones becoming relatively even more inefficient as they age?

The point at which obsolescence has rendered the building and site together worth less than the site alone is the point at which redevelopment or refurbishment takes place' (Dubben and Sayce 1991: 57). Figure 15.1 shows this relationship between decline in present value and redevelopment. The total value of a property is the sum of the present values of the building and the land on which it stands. Using straight-line depreciation, the value of the building reduces each year over its expected useful life. Once the total value equates to the site value, and thus there is no value to the building, the property should be redeveloped. Indeed, redevelopment may still offer an enhancement to the total value, even after deducting the costs involved, when total value is only marginally above site value.

The annual rate of depreciation will vary between different property categories. From a ten-year sample size analysis of UK data, Crosby (2006) concluded that shops experienced low rental depreciation, industrial more, but offices were the most vulnerable. Askham (1989c: 52) found that new office building rental values 'depreciate at a relatively low rate over the first five years of the building's life, gathering momentum between years 5 and 25 thereafter levelling out but rising again as it ultimately enters the final stage of total obsolescence'.

The effects of obsolescence may be particularly difficult to recognise, isolate and measure in falling markets where all property values are declining. Weatherhead (1997: 126) defined the characteristics of building obsolescence in such circumstances as where 'the rental growth potential of the building is reduced in comparison with trends in full market rents or, alternatively, in a period of falling real estate values, the fall in rental value is faster than the trend in the fall of full market rents'.

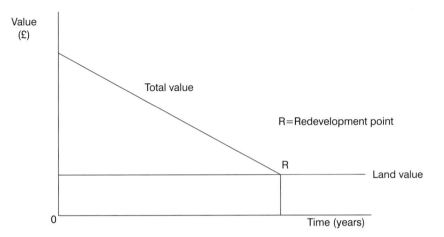

Figure 15.1   Decline in present value and redevelopment point (adapted from Dubben and Sayce 1991: 58)

15.3 Format and content of calculations

The contractor's basis, as used in rating valuation, involves five stages, following the precedent established by the Lands Tribunal in *Gilmore (VO)* v. *Baker Carr (No 2)* [1962] 3 All ER 230. A sixth stage to cover 'the negotiations' was also suggested in *Imperial College of Science and Technology* v. *Ebdon (VO) and Westminster City Council* (1985) 273 EG 81 but usually this aspect is considered as part of the other five, which are (Valuation Office Agency 2006c: 3):

- Stage 1: Estimation of the replacement cost (ERC)
- Stage 2: Estimation of the adjusted replacement cost (ARC)
- Stage 3: Estimation of the site value
- Stage 4: Application of the statutory decapitalisation rate
- Stage 5: Consideration of the results of Stage 4 to see whether it represents the correct answer on the statutory hypothesis

The DRC method requires each of the following to be undertaken (RICS 2007e: 7–20):

- assessing replacement cost;
- estimating the site value of a specialised property;
- calculating the cost of the buildings and site improvements of a specialised property;
- assessing depreciation;
- other considerations; and
- final reconciliation.

It will be seen that both methods require similar basic calculations of:

- total replacement costs of construction including fees
- *less* allowances for depreciation

- *plus* site value (usually from comparables)
- = capital value of existing property
- (and when required) multiply by a percentage = estimated market rent per annum.

There is also a need to review the final values to ensure they appear logical, consistent and appropriate. Each of these stages will now be considered in more detail.

Estimating the costs of construction

Costs of construction of the building are estimated in a modern style and materials, if using the renewal approach, or as it stands if direct replacement is required. They should include everything necessary to complete the construction from a 'green field' site to completed buildings, fit for and capable of being occupied and utilized for the current use. For rating, the costs are taken as relevant at the date of valuation. With DRC, there is an assumption the building is completed by the valuation date and costs are adjusted by backdating to the assumed tender date that would enable completion by then. In all cases, an allowance needs to be made for geographical regional variations in costs.

The best evidence of costs could be provided by actual estimates, quotations or tender prices from contractors directly relating to the subject property or a similar construction. Alternatively, costing data can be obtained from the Building Cost Information Service (2008b and BCIS Wessex 2008), a quantity surveyor or other sources such as *Spon's* (Davis Langdon 2008). Due allowance should be made for all appropriate professional and statutory fees. A provision for short-term finance over half of the construction period is usually made within a DRC valuation.

Deductions for depreciation

Differences between a new building and that actually existing must be taken into account. Deductions are made to reflect age and state of obsolescence and in some instances for over-ornamentation or excess detail, where the current building is more elaborate than a new one would be, or for surplus accommodation. It is accepted 'that just because a certain amount of cost has been incurred it does not necessarily mean that such expenditure is reflected in the value of the property' (Bond and Brown 2006: 173).

Britton *et al.* (1991: 72) warn against the 'danger of duplicating allowances' at this stage in that where the construction costs are based on modern design and materials then surplusages, poor layout, and so on will already be 'designed out' in those costs and 'the main further allowance will be for the physical effects of age' only.

Whatever method for calculating depreciation is adopted, two elements are needed. These are age and future life. Britton *et al.* (1991: 74–6) concluded that while age is factual, 'future life is a pure estimate and should be the best guess of the period over which the building, plant or machinery will remain suitable for its existing purpose. It is not a simple estimate of physical life' and 'the older the building the more sensitive the value is to errors of estimation of future life'.

Assess site value

The land on which the buildings stand is valued as if restricted to its current use for the contractor's basis, but for DRC it is 'based on price which would have to be paid for an equivalent site if the premises had to be replaced' (College of Estate Management 2004: 6).

Care needs to be taken when adjusting evidence of market transactions from other types of development land. It is usually assumed that the land has the benefit of planning permission for the replacement of the existing buildings or to erect modern substitute buildings, whichever is applicable. Providing the use to which the buildings are put is largely similar to one for which there is a market, it should be possible to find comparables. The main difficulty is when the land is being put to a use for which there may be no market. Then planning permission for a use, or a range of uses, prevailing near the actual site and for which there is reasonable expectation consent would be granted, could be assumed. However, the actual use of the property may be so special that it may prove impossible to categorize it in general market terms.

When the land value is added to the depreciated value of the buildings, the resultant figure is the effective capital value (ECV) in rating or gross replacement cost for DRC valuations.

Decapitalise to find estimated market rent

This will not be required for asset valuations, but will for rating and other circumstances where a rental value needs to be allocated. The decapitalisation interest rate to be used has been the subject of much discussion and litigation. It could be the all risks yield considered appropriate for the property type, but this is not universally accepted. For rating assessments in England for the 2005 valuation list, statutorily prescribed figures are used of 3.33 per cent for educational property and hospitals and 5 per cent for anything else. However, in several decided cases the effective capital value figure has been subjected to adjustments or an amended decapitalisation rate used, to reflect such factors as variations in demand.

Finally, the valuer is recommended to 'stand back and look' at the calculated rental figure to see if it fairly reflects what the tenant would be willing or able to pay.

Example

The following shows the valuation determined by the Lands Tribunal in the case of *Eastbourne Borough Council & Wealden District Council* v. *Paul Stuart Allen (VO)* [2001] RA 273. This was an appeal on the rateable value assessed on The Sovereign Centre, The Foreshore, Eastbourne, East Sussex for the 1995 revaluation. The subject property was a local authority leisure centre. The Tribunal rejected a profits method approach and adopted the contractor's basis.

Example continued overleaf ...

SOVEREIGN CENTRE, EASTBOURNE
1995 RATING LIST
CONTRACTOR'S BASIS VALUATION BY THE LANDS TRIBUNAL

Stage 1 Estimated replacement cost (ERC)
Adopt simple substitute building
Mr Allen's (the Valuation Officer) area =   5,712m²
Less diving pool   126
    competition and teaching pool hall   39
    dry changing area   77
    viewing areas   41
    reception area   90
    internal walls   82
                            =   455m²
                            =   5,257m²

Total floor area
5,257 m² @ £784 per m² =   £4,121,488
Reduction for external envelope
5,257 m² @ £46 per m²   − £ 241,822
Reduction for internal finishes
5,257 m² @ £16.50 per m²   − £   86,740
Total building costs   £ 3,792,926
Add external works agreed   £   504,510
                                = £4,297,436
Professional fees @ 10% - agreed = £ 429,744
                                ERC = £ 4,727,180

Stage 2 Adjustment of ERC to reflect age and obsolescence
Age allowance   8.5 %
Functional/technical obsolescence   4.0%
Total allowances   12.5% £ 590,898
                                =   £4,136,282

Stage 3 Land value agreed   = £ 225,000
                                = £4,361,282
Stage 4 Statutory decapitalisation rate 5.5%   = £ 239,870
Stage 5 Adjustment   NIL = £ 239,870
RV as at 1 April 1995   SAY   £240,000

Full text of case available from http://www.landstribunal.gov.uk/decisions/decisions.htm

15.4 Costs in use

Aside from the cost method of valuation, costs can sometimes be used in another way. Although seldom employed by valuers, this approach is known as costs in use and is a discounted cash flow technique that can be used to express the total costs incurred from the use of a building. It can be calculated as a capital sum, being the total of the initial capital cost and the present values of recurring costs in connection with the use. Alternatively, it can be expressed as an annual cost, which is found by calculating the annual equivalent of the initial capital cost and any recurring capital expenditures, added to the regular annual costs.

Progress check questions

- When and why would you use either the contractor's basis or DRC method of valuation?
- How does the depreciated replacement cost approach to valuation differ from an investment method or comparables approach?
- Why is this cost method not always reliable?
- What can be done to make the valuations found from the cost method more accurate?
- How is it decided whether to assess costs of construction on a renewal or replacement basis?
- Why would a DRC valuation not normally be acceptable for secured lending purposes?
- What variations of obsolescence are there?
- How may deductions to reflect age and obsolescence of the buildings be calculated?
- Why can it be difficult to assess the site value and how is it done?
- When it is required, how can a rental value be found from a costs method of valuation?

Chapter summary

The cost approach to valuation is mainly used when value cannot be arrived at by any of the other methods. This will be on a property for which there is no market, or for which there is insufficient direct comparable market evidence or which cannot be valued with the profits method. It is primarily used in rating or asset valuation for the assessment of the value of unusual or specialist properties. It can also be used in support of a value arrived at by another technique. In the UK, there are two variations of the cost approach; the contractor's basis and the depreciated replacement cost (DRC) method. Both equate cost with value, being based on current cost of constructing the property plus the site value. Allowances for the depreciation in the value of the buildings due to age and physical, functional, economic and other forms of obsolescence must be deducted from the construction costs. When required, rental value is found by taking a percentage of the final capital value. A separate technique that is more based on an accountancy basis rather than valuation is costs in use, which assesses the present value of all costs incurred from the use of a building.

Landlord and tenant valuations

In this chapter …

- When premiums may need to be paid for leasehold interests and how they can be calculated.
- Lease surrender and renewal and how to value the rent for the new lease.
- Sale and leaseback agreements to release capital whilst retaining occupation of a property.
- Releasing latent value through joining together one or more property interests ('marriage') or splitting them up ('divorce').
- Effects of some lease terms on rents.
- Calculating the rental value when a lease is on terms that are different to other market comparables.
- Leasehold term and reversion valuations, the inherent mathematical error they contain and how to resolve it.
- Whether valuing leasehold interests using dual rate tax adjusted formulae and tables is necessary or realistic.
- Premium payments paid in the 'reverse' or opposite direction to that normally expected and how to value property let at more than its current market rent.
- Types of lease inducements that may be offered to tenants and how to analyse their effects on value.

There is often a need for valuations to be undertaken on properties subject to lease agreements. This may be to agree a rent or premium or to negotiate a settlement or other transaction involving landlord and tenant. The principal needs for valuations and how they can be calculated are covered in this chapter. The basic techniques described in previous sections are applied to practical situations. However, in addition to the financial considerations, there will often be legal and management issues that the valuer will need to consider. For instance, statutory legislation may regulate the level of rent that can be sought or provide compensation to the tenant for disturbance or improvements. These factors will also need to be considered in professional practice alongside the valuation considerations, but are topics beyond the scope of this text. The further reading list at the end of this chapter provides suggestions of sources that supply information on such issues.

16.1 Calculation of a premium on the grant of a lease at a reduced rent

'A premium is the price paid by an actual or prospective lessee to a lessor or previous lessee, in consideration for the rent being below that which would otherwise be payable as the full rental value' (Valuation Office Agency 2006a: part 9).

The problem with lease premiums is that they can comprise payments for other elements in addition to the rental reduction. These have been identified by the VOA (Valuation Office Agency 2006a: part 9) as including:

- goodwill
- tenant's fixtures and fittings
- stock
- residual value of improvements ignored on review or renewal, and/or
- capitalised profit rent/key money.

This makes it difficult to analyse comparables unless the premium paid can be identified into its constituent parts.

When a lease is granted at less than its market rent, it can be because the landlord is prepared to forgo some of the annual rental income in return for a capital payment at the start of the lease. The amount of this premium will depend on the parties, but generally, it is intended to equate with the reduction in rent. To calculate the sum involved requires a 'before and after' comparison of each party's position to check their respective positions if the deal proceeds on the suggested basis rather than as a straightforward letting at market rent.

Example

The freeholder agrees to grant a 15-year lease with five-yearly reviews on a property worth £100,000 per annum on FRI terms. A premium of £250,000 has been agreed with the prospective tenant on the basis that, on review, the rent will be increased to market rent *less* the annual equivalent of the initial premium only. Current freehold all risks yield for this property is 8 per cent. What commencing rent should the tenant pay?

Valuation from landlord's viewpoint:
Freehold value if let at MR:

	£	
Net income =		100,000 p.a.
× YP perp @ 8%		12.5
	Capital value =	£1,250,000

Proposed freehold valuation:
Term

	£	
Net income =		x p.a.
× YP 15 yrs @ 7% *		9.108
		= £ 9.108x

Reversion to MR

	£	
Net income =	100,000 p.a.	
× YP perp def 15yrs @ 8%	3.941	
	= £394,100	

Add agreed premium = £250,000

Capital value = £644,100 + 9.108x

* = lower yield as term net income considerably lower and thus 'more secure' than reversion

Value let at MR should equal proposed interest:

Therefore £1,250,000 = £644,100 + 9.108x
x = £605,900/9.108 = £66,520

In other words the freeholder requires a rent of £66,500 per annum

Valuation from tenant's viewpoint:

	£	
Market rent =	100,000 p.a.	
Less annual equivalent of premium:		
£250,000/YP 15yrs @ 9%		
+ 4% (7.1459)	=(£ 34,985 p.a.)	
	Rent payable = £ 65,015 p.a.	

That is, tenant can afford to offer a rent of £65,000 per annum.

Notes

- leasehold MR yield taken at 1 per cent above freehold ARY to reflect reduced security of investment;
- no tax taken in decapitalising the premium to approximately allow for tax relief claimable by tenant on premium.

Conclusion

Tenant cannot quite offer the rental the freeholder is seeking, but the figures are very close. It depends on each side's position and viewpoint, but if the freeholder would prefer a capital payment and accept a reduced income (possibly for business cash flow reasons) then a negotiated settlement may be reached. This could be on the basis that some of the £1,500 difference between the rentals would be forgone and the rent payable for the first five years of the new fifteen-year lease will be SAY £65,750 per annum. Based on this, reviews at the 5th and 10th years could be geared to 65.75 per cent of market rent (£65,750 × 100/£100,000).

16.2 Capital payment on assignment of a lease

A payment will normally be sought when a lease is assigned at a passing rent that is less than the rack rental value. The amount paid will reflect the value of the potential profit rent that the assignee will acquire. Even when the rent payable is the market rent, there may still be a premium paid for other elements such as tenant's fixtures and fittings that will pass to the assignee, as mentioned in Section 16.1 above.

Example

A twenty-year lease of an industrial unit is to be assigned. The lease provides for only one rent review, after ten years, and has an unexpired term of four years. The tenant is responsible for internal repairs only. Current rent payable is £65,000 per annum, and from comparables you estimate the market rent to be £100,000 per annum if let on FRI terms. Nearby properties let at market rent show a leasehold yield of 12 per cent. Lessee pays tax at 30 per cent and can obtain a net annual sinking fund rate of 3 per cent. What sum is payable on assignment?

Market rental value on FRI terms =		£100,000 p.a.	
Adjust to gross terms as present lease:			
Add external/structural repairs @ say 5% MR=	£ 5,000 p.a.		
Add insurance @ say 3.5% MR =	£ 3,500 p.a.		
Add management charge @ say 3% MR=	£ 3,000 p.a.		
MR on internal repairing and insuring basis (rent receivable) =		£111,500 p.a.	
Less rent payable =		£ 65,000 p.a.	
Net profit rent=		£ 46,500 p.a.	
× YP 4 yrs @ 12%+3% (tax 30%)		2.167	
Capital value=		£100,765	
SAY price payable=		£100,000	

(ignoring value of fixtures and fittings, goodwill and/or 'key money')

'Outgoings' are added back to bring comparable value in line with internal repairing and insuring terms as under existing lease.

16.3 Lease surrender and renewals

A surrender and renewal is where the tenant reaches an agreement with the landlord to surrender the existing lease back to the landlord before the expiry of the term, in return for the grant of a new lease on part or whole of the same property. It is a voluntary arrangement; neither party can be compelled to agree to the transaction. Thus, the terms of the agreement must be mutually acceptable to both sides. Apart from the rental, the landlord may also take the opportunity to revise the terms of the occupation within the new lease.

The main reason why tenants would seek surrender and renewal is that the existing unexpired term of the lease offers insufficient security. The tenants would normally look to improve this security before carrying out substantial improvements to the property, or trying to raise finance using their interest as collateral. Alternatively, the tenants may be looking to sell their business, the value of which would be enhanced by a longer unexpired lease term. The landlord may benefit from an immediate increase in rental income and/or from a premium and/or from improvements to the property. Alternatively, the landlord may cooperate for the sake of maintaining good relationships with the tenant.

The value of the existing unexpired term of the lease must be taken into account. Therefore, four valuations are required:

- existing freehold subject to the existing lease;
- proposed freehold after the existing lease has been surrendered and the new lease granted;
- existing leasehold; and
- proposed leasehold under the new lease.

The basic calculation then required is that, whatever premium or rent is agreed for the new lease, neither party should be in a worse position after the surrender and renewal than they were before. It is a comparison between their 'before and after' positions.

Thus, the following must be true:

- Existing tenant's interest = Proposed tenant's interest
- Existing landlord's interest = Proposed landlord's interest

The appropriate new rent and/or premium payable is found from this balancing of 'existing must equal proposed' calculations. This will indicate the maximum figure the tenant can afford to offer and the minimum the landlord would be prepared to accept. Providing a suitable compromise between these figures can be reached through negotiation, the deal can proceed. When both parties would be better off after the surrender and renewal it should be easy to reach a settlement. When only one party gains, it may still be possible for a deal to be effected providing the party that is better off can adequately compensate the other party's loss from out of their own profit. Agreement will not be possible though when both sides would be worse off afterwards than they were beforehand.

There are two alternative approaches to balancing the 'present v. proposed' equation to enable a settlement to be reached between the parties:

- the rent of the new lease is adjusted; either for the initial term of the new lease until the first rent review; or for the entire term of the new lease; or
- a premium is paid where the new lease rent is already agreed.

Blank pro-forma worksheets are provided in Figures 16.1, 16.2, 16.3 and 16.4 to assist in preparing surrender and renewal valuations, with and without allowances for outgoings.

Surrender and renewal
VALUATION ANALYSIS OF:

Freehold Valuation of Present Interest
Term

Net income		£	pa	
× YP for	yrs @	%		
			= £	

Reversion

Net income		£	pa
× YP perp def	yrs @	%	
		= £	
	Capital value = £	
	SAY = £	

Freehold Valuation of Proposed Interest
Term

Net income		£	pa	
× YP for	yrs @	%		
			= £	

Reversion

Net income		£	pa
× YP perp def	yrs @	%	
		= £	
Less costs		= £	
	Capital value = £	
	SAY = £	

Present=Proposed analysis:
£   =£

Figure 16.1 Freehold valuation pro-forma worksheets for surrender and renewal valuation

Surrender and renewal
VALUATION ANALYSIS OF:

Leasehold Valuation of Present Interest

Rent receivable		£	pa
Less rent payable		£	pa
Net profit rent		£	pa
× YP for	yrs @	%	
+	% & tax @	%	
	Capital value = £	
	SAY = £	

Leasehold Valuation of Proposed Interest

Rent receivable		£	pa
Less rent payable		£	pa
Net profit rent		£	pa
× YP for	yrs @	%	
+	% & tax @	%	
		= £	
Less costs		= £	
	Capital value = £	
	SAY = £	

Present=Proposed analysis:
£   =£

Figure 16.2 Leasehold valuation pro-forma worksheets for surrender and renewal valuation

Surrender and renewal
VALUATION ANALYSIS OF:

Freehold Valuation of Present Interest
Term

Gross income		£		pa
Less outgoings				
Ext/struct reps @	£	pa		
Insurance @	£	pa		
Management @	£	pa		
		= £	pa	
Net income		£	pa	
× YP for	yrs @	%		
			= £	

Reversion

Net income		£	pa	
× YP perp def	yrs @	%		
			= £	
	Capital value = £			
	SAY = £			

Freehold Valuation of Proposed Interest
Term

Net income		£	pa	
× YP for	yrs @	%		
			= £	

Reversion

Net income		£	pa	
× YP perp def	yrs @	%		
			= £	
Less costs		= £		
	Capital value = £			
	SAY = £			

Present=Proposed analysis:
£   =£

Figure 16.3   Freehold with outgoings valuation pro-forma worksheets for surrender and renewal valuation

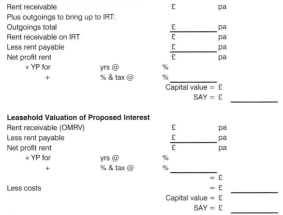

Surrender and renewal
VALUATION ANALYSIS OF:

Leasehold Valuation of Present Interest

Rent receivable		£		pa
Plus outgoings to bring up to IRT:				
Outgoings total		£	pa	
Rent receivable on IRT		£	pa	
Less rent payable		£	pa	
Net profit rent		£	pa	
× YP for	yrs @	%		
+	% & tax @	%		
	Capital value = £			
	SAY = £			

Leasehold Valuation of Proposed Interest

Rent receivable (OMRV)		£	pa	
Less rent payable		£	pa	
Net profit rent		£	pa	
× YP for	yrs @	%		
+	% & tax @	%		
		= £		
Less costs		= £		
	Capital value = £			
	SAY = £			

Present=Proposed analysis:
£   =£

Figure 16.4   Leasehold with outgoings valuation pro-forma worksheets for surrender and renewal valuation

Example

Surrender and renewal example 1: find the premium to be paid

Acting between the parties, assess a fair premium to be paid for the renewal of Jackson's lease of a shop unit. The shop needs modernisation and refurbishment. It is let at a current rental of £50,000 per annum on FRI terms; but with an expenditure of £200,000 on improvements, its rental value could be increased by £30,000 per annum. Current market rent is estimated at £80,000 per annum (that is, after improvements this would rise to £110,000 per annum). Jackson holds a lease from Edwards, which has eight years left unexpired without rent review. Jackson wants to take a new lease for fifteen years at £80,000 per annum, without review, on the basis that he will undertake the improvement work. Edwards, the freeholder, is agreeable in principle to the proposal. What premium is payable?

Valuations

Assumptions derived from comparable market evidence:
- freehold equivalent yield is 7 per cent and leasehold yield 8.25 per cent;
- tenant pays tax @ 20 per cent; and
- net annual sinking fund rate is 3 per cent.

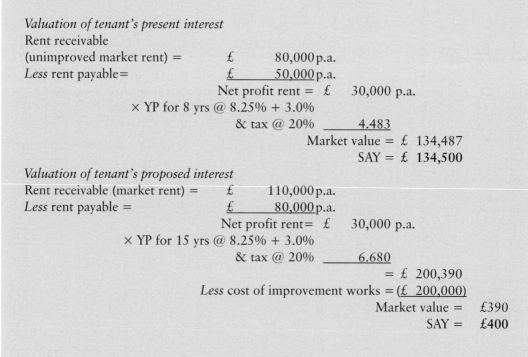

Valuation of tenant's present interest
Rent receivable
(unimproved market rent) =   £   80,000 p.a.
Less rent payable =   £   50,000 p.a.
                                        Net profit rent = £   30,000 p.a.
                   × YP for 8 yrs @ 8.25% + 3.0%
                                    & tax @ 20%   4.483
                                             Market value = £ 134,487
                                                      SAY = £ 134,500

Valuation of tenant's proposed interest
Rent receivable (market rent) =   £   110,000 p.a.
Less rent payable =   £   80,000 p.a.
                                        Net profit rent = £   30,000 p.a.
                   × YP for 15 yrs @ 8.25% + 3.0%
                                    & tax @ 20%   6.680
                                                      = £ 200,390
                        Less cost of improvement works = (£ 200,000)
                                             Market value =   £390
                                                      SAY =   £400

Therefore for tenant's present and proposed interests to be equal, a 'reverse' premium (paid by the landlord to the tenant) must be paid.

If sum of reverse premium = x

Then x = £134,500 – £400

Thus x = £134,100

Therefore the tenant requires a landlord's contribution of at least £134,100 towards the cost of the improvement works in order to take the new lease on the terms agreed.

Valuation of landlord's present interest

Term

Net income	£	50,000 p.a.
× YP for 8 yrs @ 7%		5.971
		£298,565

Reversion to market rent

Net income	£	80,000 p.a.
× YP perp def 8 yrs @ 7%		8.314
		£665,153
	Market value	=£963,718
	SAY	=£964,000

Valuation of landlord's proposed interest

Term

Net income	£	80,000 p.a.
× YP for 15 yrs @ 7%		9.108
		£728,633

Reversion to market rent

Net income	£	110,000 p.a.
× YP perp def 15 yrs @ 7%		5.178
		£569,558
	Market value	=£1,298,191
	SAY	=£1,300,000

Freeholder's interest significantly increases in value if the surrender and renewal goes ahead. Should present interest = proposed interest, landlord can afford to pay tenant up to: £1,300,000 – £964,000 = £336,000.

Conclusions

The landlord could comfortably pay the contribution towards the improvement costs that the tenant is seeking. In fact, the landlord could afford to pay even more. The tenant's representative should be aware of this and will probably try to negotiate a higher settlement figure. Suppose the landlord offers to pay the tenant 75 per cent of the cost of the improvements (£150,000), the surrender and renewal could proceed on the agreed terms and both parties will benefit from the transaction. The landlord will also ultimately benefit from the reversion on the improved property.

The position of both parties will be:

- Landlord pays tenant a 'reverse premium' of £150,000 as a contribution to the costs of the improvements. This reduces the proposed freehold value from £1,300,000 to £1,150,000, which still leaves the landlord better off than with the present lease by £186,000 (£1,150,000 – £964,000).
- Tenant receives £150,000, making proposed interest value £150,400 compared to the present interest of £134,500 and a gain of £15,900 plus an improved property and a new longer lease with no rent reviews.

Example

Surrender and renewal example 2: finding an initial rent

A tenant of a retail unit has two years left to expiry of the existing lease, without review, and wishes to surrender this agreement and take a new lease for fifteen years. The present lease is on internal repairing terms only at a current rent of £200,000 per annum. The landlord is willing to grant the new lease provided that it is on full repairing and insuring (FRI) terms and that the rent payable can be reviewed to open market rental value every three years. The present estimated rental value (ERV) is £250,000 per annum on FRI terms and market freehold all risks yield at this rental level is currently 5 per cent. Acting between the parties, show what rent the tenant can expect to pay for the initial three years of the proposed new lease.

Assumptions:

- leasehold all risks yield is 1 per cent higher than freehold;
- tenant pays tax at 30 per cent; and
- net annual sinking fund rate is 3 per cent.

Valuations

Existing freehold valuation
Term

Gross income =			£200,000 p.a.
Less outgoings:			
Repairs & decs @ say 5% MR =	£	12,500	
Insurance @ say 3% MR =	£	7,500	
Management @ say 4% rent =		£8,000	
		=£ 28,000	p.a.
	Net income=	£ 172,000 p.a.	
	× YP for 2 yrs @ 4%	1.886	
	=	£324,408	

Reversion to market rent

Net income =	£	250,000 p.a.
× YP perp def 2 yrs @ 5%		18.141
	=	£4,535,147
	Market value =	£4,859,556
	SAY =	£4,860,000

Proposed freehold valuation

Term

Net income =	£	*x* p.a.
× YP for 3 yrs @ 4%		2.775
		£2.775*x*

Reversion to market rent

Net income =	£	250,000 p.a.
× YP perp def 3 yrs @ 5%		17.277
		£4,319,188
	Market value =	£4,319,188 + £2.775*x*

Note to both freehold valuations: term yield reduced by one per cent to reflect increased security of term income compared to market rent.

Solve the equation of present interest = proposed interest to find *x*

$$£4,860,000 = £4,319,188 + 2.775x$$

$$x = (£4,860,000 - £4,319,188)/2.775$$

$$x = £194,728$$

Therefore the minimum rent the landlord will accept is £194,887 p.a.

Existing leasehold valuation

Rent receivable (market rent on FRI terms) =	£	250,000 p.a.
Add Outgoings to make IRT	£	28,000 p.a.
Market rent on existing lease terms =		£278,000 p.a.
Less rent payable =		£200,000 p.a.
Net Profit Rent =		£ 78,000 p.a.
× YP for 2 yrs @ 6% + 3% & tax @ 30% =		1.309
	Market value =	£102,130
	SAY =	£102,000

Proposed leasehold valuation

Rent receivable (market rent) =	£	250,000 p.a.
Less rent payable =	£	*x* p.a.
Net profit rent =		£250,000 − *x* p.a.
× YP for 3 yrs @ 6% + 3% & tax @ 30%		1.915
	Market value =	£478,750 − 1.915*x*

Again, '*x*' is used as proposed rent payable

Present = Proposed
£102,000 = £478,750 − 1.915x
x = (£478,750 − £102,000)/1.915
x = £196,736

Thus the maximum rent the tenant can afford is £196,736 per annum.

Conclusion

The tenant can afford to pay what the landlord is seeking. Usual settlement is likely to be around mid-way between both parties' figures, which in this case are extremely close anyway. Therefore the suggested settlement of rent for the first three years of the new lease is **£195,500 per annum**, which will be acceptable to both sides.

16.4 Premium on variation of a condition of a lease

Most lease covenants restricting the use of a property, or preventing alteration, assignment or subletting, can be varied with the landlord's consent, such consent not to be unreasonably withheld. Where such a covenant is more restrictive, however, then sometimes the landlord can realise some value. The basis of the valuation is that the landlord should suffer no loss to his/her interest and can expect to share in any gain in the tenant's interest.

The gain in the value of the tenant's interest is derived from:

Value of lease without the condition = x
Value of lease with the condition imposed = y
Gain in value = x − y

Example

Plum has a leasehold interest in a small office building situated in a retail high-street location. The lease has seven years left to run, without further review, at a rent of £30,000 per annum on FRI terms. The user covenant in the lease absolutely prohibits usage other than as offices. The current estimated MR for office purposes is £40,000 per annum

Plum has now approached his landlord, Mustard (the freeholder), for consent for a change of use to retail purposes, for which the current estimated MR is £60,000 per annum. The local planning authority has indicated a planning application for such a change would probably be approved. Plum pays tax at 40 per cent and can obtain 3 per cent net on his annual sinking fund instalments. On the basis he seeks a yield of 7 per cent on his interest, what premium should Plum pay?

The landlord is assumed to suffer no loss to his interest if he/she agrees to relax the condition, as the rent paid by the tenant will not alter for the unexpired term, and will be at least as much on review for retail purposes as it would have been for offices. The landlord will thus look to any gain in the tenant's interest to calculate the premium payable rather than any adjustment of the freehold value.

Existing value with condition imposed from the tenant's viewpoint:

Rent receivable (as offices) =	£	40,000 p.a.	
Less rent payable =	£	30,000 p.a.	
	Net profit rent =	£	10,000 p.a.
× YP 7yrs @ 7% + 3% (tax 40%)			3.478
		Capital value =	£34,780
		SAY =	**£35,000**

Tenant's proposed value without condition:

Rent receivable (as retail) =	£	60,000 p.a.	
Less Rent payable =	£	30,000 p.a.	
	Net profit rent =	£	30,000 p.a.
× YP 7yrs @ 7% + 3% (tax 40%)			3.478
		=	£104,340
Less costs of approvals and conversion to shop say =			(£25,000)
		Capital value =	£79,340
		SAY =	**£79,000**

Therefore: gain in value of tenant's interest = £44,000 (£79,000 – £35,000)

How this is divided between landlord and tenant will depend on the negotiating power and ability of both sides. In extreme cases, a landlord could demand all, or the majority of any gain in value, since it is being directly created from his/her consent to the change of the lease condition. However, in this instance, if both sides agree to equally share the gain, a premium of £22,000 will be payable (£44,000/2) by the tenant.

Alternatively, a revised rent for the remaining seven years of the lease can be obtained as follows:

Annual equivalent of premium
= £22,000/YP 7yrs @ 7%+3% (tax 40%)
= £6,325 per annum
Therefore: revised rent = existing rent + AE of premium
= £30,000 + £6,325
= **£36,325 per annum**

16.5 Sale and leaseback

A sale and leaseback can be undertaken by a freeholder (or the owner of a very long leasehold interest) as a means of raising capital and yet retaining possession of the property. The principle is that the freehold or long leasehold interest is sold for a capital sum on condition that the new owner grants the previous owner an occupational lease on part or whole of the property at the market rent, or a lower rent which is linked to the purchase price. Clearly, the current owner becomes a tenant and has to pay an annual rent, but has released a substantial capital sum and can remain in occupation.

A US perspective on this type of transaction was provided by the Vice President of United Trust Fund:

> Sale/leasebacks give businesses additional funds that can be used to enhance liquidity, expand operations, retire high-cost debt or finance expansions. This financing method allows companies in need of cash to realize 100% of the property's fair market value, compared to 60 to 80% with conventional debt sources ... a typical lease is structured with an initial term of 15 to 25 years plus options, which permit the company to control the property for its entire useful life.

> (Domb 2000)

Example

A freeholder owns a property that has a current ERV with vacant possession (VP) on the open market of £100,000 per annum. The market all risks yield of the investment is deemed to be 8 per cent. The owner now wishes to dispose of her freehold interest in return for a 50-year FRI lease on the property at an initial rent of £80,000 per annum and with five-yearly rent reviews to market rent. What sale price can the owner expect to achieve for the property on this basis?

Valuation of freehold interest with VP:

Net income =	£	100,000 p.a.
× YP perp @ 8%		12.5
	Capital value =	£ 1,250,000

This would be the price the owner could expect to achieve on the open market if the property is vacated and sold. However, on the proposed basis of sale and leaseback, the valuation which would be undertaken by a prospective purchaser would be:

Valuation of proposed freehold interest after sale and leaseback:

Term

Net income =	£	80,000 p.a.
× YP 5yrs @ 7%		4.1
	= £	328,000

Reversion

Net income =	£	100,000 p.a.
× YP perp def. 5yrs @ 8%		8.507
	=	£ 850,700
	Capital value =	£1,178,700
Less costs of sale and arranging		
leaseback @ say 5% =		(£58,935)
		= £1,119,765
	SAY	= £1,120,000

This would be the expected value realisable on the proposed sale and leaseback. Of course, the above valuation could have been carried out on a hardcore basis if preferred. The expected sale price would represent 89.6 per cent of the vacant possession value ($£1.12m \times 100/£1.25m$).

16.6 Marriage valuations

'Marriage value' occurs when the value of the whole interest exceeds the sum of its parts. It can be more fully defined as: 'Latent value which would be released by the merger of two or more interests in land. For example, two adjoining parcels may be worth more as one property than the aggregate of their separate values. Similarly, two interests in the same property may have a greater value when merged than the sum of their individual values' (Parsons 2004: 168).

The most frequent occasion when marriage value occurs is where the value of the unencumbered freehold interest in a property exceeds the combined value of that same freehold, subject to a tenancy, and the value of the leasehold interest held by the occupying tenant. In such a situation, the tenants could bid to buy out the freeholder's (their landlord's) interest, or vice versa, to realise the additional inherent value. Sometimes statutory legislation empowers tenants to obtain such a purchase, for example the Leasehold Reform Act 1967 in UK.

In these circumstances, the marriage value is equal to:

Freehold value not subject to existing lease (that is VP valuation basis)
Less
The *sum* of the existing freehold value + existing leasehold value.

There will thus be three valuations required:

1. freehold value on vacant possession (VP) basis;
2. existing freehold value subject to current tenancy; and
3. existing leasehold value.

and the marriage value (disregarding costs of sale or purchase) will be: $1 - (2+3)$.

Marriage value can also occur when two adjoining properties would be worth less if offered on the market separately by separate owners than if they were offered together, with potential conversion possibilities, to make one large unit. In addition, it can exist when the sum total value of separate development sites, valued and purchased individually, is lower than the value of the resulting large assembled site.

A blank pro-forma worksheet is provided in Figure 16.5 to assist in preparing marriage valuations.

Marriage value
VALUATION ANALYSIS OF:

Freehold Valuation of Present Interest
Term

Net income   £   pa
   × YP for   yrs @   % _____
                                     = £

Reversion

Net income   £   pa
× YP perp def   yrs @   % _____
                                      = £   _____
              Capital value = £ _____
                    SAY = £ _____

Freehold Valuation of Proposed Interest

Net income   £   pa
   × YP for   yrs @   % _____
              Capital value = £
                    SAY = £ _____

Leasehold Valuation of Present Interest

Rent receivable   £   pa
Less rent payable   £   pa
Net profit rent   £   pa
   × YP for   yrs @   %
     +   % & tax @   % _____
              Capital value = £
                    SAY = £ _____

Sum of present freehold + leasehold interests   = £   + £
Proposed freehold – above sum = £   – £
Marriage value   =   £

Split of marriage value agreed =   % tor freeholder
Present freehold value + aggreed marriage value split   =   £   + £
                                      = £

Say price to be paid by tenant to purchase frehold   = £ _____

Figure 16.5   Marriage valuation worksheet

Example

Marriage valuation example 1

A UK residential tenant, occupying under a regulated tenancy protected by the Rent Acts, wishes to acquire the freehold interest of the house she occupies. Vacant possession value of the freehold is estimated at £200,000, and its value subject to the tenancy as an investment property is £80,000. The gain, or marriage value, will be £120,000 from the 'marriage' of the leasehold and freehold interests.

The split of this marriage value would depend on the outcome of negotiations between the parties, but if they agreed an equal division of the additional value created, the price to be paid by the tenant to acquire the freehold would be:

Existing freehold value + agreed split of marriage value

= £80,000 + £120,000/2

= £140,000 to purchase the freehold interest

As a result of paying this figure, the former tenant would be in possession of a freehold interest worth £200,000 on the open market, and could if desired realise an immediate capital gain of £60,000 less costs by vacating and selling.

Example

Marriage valuation example 2

A leaseholder occupies an office property under a ground lease with an unexpired term of twenty years at a fixed ground rent of £10,000 per annum. From comparables, current market rent of the office building is estimated at £200,000 per annum. The lessee wishes to buy the freehold interest from the landlord. Freehold all risks yields are 8 per cent. The lessee pays tax at 40 per cent and can obtain 3 per cent net annual return on a sinking fund. What price is likely to be agreed for this purchase?

Valuation of existing freehold (with differential term and reversion yields)
Term

	£	
Net income =	10,000 p.a.	
× YP 20yrs @ 5%	12.4622	
		= £124,622

Reversion to MR

	£	
Net income =	200,000 p.a.	
× YP perp def 20 yrs @ 8%	2.6818	
		= £536,360
	Capital value =	£660,982
	SAY =	**£661,000**

- Low yield used in term as income very 'secure' being only 5 per cent of the market rent;
- however, could reasonably be argued a low income, fixed for twenty years is a poor investment and should be capitalised at a high yield, such as 12 per cent;
- accordingly, use of equivalent yield would be preferable.

Valuation of existing freehold using equivalent yield
Term

Net income =	£	10,000 p.a.		
× YP for 20 yrs @ 8%		9.818		
		= £	98,181	

Reversion to market rent

Net income =	£	200,000 p.a.		
× YP perp def 20 yrs @ 8%		2.682		
		= £	536,371	
	Capital value =	£	634,552	
	SAY =	£	634,500	

Valuation of freehold with VP

Net income =	£	200,000 p.a.
× YP perp @ 8%		12.5
Capital value =	£2,500,000	

Valuation of present leasehold interest

Rent receivable =	£	200,000 p.a.
Less rent payable =	£	10,000 p.a.
Net profit rent =	£	190,000 p.a.
× YP for 20 yrs @ 9.5%		
+ 3% & tax @ 40%		6.368
Market value =	£1,209,989	
SAY =	£1,210,000	

Leasehold yield assumed at 1.5% above freehold ARY.
Marriage Value
= proposed freehold – (present freehold + present leasehold)
= £2,500,000 – (£634,500 + £1,210,000)
= £655,500

Conclusions

- The lessee could afford to pay up to £1,290,000 (£634,500 + £655,500) to buy out the freeholder's interest;
- however, the lessee would normally expect to retain part of the marriage value;
- if the parties agreed a 50/50 split of this, the lessee would pay £962,250 to buy the freehold, namely existing freehold + half marriage value = £634,500 + £655,500/2.

Example

Marriage valuation example 3

Titania Building is occupied on a 25-year lease by Lysander Bank from the freeholder, Oberon Estates. The lease was granted eight years ago on full repairing and insuring terms with upwards-only rent reviews every five years. The current rent passing is £62,500 per annum. The bank modernised the building at the start of the lease term at a cost of £80,000. It was a condition of the lease that this modernisation be undertaken and it was agreed that the value of this work could be rentalised at review. Four years ago, the bank extended the ground floor by 50 sq.m. at a cost of £60,000. Oberon Estates consented to the extension and agreed that the value of this improvement work would be disregarded at rent reviews for the remainder of the lease.

Titania Building originally provided 387 sq.m. net internal area, equally distributed between three floors. There is no lift, but the building does have central heating and double-glazing. There have been few market transactions recently, but what evidence exists indicates that a building of this size and nature would expect to let on a standard 'institutional lease' on the open market for £180 per sq.m. per annum

Lysander Bank occupies the ground and first floors but has recently sublet the second floor for five years without review to Quince Enterprises for £27,000 per annum on internal repairing terms only. The bank is interested in purchasing the freehold interest in Titania Building. Acting between the parties, and assuming Oberon Estates are agreeable to a sale, advise what price they should expect to pay.

Assumptions made (based on market comparable evidence):

- freehold all risks yield = 7 per cent;
- leasehold all risks yield = 8 per cent;
- net sinking fund rate = 3 per cent; and
- bank's tax rate = 30 per cent.

Subletting: £27,000/129sq.m = £209.30 per sq.m p.a.
Present rent on subject: £62,500/387sq.m = £161.50 per sq.m p.a.
Current market rents = £180 per sq.m p.a.

Conclude: existing rent on subject is lower than MR, but subletting appears higher, although it is not on FRI terms as would be expected on an institutional lease.

Sublessee only pays for internal repairs.
Lysander Bank pay for:
External and structural repairs say = 5 per cent of MR
Insurance say = 3.5 per cent of MR
Management say = 4 per cent of rent passing.

So sublessee's rent should be *higher* than FRI rent to allow for costs of outgoings to be met by landlord out of rent collected.

Adjusting for these outgoings:

- £209.30 per sq.m. less 4 per cent management charge = £209.30 – £8.37 = £200.93 per sq.m.
- Repairs and insurance = 7.5 per cent of MR = 0.075 × £180 = £13.50 per sq.m.
- Rent collected less management less repairs and insurance = £200.93 – £13.50 = £187.43 per sq.m. per annum if let on FRI terms
- This is above the market rent level that would apply to the whole building, but this is a subletting of one floor only, so quantum adjustment for size appears to have been made of 4.1 per cent: (£187.43 – £180) × 100/£180. Thus subletting rent is in line with current market rent values.

Valuations now required:

- present freehold (subject to letting to the bank); and
- present leasehold (including subletting of top floor); and
- proposed freehold (still subject to subletting of top floor, but with VP assumed for lower two floors).

Valuation of present freehold

- Term (next two years until bank's rent review) @ £62,500 per annum
- Reversion in two years' time to MR disregarding 20 sq.m. extension (MR = £180 × 387 sq.m. = £69,660 per annum)
- Final reversion to MR including the extension in seventeen years' time when lease ends (MR = £180 × 437 sq.m. = £78,660 per annum)

Valuation of present freehold using hardcore method and equivalent yield

Assume equivalent yield = 7% (found from comparables)

Hardcore income =	£	62,500 p.a.
× YP perp @ 7%		14.285
		= £ 892,857
Middle slice income =	£	7,160 p.a.
× YP perp def 2 yrs @ 7%		12.477
		= £ 89,340
Top slice income =	£	9,000 p.a.
× YP perp def 17 yrs @ 7%		4.522
		= £ 40,698
		Market value = £1,022,895
		SAY = £1,025,000

Freehold valuation: equated yield approach

Assumed annual rental growth rate = 2.5 per cent (based on market evidence); equated yield = 9.2 per cent (found from formula in Section 12.4 above, where ARY is 7 per cent); rental income will be current rent for next two years, then unimproved market rent until end of year 17 and improved market rent thereafter.

Years	Income	PV factor	DCF
1 to 2	£62,500	1.754	£109,647
3 to 7	£73,187	3.245	£237,491
8 to 12	£82,804	2.090	£173,042
13 to 17	£93,685	1.346	£126,084
18 to 22	£119,690	0.867	£103,771
23 to perp*	£135,419	2.061	£279,099
		GPV =	£1,029,134
		SAY =	£1,025,000

* = income capitalised @ ARY to reflect continued future rental growth but then discounted at equated yield

Valuation of present leasehold

Present rent receivable will be the net rent actually collected from the subtenant plus the market rent value of the two floors occupied by the bank itself (notional rent receivable or rent that would otherwise be paid for a similar building). When the sublease ends in five years' time, it can be assumed that the lease would be renewed or another subletting arranged and this would continue through until end of head lease 17 years from now.

Total present rent receivable

Net rent from subletting of second floor
@ £187.43 × 129 sq.m. = £ 24,178 p.a.
Market rent value of remainder
(including extension)
@ £180 × 308 sq.m. = £ 55,440 p.a.
Total = £ 79,618 p.a.

Term
Rent receivable= £ 79,618 p.a.
Less rent payable = £ 62,500 p.a.
Net profit rent £ 17,118 p.a.
× YP 2 yrs @ 8% + 3%
& tax @ 30% 1.275
=£21,825

Reversion

Rent receivable =	£	79,618 p.a.	
Less rent payable =	£	69,660 p.a.	

$$\text{Net profit rent} = £ \quad 9,958 \text{ p.a.}$$

× YP 15 yrs @ 8% + 3%	
& tax @ 30%	6.377
× PV 2 yrs @ 8%	0.857
	= £54,421
Market value =	£76,246
SAY =	**£76,250**

Valuation of proposed freehold

Assume building will again be let as a whole to one tenant, thus overall present value of market rent is 437 sq.m. at £180 = £78,660 per annum.

Net income (MR) =	£	78,660 p.a.
× YP perp @ 7%		14.285
Market value =		£1,123,658
SAY =		**£1,124,000**

Marriage value = proposed freehold less (present freehold + present leasehold)
= £1,124,000 − (£1,025,000 + £76,250)
Marriage value = £22,750

As marriage value is a relatively small amount compared to the overall value, the bank can expect to pay present freehold value plus whole or most of the marriage value to buy out the freeholder. Expected purchase price will thus be say **£1,045,000**.

16.7 Divorce valuations

'Divorce value' is the opposite of 'marriage value' and would occur when the interest in a property is worth less if offered as a single entity than it would be if broken up and offered as several smaller interests. In other words, the sum of the parts is greater than the value of the whole. To achieve a divorce value, a property may be split up *physically* (into smaller parcels) or *legally* (into freehold and leaseholds).

Example

Divorce valuations example 1

A freehold site of 0.33 hectare has planning permission for the development of fourteen detached houses. From comparables, such sites are selling for £1,500,000 per hectare, although individual building plots for detached houses are selling at £50,000 each. Instead of selling the complete site to a single developer, it could be broken up and sold as fourteen individual plots to separate persons wishing to build their own houses, or a self-build group. The result would be as follows:

Site sold complete = 0.33 ha @ £1,500,000 per ha = £500,000
Site sold as 14 plots each @ £50,000 = £700,000
Therefore: Gain in value (that is divorce value) = **£200,000**

Example

Divorce valuations example 2

The owner of a small freehold office building wishes to obtain the maximum capital value of his investment. The building has a ground plus three upper floors. Each floor comprises 300 sq.m. NIA and there is a lift with a twelve person capacity which serves all floors. Similar sized and quality buildings in the town are letting at £250 per sq.m., whereas small suites of around 250 to 300 sq.m. are achieving rentals of £280 per sq.m., both on FRI terms. Market evidence also indicates that freehold all risks yield is 6 per cent.

Thus, divorce value can be found as follows:

Valuation of freehold interest let to a single tenant for the whole building
Net income: 1,200 sq.m. @ £250 = £   3000,000 p.a.
× YP perp @ 6%   16.667
                                Capital value SAY = £5,000,000

Valuation of freehold if each floor let separately
Net income: 300 sq.m. @ £280 × 4 =   £336,000 p.a.
× YP perp @ 6.25%   16
                        Capital value = £5,376,000

Note: all risks yield increased by 0.25 per cent where separate lettings arranged to reflect potentially smaller tenants that do not represent quite as good an investment risk compared to one larger company occupying whole building.

Therefore: Divorce value = £5,000,000 – £5,376,000 = £376,000

16.8 Effect of improvements and restrictive user clauses on valuations

When a lease contains an express clause requiring certain tenant's improvements to be disregarded when assessing the rental value at review, or a similar principle applies at lease renewal, thought needs to be given to the following when assessing the rental value:

- What are these improvements? Alterations to the demised premises? Additions to the demised premises? Provision of fixtures to the demised premises?
- And are any of these definable as repairs not improvements, or equipment that is a fitting or a tenant's fixture?
- Do any of these have no effect or an adverse effect on the rental value at review?

How is the value of improvements to be assessed? In *GREA Real Property Investments Ltd* v. *Williams* (1979) 250 EG 651 the court stated: 'The tenant must be credited with the rental equivalent of his/her expenditure upon review, which may be affected by inflation, but this was to be the sole benefit to him/her ... Improvements are paid for at the time they are carried out and at the end of the lease tenure for the benefit of the landlord, therefore, the tenant should consider them as wasting assets ... The improvements should be valued in their existing state as at the review date.'

There are three possible approaches to valuing improvements. The usual method is by direct rental comparison. An alternative may be:

- calculate the rental value of the premises as they stand with the improvements (a);
- establish the amount of increased turnover generated by the improvements, expressed as a percentage (b); and
- reduce rental value in (a) by the percentage in (b).

The third option is the cost method, with the annual equivalent of the expenditure being taken over the useful life of the improvement.

At times, it has been considered good estate management to include a restrictive user clause in a lease. With properties where a balanced mix of users is desired, such as in a shopping centre, this can still be a valid consideration. However, when a property can be used only for a certain purpose, then its market rent value is likely to be adversely affected and a reduction applied – see *UDS Tailoring Ltd* v. *BL Holding Ltd* (1982) 261 EG 49 and *Plinth Property Investments Ltd* v. *Mott Hay & Anderson* (1978) 249 EG 1167.

No effect on rent is likely if a change of use is permitted within a fairly wide class, with the proviso that the landlord will not unreasonably withhold consent. Also check that the clause is not restrictive but merely *permissive*, that is allowing something which would otherwise be considered a breach: *Sydenham* v. *Enichem Elastomers* [1986] NPC 52 and *The Faucet Inn Pub Co plc* v. *Ottley Corporation* [2006] 14 EGCS 174. When a discount is warranted, it must be deducted from a true 'open user' market rent, not from a restrictive user comparable.

Example

A pension fund requires an open market capital valuation of its freehold interest in a warehouse property known as Unit A. They would also like a calculation of worth to them of the investment, based on a 12 per cent target equated yield.

The building is situated on a plot of 1.3 hectare and is part of a modern trading estate near a motorway junction, five miles from a regional airport and three miles from major docks. The original building is twelve years old and provides 4,000 sq.m. of warehouse/distribution space and has an eaves height of 7 metres. An ancillary office block section is at the front elevation.

It is on two storeys and measures 550 sq.m. All floor areas have been measured on a gross internal area basis.

The property has been let to a public limited company on a 25-year full repairing and insuring lease that commenced twelve years ago. The lease is subject to five-yearly upwards-only rent reviews. As a condition of the lease, the tenants installed heating, lighting and sprinklers to the building. With the freeholder's consent, the tenants also built an extension on to the original building eight years ago which increased the floor area of the distribution space by 800 sq.m. The lease contains a clause stating that this improvement is to be disregarded at subsequent rent reviews or other reassessment of the current occupier's rent for a period of 21 years. The tenant's current rent, agreed as from the last review two years ago, is £200,000 per annum.

The following open market comparable evidence of other buildings on the same trading estate has been obtained.

Unit B is a distribution unit of 5,000 sq.m. GIA constructed twelve years ago. The eaves height is 6 metres. The total area includes an office content of 10 per cent. The unit was let two months ago at £250,000 per annum with a six-month rent-free period. The new lease was for fifteen years with five-yearly upwards or downwards rent reviews on FRI terms. The lease contains no unusual or unduly onerous terms.

Unit C is a warehouse measuring 4,500 sq.m. GIA and includes offices of 500 sq.m. Its eaves height is 5 metres. A new lease for ten years was granted six months ago at a rent of £210,000 per annum on FRI terms and with an upwards-only rent review after five years. The first four months of the lease are rent-free. The building was constructed eleven years ago and the lease is on standard institutional terms. The building, as let and occupied by the tenants, has just been sold as an investment to an insurance company for the sum of £2.2 million.

Analysis of comparables

Factor	Subject	Unit B	Unit C
Age	12 yrs	12 yrs	11 yrs
GIA total (sq.m)	4550	5000	4500
Office GIA	550	500	500
Office %	12%	10%	11%
Lease term unexpired	13 years	15 years	10 years
Rent reviews	5 yearly	5 yearly	5 yearly
Review basis	Up only	Up/down	Up only
FRI	yes	yes	yes
Other	T imps*	Rent-free	Rent-free
Current rent	£200,000	£250,000	£210,000
Rent per sq.m	£43.96	£50.00	£46.67
Eaves height	7m	6m	5m

*Improvements disregarded until 13 years from now, or in other words until end of current lease.

Effect of rent frees over whole lease term:

Unit B: 6 months on 15 yr lease
Unit C: 4 months on 10 yr lease
Unit B: (YP 14.5 yrs × PV 0.5 yrs)/YP 15 yrs
Unit C: (YP 9.67 yrs × PV 0.33 yrs)/YP 10 yrs

ARY derived from sale of unit C: £210,000/£2.2m = 0.0955 = 9.55%
say adopt 9.5 per cent for rent-free analysis and reversionary yield for main valuation.

Unit B: effect of rent free
YP 14.5 yrs @ 9.5% =   7.7029
PV 0.5 yrs @ 9.5% =   0.9556
YP 15 yrs @ 9.5% =   7.8281
Adjustment factor
=(7.7029 × 0.9556)/7.8281= 0.9403
1 − 0.9403 = 0.0597 = 5.97%;
say 6%

Unit C: effect of rent free
YP 9.67 yrs @ 9.5% =   6.1496
PV 0.33 yrs @ 9.5% =   0.9704
YP 10 yrs @ 9.5% =   6.2787
Adjustment factor
=(6.1496 × 0.9704)/6.2787 = 0.9505
1 − 0.9505 = 0.0495 = 4.95%;
say 5%

Rent review basis on Unit B of upwards or downwards is more favourable to tenants than upwards only reviews on Units A and C. This is likely to be reflected in a slightly higher tenant rental bid. Comparatively low eaves height of Unit C is a disadvantage and probably will result in a lower rent being achieved.

Percentage adjustment of comparables to bring in line with subject building

Factor	Unit B	Unit C
Age	nil	nil
GIA total (sq.m)	nil	nil
Office %	nil	nil
Lease UT	nil	nil
RRs	nil	nil
RR basis	say −3%	nil
FRI	nil	nil
Other (rent frees)	−6%	−5%
Eaves height	nil	say +5%
Total adjustments	−9%	nil

Adjusted rents:

Unit B = £50.00 less 9% = £45.50 per sq.m.
Unit C = £46.67 less nil = £46.67 per sq.m.
SAY adopt market rent of £46 per sq.m for subject property.

Applied to Unit A floor area:

Excluding extension = 4,550 sq.m. @ £46 = £209,300 p.a.
= **SAY £209,000 p.a.**
With the extension = 5,350 sq.m. @ £46 = £246,100 p.a.
= **SAY £246,000 p.a.**

Freehold valuation: equivalent yield approach

Term
Net income =   £   200,000 p.a.
× YP for 3 yrs @ 9.5%   2.509
                                  = £   501,781

Reversion to intermediary rent
Net income =   £   209,000 p.a.
× YP 10 yrs @ 9.5%   6.279
× PV 3 yrs @ 9.5%   0.762
                                  = £   999,495

Reversion to market rent
Net income =   £   246,000 p.a.
× YP perp def 13 yrs @ 9.5%   3.235
                                  = £   795,844
                         Market value = £ 2,297,120
                             SAY = £ 2,300,000

Nominal equivalent yield = 9.5 per cent (annually in arrear); true equivalent yield = 10.09 per cent (quarterly in advance).

Freehold valuation: DCF approach

Assumed annual rental growth rate = 3 per cent (based on market evidence); equated yield (client's target rate) = 12 per cent.

Example continued overleaf …

Calculation of worth to client

Years	Income	PV factor	DCF
1 to 3	£200,000	2.402	£480,366
4 to 8	£228,380	2.566	£585,979
9 to 13	£264,755	1.456	£385,459
14 to 18	£361,259	0.826	£298,444
19 to 23	£418,799	0.469	£196,318
24 to perp*	£485,502	0.777	£377,097
		GPV =	£2,323,663
		SAY =	£2,300,000

*Income capitalised @ ARY to reflect continued future rental growth but then discounted at equated yield.

No holding period specified by pension fund, so worth calculated in perpetuity. This confirms the objective valuation figure is reasonable from this client's perspective.

16.9 Constant rent

When a property is let with rent reviews that do not conform to the standard or most frequently found period in the market, should an adjustment be made to the rent at each assessment and, if so, by how much? There is no universal agreement between valuers on whether any adjustment should be made. However, an argument can be advanced that some adjustment is justified, either to compensate the tenant for having more frequent than normal reviews or the landlord for having to wait longer for a review of the rent.

Supposing comparable properties all have five-yearly rent reviews and the subject property to be valued has seven-yearly reviews. Then the landlord has to wait an additional two years before the rental income could be increased compared to investments owned by other landlords. Over a 21-year period, there will only be three reassessments of rent, compared to four with the five-yearly reviews. A slightly higher rental at each review, compared to those on the five-year pattern, would help to offset this disadvantage. In a recessed market, however, it is questionable whether an increased rent at the start of a lease term or review period could be obtained from a tenant to reflect longer-than-usual review periods in the lease. Conversely, if the subject property has three-yearly reviews, the tenant would argue for a discounted rent at each review to reduce the impact of the more frequent increases.

An argument against making any adjustments to rents is advanced by the Valuation Office Agency who advise that 'VOs should take steps to determine the local "norm" for rent review patterns for each class of property within their local office area. Once the norm is established it can be argued that a balance has been struck in the real world between the interests of landlords and tenants and therefore changes in rent would not normally be expected at less than the norm interval' (VOA 2006a: part 15).

An alternative term for increasing rents to reflect longer rent review periods is *overage*, which 'means the enhancement of a rent because the period between the reviews specified in the lease is greater than normal' (VOA 2006a: part 15).

Should adjustments be considered appropriate, how they can be calculated has been addressed in practice by two alternative approaches:

- a mathematical basis, using formulae such as Jack Rose's 'Constant Rent' or Philip Bowcock's 'REAT' equations; or
- a rough 'rule-of-thumb' using a percentage.

Rose's Constant Rent 'Uplift' Formula (Bowcock and Rose 1979: 9):

$$k = \frac{(1+r)^n - (1+g)^n}{(1+r)^n - 1} \times \frac{(1+r)^t - 1}{(1+r)^t - (1+g)^t}$$

where:
k = uplift factor, or multiplier to be applied to market rent with normal reviews to compensate for abnormal review pattern
r = equated yield on property
n = number of years to next review in actual lease
t = number of years between normal reviews for that type of property
g = anticipated annual rental growth

Figure 16.6 provides an example of how the constant rent formula could be entered into a spreadsheet so that calculations of the adjustment factor and the adjusted rental can be quickly undertaken. The code to be entered into the relevant cells in Figure 16.6 to carry out calculations is shown in Figure 16.7

Whilst the constant rent formula provides a mathematical solution to rental uplift, in practice a much simpler 'rule of thumb' is often adopted:

- rent review period of subject property *minus* rent review period of comparable
- multiplied by a percentage from 1 to 3 per cent depending on prevailing general annual rental growth rates.

In deciding on the percentage multiplier for the uplift or downlift factor, the following is an approximate guide:

- 1 per cent if average market rental growth rate per annum is 1 to 3 per cent
- 2 per cent if average market rental growth rate per annum is 4 to 6 per cent
- 3 per cent if average market rental growth rate per annum is 7 per cent or more.

	A	B	C
1	Constant Rent 'Uplift' Calculator		
2			
3	Equated Yield =	10	%
4	Annual Rental Growth =	3	% p.a.
5	Years between Rent Reviews on subject property =	7	years
6	Years between Rent Reviews on comparables =	5	years
7			
8	Uplift Factor (multiplier applied to comparable MR) =	1.0251	
9	i.e. Market Rent adjusted by	2.51	%
10			
11	Market Rent on same terms as comparables =	£1,000	
12	Adjusted or 'Constant' rent on actual lease terms =	£1,025	

Figure 16.6   Using a spreadsheet to calculate constant rent adjustment

Constant rent 'uplift' calculator

Equated yield =	10 %
Annual rental growth =	3 % pa
Years between rent reviews on subject property =	7 years
Years between rent reviews on comparables =	5 years

Uplift factor (multiplier applied to comparable MR =   [Formula A]
ie market rent adjusted by [Formula B] %

Market rent on same terms as comparables =	£1,000
Adjusted or 'constant' rent on actual lease terms =	[Formula C]

Code for Formula A
$= ((((1+B3/100)\wedge B5)-((1+B4/100)\wedge B5))/(((1+B3/100)\wedge B5)-1))*((((1+B3/100)\wedge B6)-1)/(((1+B3/100)\wedge B6)-((1+B4/100)\wedge B6)))$

Code for Formula B
$= (B8-1)*100$

Code for Formula C
$= B11*B8$

Figure 16.7   Programming code to be entered into spreadsheet cells to undertake calculations shown in Figure 16.6

Example

The full rental value of a factory is to be assessed. It is let on a 30-year lease with ten-yearly rent reviews. Comparable evidence from similar properties let on ten to fifteen year terms with five-yearly reviews indicates a market rent of £30,000 per annum. Allowing for annual rental growth of 3 per cent and equated yield of 14 per cent estimate the rental value on the subject property to allow uplift for the longer review periods.

Using a basic 'rule of thumb' approach:

Rent review period on subject *minus* rent review period on comparables

$= 10 - 5 = +5$

$\times$ 1% per year (as rental growth quite low at 3% per annum)

$= +5\%$ uplift

Thus market rent should be increased by 5 per cent to reflect the ten-year rent review pattern: £30,000 $\times$ 1.05 = £31,500 per annum

Alternatively, a very mathematical approach can be taken, using Rose's Constant Rent formula where:

$r = 0.14$ (equated yield on property)
$n = 10$ (number of years to next review in actual lease)
$t = 5$ (number of years between normal reviews for that type of property)
$g = 0.03$ (anticipated annual rental growth)

$$k = \frac{(1+0.14)^{10} - (1+0.03)^{10}}{(1+014)^{10} - 1} \times \frac{(1+0.14)^5 - 1}{(1+014)^5 - (1+0.03)^5}$$

$$k = \frac{3.2072 - 1.3439}{3.2702 - 1} \times \frac{1.9254 - 1}{1.9254 - 1.1593}$$

$$k = \frac{2.3633}{2.7073} \times \frac{0.9254}{0.7661} = 0.08729 \times 1.2079$$

$$k = 1.0544$$

Therefore uplift factor is 5.44%: ((1.0544 – 1) $\times$ 100)
£30,000 plus 5.44% = £30,000 $\times$ 1.0544 = £31,632 per annum
Conclusion: from the two approaches, market rent allowing for 10 year rent review SAY = **£31,600 per annum**

Alternatively, suppose the property was to be offered on lease terms with three yearly rent reviews. What reduction for the shorter review periods may be justified?

Using a basic 'rule of thumb' approach:

Rent review period on subject *minus* rent review period on comparables
$= 3 - 5 = -2 \times$ 1% per year (as rental growth quite low at 3% per annum)
$= -2\%$ reduction

Thus market rent should be decreased by 2 per cent to reflect the three-year rent review pattern:

£30,000 × 0.98 = £29,400 per annum

Alternatively, using Rose's Constant Rent formula where:
r = 0.14 (equated yield on property)
n = 3 (number of years to next review in actual lease)
t = 5 (number of years between normal reviews for that type of property)
g = 0.03 (anticipated annual rental growth)

$$k = \frac{(1+0.14)^3 - (1+0.03)^4}{(1+014)^3 - 1} \times \frac{(1+0.14)^5 - 1}{(1+014)^5 - (1+0.03)^5}$$

$$k = \frac{1.4815 - 1.0927}{1.4815 - 1} \times \frac{1.9254 - 1}{1.9254 - 1.1593}$$

$$k = \frac{0.3888}{0.4815} \times \frac{0.9254}{0.7661} = 0.8074 \times 1.2079$$

$$k = 0.975$$

This indicates the reduction factor is 2.25%: ((1 − 0.975) × 100)
£30,000 less 2.25% = £30,000 × 0.975 = £29,250 per annum
Conclusion: from the two approaches, market rent allowing for three-yearly rent review SAY
= **£29,300 per annum**

16.10   Rental equivalent on an annual tenancy

The Rental Equivalent on an Annual Tenancy (REAT) is a formula devised by Philip Bowcock (Bowcock and Rose 1979: 9). It may be used to calculate the rental under an annual tenancy where rent could be reassessed each year and where comparables are let on longer terms and with less frequent rent reviews. The formula is:

$$\frac{i-g}{1 - \left(\frac{1+g}{1+i}\right)^z} \times \text{YP } z \text{ years S/R @ } i\%$$

where:
i = equated yield percentage expressed as a decimal;
g = estimated annual rental growth percentage expressed as a decimal; and
z = number of years between rent reviews on comparable properties.

The REAT formula can be used to adjust comparable evidence obtained on a standard rent review pattern (for example five-yearly) to an equivalent rental on a year-to-year tenancy basis. Alternatively, the 'rule of thumb' approach can be used to find the reduction factor.

Again, there is no universal agreement between valuers whether such a decrease is always correct or justified. In relation to rating valuation where a tenancy from year to year is assumed, the VOA state that 'there is no precedent to suggest that the rent must be reviewed at the end of each year. On the contrary the Lands Tribunal has indicated that the rent may be varied as much on the initiative of the tenant as on that of the landlord' – as per *W. A. Rawlinson & Co Ltd* v. *Pritchard (VO)* (1959) 52 R & IT 182. 'There are also arguments as to why rents from year to year should not be adjusted, or in fact may be higher than those fixed for a term of years ... Furthermore annual tenancies are not normally attractive to landlords because of the high cost of annual rent reviews, the uncertainty of income and the potential loss of value on the capital investment' (Valuation Office Agency 2006a: 38).

Example

A property is occupied on a yearly tenancy. Comparable evidence obtained from properties let for terms of twenty years with five-yearly rent reviews indicates the market rent is £1,000 per sq.m. Using the REAT formula, what rental would be appropriate on the subject property to reflect the nature of the lease? Annual rental growth is estimated at 8 per cent and the equated yield at 16 per cent.

In the formula: $i = 0.16$; $g = 0.08$; $z = 5$; YP single rate 5 years @ 16% = 3.2743

$$REAT = 0.26628 \times 3.2743 = 0.872$$

Therefore comparable rental should be reduced by 12.8 per cent $(1 - 0.872 = 0.128)$ to reflect annual tenancy terms. Thus appropriate rent = £1,000 × 0.872 = £872 per sq.m. per annum
Alternatively, using the 'rule of thumb' approach:

Rent review period on subject *minus* rent review period on comparables

$= 1 - 5 = -4$
× 3% per year (as rental growth high at 8% per annum) $= -12\%$ reduction

Thus market rent should be decreased by 12 per cent to reflect the yearly rent review pattern:

£1,000 × 0.88 = £880 per sq.m. per annum

Conclusion: from the two approaches say adjusted rent = **£875 per sq.m. per annum**

16.11 Problems with leasehold valuations, including double sinking fund error theory and relevance of 'dual rate' calculations

There are several problems with traditional leasehold valuation methods. First, the annual sinking fund is calculated only to replace the historic purchase price. While the same nominal sum of money should be recovered by the leaseholder at the end of the lease, its real value will have declined and will be insufficient to purchase a similar leasehold interest to the one that has just expired. While in theory the use of a sinking fund should allow the purchase of new leasehold interests on expiry of old ones in perpetuity, in practice this is difficult to achieve.

Second, the assumption made when calculating the sinking fund for a dual rate years purchase is that the interest rate and the tax payable will remain the same throughout. In reality, this seldom happens, particularly over longer terms. Should the interest rate fall and/or tax rate increase, there will be a shortfall in the accumulated funds at the end of the lease term.

Third, investing in a sinking fund requires forgoing some of the 'profit rent'. When the leaseholders occupy their own property, this profit is only a notional sum anyway. It may be 'money saved', but there is no actual cash flow from which sinking fund instalments can be deducted. Accordingly, and having consideration to the first two shortcomings in using the sinking fund approach, many leaseholders decide to accept they own a wasting asset that will reduce to nil value. They dispense altogether with setting up and investing in a sinking fund and 'write off' the expenditure completely over the lease term in the same way as allowing for depreciation on fixed assets considered in Section 14.1.

Fourth, when analysing market comparables, how can leasehold transactions be accurately assessed unless the net annual sinking fund and the tax rate used in the valuation are known? Moreover, prices or premiums paid for leases often comprise other payment elements apart from a capitalisation of the profit rent (see Section 16.1). Accurately deriving leasehold all risks yields as would be used in a dual rate tax adjusted basis from market evidence is therefore problematic.

Fifth, there is the effect of taxation in a dual rate years purchase as the unexpired term lengthens. Over longer periods, making deductions for tax has less effect on the discounting factor. This point is illustrated by Isaac and Steley (1999: 116) who prove the effect of tax on the YP number diminishes as the term increases:

YP 4 yrs @ 9% & 3% =   3.039
YP 4 yrs @ 9% & 3% (tax 40%) =   2.048
1 − (2.048/3.039) =   0.326
× 100 =   32.6
Therefore, effect of tax =32.6% reduction in YP
YP 60 yrs @ 9% & 3% =   10.402
YP 60 yrs @ 9% & 3% (tax 40%) =   9.978
1 − (9.978/10.402) =   0.04
× 100 =   4
Thus effect of tax = 4% reduction in YP

The conclusion that can be drawn from the above is that with longer lease terms the effect of tax diminishes and thus it is arguable whether tax-adjusted years purchases need to be used when valuing this type of interest. Indeed, many would argue that the whole dual rate concept is flawed anyway, so why not value leaseholds using a single rate approach, as used with freeholds? The VOA in relation to rating valuation provides a precedent for this approach. In assessing the annual equivalent of premiums, they had long calculated them using dual rate tables at 6 and 3 per cent, with no allowance for tax, regardless of the type of property. However, they replaced this by the use of a single-rate approach for the 1990 rating lists and have retained this method thereafter. Different percentage rates are recommended for each class of property, which are intended to 'represent average national yields' (Valuation Office Agency 2006a: part 12).

Example

Value the leasehold interest in an office property held on ground lease with an unexpired term of 80 years without review. The estimated MR is £100,000 per annum and the ground rent is £2,000 per annum. Assume an ARY of 8 per cent and net asf rate of 2.5 per cent with tax paid at 28 per cent.

Conventional valuation:

Rent receivable =	£	100,000 p.a.
Less rent payable =	£	2,000 p.a.
	Net profit rent = £	98,000 p.a.
× YP 80yrs @ 8% + 2.5% (tax 28%)		11.6834
	Capital value =	£1,144,973
	SAY =	£1,145,000

However, with a long unexpired term, allowing for tax has a negligible effect mathematically. In addition, with such a long period, it is similar to a freehold and so a single rate no tax approach could be used to find the answer more easily. To compensate for not adjusting for tax and annual sinking fund the all risks yield should be increased slightly as, unlike a freehold, it is still a wasting asset despite the length of term and will have a nil value in 80 years' time.

Net profit rent =	£	98,000 p.a.
× YP perp @ 8.55%		11.6959
	Capital value =	£1,146,199
	SAY =	£1,145,000

The sixth and final problem with traditional methods is the 'double sinking fund error' in leasehold term and reversion valuations. The error occurs due to allowing for one sinking fund on the term capital value and another on the reversion. The amount invested by the end of the term will continue to earn interest during the reversion, but this is not included in 'conventional' dual-rate calculations. However, only a single sinking fund for the whole investment is needed anyway, which should be calculated over the term *plus* reversion periods.

To find the single sinking fund the final capital value needs to be known. However, until the sinking fund figure is fixed, the years purchase numbers cannot be calculated and without these the capital value cannot be found! It is a circular argument. The solution is to use an algebraic equation to find the answer.

The other aspect to take into consideration is that the sinking fund instalments over the term plus reversion will accumulate to a capital sum equivalent to the current market value of the lease. The present value of this sum needs to be included in the assessment of the value of the interest. It is discounted at the all risks yield, as is the case with all single-rate appraisals (such as freehold valuations) where future sums receivable are multiplied by the present value formula at the ARY.

Thus if 'x' represents the capital value of the investment, the sinking fund can be found accordingly, expressed in terms of 'x'. For example, suppose a leasehold interest has a profit rent of £10,000 per annum for the next four years rising to £20,000 per annum for the following eleven years. The net annual sinking fund rate is expected to be 3 per cent and tax is to be allowed at 40 per cent.

Without eliminating the double sinking fund error the valuation would be:

Term

Net Profit Rent =	£	10,000 p.a.
× YP for 4 yrs @ 10% + 3%		
& tax @ 40%		2.007
		= £20,065

Reversion

Net Profit Rent =	£	20,000 p.a.
× YP for 11 yrs @ 10% + 3%		
& tax @ 40%		4.345
× PV 4 yrs @10%		0.683
		= £ 59,359

$$\text{Market value (assignment price)} = £ \quad 79,424$$
$$\text{SAY} = £ \quad 79,500$$

The problem with this is that effectively two sinking funds are being set up; one for the first four years and another for the subsequent eleven years. At the end of the term period, the fund will have accumulated to £20,065 (to replace the capital sum up to that point). This sum of money will, however, remain invested for the next eleven years and continue to earn interest. This is not taken into account in the conventional calculations. A second sinking fund then runs for the last eleven years to replace the capital amount of £59,359 attributable to the reversion.

It is the 'missing interest' on the term sinking fund that the double sinking fund adjusted valuation makes allowance for. Using the tax adjustment factors table from *Parry's Tables*, the required calculations become:

Capital value =	£	x
× ASF 15 yrs @ 3%		0.0538
× Gross tax factor @ 40%		1.667
		= 0.0897
	ASF =	0.0897x

Valuation:

Term

Profit rent =	£	10,000 p.a.
less ASF =		0.0897x
		= £10,000 – 0.0897x
	× YP 4 yrs @ 10%	3.17
		= £31,700 – 0.284x

Reversion

Profit rent =	£	20,000
less ASF =		0.0897x
		= £20,000 – 0.0897x
	× YP 11 yrs @ 10%	6.495
	× PV 4 yrs @ 10%	0.683
		= £88,720 – 0.3983x

Capital replaced at end of lease by sinking fund (x)
$\times$ Present value 15 yrs @ ARY of 10% (0.2394) = $\qquad\qquad$ $0.2394x$
$$\text{Capital value} = \pounds120,420 - 0.4429x$$

$x = 120,420 - 0.4429x$
$x + 0.4429x = 120,420$
$x = 120,420/1.4429$
$x = 83,457$

Conclusion is that capital value adjusted for double sinking fund error = say £83,500
To prove this works, insert the figures back into the full valuation:

Annual sinking fund for 15 yrs @ 3% =0.05376658
$\times$ Gross tax factor @ 40% = $\qquad$ 1.666666667
$\qquad\qquad$ Annual sinking fund multiplier = 0.089610967
$\qquad\qquad\qquad\qquad$ $\times$ £83,500 = £ 7,482.51 p.a.

Term
Net profit rent = $\qquad$ £ $\qquad$ 10,000 p.a.
Less annual sinking fund = $\qquad$ £ $\qquad$ 7,482 p.a.
$\qquad\qquad\qquad\qquad$ = £ $\qquad$ 2,518 p.a.
$\qquad\qquad$ $\times$ YP for 4 yrs @ 10% $\qquad$ 3.170
$\qquad\qquad\qquad\qquad\qquad$ = £7,980

Reversion
Net profit rent = $\qquad$ £ $\qquad$ 20,000 p.a.
Less annual sinking fund = $\qquad$ £ $\qquad$ 7,482 p.a.
$\qquad\qquad\qquad\qquad$ = £ $\qquad$ 12,518 p.a.
$\qquad\qquad$ $\times$ YP for 11 yrs @ 10% $\qquad$ 6.495
$\qquad\qquad\qquad$ $\times$ PV 4 yrs @ 10% $\qquad$ 0.683
$\qquad\qquad\qquad\qquad\qquad$ =£ 55,530

Capital replaced at end of lease by
sinking fund = $\qquad$ £ $\qquad$ 83,500
$\times$ present value @ ARY of 10% = $\qquad$ 0.239
$\qquad\qquad\qquad\qquad$ = £ $\qquad$ 19,989
$\qquad\qquad$ Market value (assignment price) =£83,499
$\qquad\qquad\qquad\qquad\qquad$ SAY =£83,500

Eliminating the double sinking fund error has shown that the property was undervalued using the conventional term and reversion approach. The corrected value of £83,500 in this instance is just over 5 per cent higher than the original figure of £79,500.

16.12 'Reverse premiums'

A 'reverse premium' can be calculated in the same way as other premiums. The difference is it is paid in the 'reverse' direction to the normal expectation. For instance, it would normally be expected that a new lease granted at less than market rent will have a potential profit rent and this would enable the tenant to offer a premium to the landlord. The profit rent would have some asset value, which the premium would represent.

However, suppose instead of a profit rent, the lease attracted a 'loss rent', that is the rent payable to the landlord is higher than the current market rent for the property. This is known as an *overrented property* and it can occur when the market has fallen since the date the current rent was agreed (also see Section 16.15 below). In this situation, the 'loss rent' could be capitalised and the premium found from this would be paid in the 'reverse' direction. Thus, the outgoing tenant would need to *pay to* the assignee this sum in order to dispose of the lease, as it has become a liability not an asset.

A landlord could make a reverse premium payment to the tenant at the commencement of a lease. This could simply be a cash incentive to take the lease or could be a contribution towards certain expenses, such as fitting out costs.

16.13 Lease inducements

The motive for granting these can vary widely between transactions, but common reasons are:

- for the landlord to secure a tenant in the face of competition and so achieve a rental income and avoid the problems of vandalism, squatting and lack of maintenance associated with vacant property; or
- the advantage of increased capital value for the landlord or easier funding; or
- cash flow advantages for the tenant; or
- taxation advantages for either party; or
- effect on a company's balance sheet.

There are many types of lease inducements that may be granted, including:

- break clauses: allowing the tenant to terminate the lease early on giving notice at a specified time, such as Year 8 in a fifteen-year lease term;
- capital payments: as a cash incentive or as a contribution towards costs;
- capping or removal of service charges: to reduce the costs burden on the tenant;
- fitting out works: to save the tenant the expense of undertaking the work;
- relocation costs: paying the tenant's costs of moving to the new property;
- rent capping: guaranteeing the rent will not be increased beyond a specified figure during a given number of years, even if market rent levels exceed this amount;
- rent-free periods: no rent paid at the start of a lease for a specified number of months (or even a year or more in depressed market conditions);
- stepped rents: rents are fixed to rise in 'steps' for a number of years so that the tenant can financially plan accordingly without the unexpected 'shock' of a large rental increase – for instance, over a ten-year lease the rent is £40,000 per annum for years 1 and 2, £50,000 per annum for years 3 and 4, £60,000 for years 5 and 6 then market rent from years 7 to 10;
- the take-back of existing premises: to save the tenant any delay and costs in disposing of their existing lease and property.

16.14 Virtual or equivalent rent and valuation analysis of lease inducements

Virtual or equivalent rent is the true annual cost of a property to a tenant. It can also be known as 'sitting rent' or 'net effective rent', which is: 'The rent that would be agreed between the parties for a letting of the premises on the relevant terms and conditions, but without incentives forming part of the transaction' (RICS 2006a: 2).

'Headline rent' is the full rental figure which the lease will be subject to at some point, and is the amount which the landlord will be keen to publicise (thus print as a 'headline' in the press) as it can give the impression the letting is more valuable than it really is. The RICS defines it as: 'The actual contracted periodic rental payment under the lease which becomes payable after all the initial incentives or concessions in the letting have ended. It is sometimes referred to as the "face rent"' (RICS 2006a: 2).

The true rental value, or 'equivalent rent', will be found after the headline sum is adjusted from an analysis of all lease inducements. 'It is the sum of the rent paid and the annual equivalent of any rent free periods, reverse premiums and capital expenditures for the use of the property' (RICS 2006a: 2). It is the equivalent rent which should be used for comparison method purposes, not the headline figure.

RICS Valuation Information Paper No. 8 provides more information, together with extensive calculated examples (RICS 2006a).

A critical aspect of analysis of lease inducements is the write-off period adopted. Devaluation should take into account the length of the benefit accruing to the tenant. However, this period could differ depending on whether seen from the tenant's or the landlord's viewpoint.

The options to choose as the write-off period of an inducement are:

- full length of the lease; or
- to first rent review; or
- to break clause; or
- useful life of incentive; or
- a specified number of years, such as ten; or
- a compromise between the above, where half the incentive is written off over one period, the other half over a different period.

Possible approaches to analysing the effects of inducements are:

- simple averaging (quick and easy but does not take into account time value of money and thus not very accurate); or
- use of years purchase and discount factors; or
- discounted cash flow (more complex but gives a fuller understanding of the effects of benefits on a yearly basis).

Example 1

A newly constructed office property has recently been let in the open market on a new twenty-year lease with upwards-only rent reviews every five years and on full repairing and insuring terms. Its net internal floor area is 1,200 sq.m. The rent agreed is £250,000 per annum, but the first year is rent-free and the landlord has also given the tenant a cash payment of £100,000 for the costs of fitting out the property. Analyse this transaction to deduce the equivalent rent for comparable purposes of such a property without the agreed rent-free and cash incentives and assuming freehold all risk yield is 8 per cent.

First, the 'write-off period' needs to be decided. The landlord will view the lease inducements as being given once over a twenty-year period, so would favour the full period of the lease. Then one year rent free is followed by nineteen years of rent, starting after one year. If the rent free was not given, then rent would have been paid for twenty years starting now. From the tenant's view though, the incentives are only effective over the first five years of the lease as once the rent review is reached the full market rent is payable thereafter. A compromise period could be adopted of say ten years, being the first half of the lease. Based on this compromise the calculations become:

Using the YP approach

£250,000 (headline rent) × YP 9 yrs def 1 yr/YP 10 yrs
£250,000 × (YP 9 yrs @ 8% × PV 1yr @ 8%)/YP 10 yrs @ 8%
= £250,000 × 6.247 × 0.926/6.710 = £215,526 per annum

Then adjust this to take account of the landlord's reverse premium of £100,000, by finding its annual equivalent:

£100,000/YP 10 yrs @ 8% = £14,903 per annum

Thus 'equivalent' or net effective rent

=	£	215,526 – £14,903
=	£	200,623
SAY =	£	**200,600** per annum
For 1,200 sq.m. =	£	167.17 per sq.m. per annum

To illustrate that this equivalent rent is correct, a discounted cash flow can be used without incorporating future rental growth as follows:

Yearly Discounted Cash Flow Analysis: *actual deal*

Year	Cash Out	Cash In	Net Cash	PV @ 8%	DCF
0	£100,000		–£100,000	1.00000	–£100,000
1		£0	£0	0.92593	£0
2		£250,000	£250,000	0.85734	£214,335
3		£250,000	£250,000	0.79383	£198,458
4		£250,000	£250,000	0.73503	£183,757
5		£250,000	£250,000	0.68058	£170,146
6		£250,000	£250,000	0.63017	£157,542
7		£250,000	£250,000	0.58349	£145,873
8		£250,000	£250,000	0.54027	£135,067
9		£250,000	£250,000	0.50025	£125,062
10		£250,000	£250,000	0.46319	£115,798
				NPV =	£1,346,039

Inserting the equivalent rent shows the same net present value if no incentives are incorporated:

Yearly Discounted Cash Flow Analysis: *without incentives*

Year	Cash Out	Cash In	Net Cash	PV @ 8%	DCF
0		£0	£0	1.00000	£0
1		£200,600	£200,600	0.92593	£185,741
2		£200,600	£200,600	0.85734	£171,982
3		£200,600	£200,600	0.79383	£159,243
4		£200,600	£200,600	0.73503	£147,447
5		£200,600	£200,600	0.68058	£136,525
6		£200,600	£200,600	0.63017	£126,412
7		£200,600	£200,600	0.58349	£117,048
8		£200,600	£200,600	0.54027	£108,378
9		£200,600	£200,600	0.50025	£100,350
10		£200,600	£200,600	0.46319	£92,917
				NPV =	£1,346,042

Example 2

A rent-free period of one year has been granted on a five-year lease. The rent payable for the other four years is £200,000 per annum. What is the 'true' or 'equivalent' annual rent being paid?

Simple averaging approach analysis

    Paying 4 × £200,000 = £800,000
    Spread over 5 years = £800,000/5 = £160,000 **per annum**

Use of years purchase and discount factors, assuming ARY of 10 per cent

'Actual value' = YP 4 yrs @ 10% × PV 1 yr @ 10% = 3.17 × 0.909 = 2.882
'Full' value = YP 5 yrs @ 10% = 3.791
So 'true' rental = actual value × £200,000/full value
that is 2.882 × £200,000/3.791 = £152,000 *per annum*

Discounted cash flow approach, assuming equated yield of 13 per cent

Year	Actual Income	PV @ 13%	DCF	Market Rent	PV @ 13%	DCF
1	£0	0.885	£0	£200,000	0.885	£176,990
2	£200,000	0.783	£156,630	£200,000	0.783	£156,630
3	£200,000	0.693	£138,610	£200,000	0.693	£138,610
4	£200,000	0.613	£122,660	£200,000	0.613	£122,660
5	£200,000	0.543	£108,550	£200,000	0.543	£108,550
		GPV =	£526,460		GPV =	£703,450

Actual value = £526,460/£703,450 of expected value = 74.83%
Thus 74.83% of MR = £149,680 *per annum* is 'true' rent.

Each of the three methods suggests a different equivalent rent, but the discounted cash flow is both the most informative and accurate, showing clearly the difference in annual cash flow. Should this letting be analysed as a comparable it is £149,680 per annum, not the headline rent of £200,000, which should be divided by the floor area to give the comparison figure per square metre.

16.15 The problem of overrented property and its valuation

Overrenting occurs when the rent payable under a lease exceeds the full open market rent value, so that the occupier is suffering a 'negative profit rent' or 'loss rent'. It is most likely to occur in a falling market, so that rent agreed at the start of the lease or at the last rent review was the market rent at that time, but values have decreased since. When a lease has an upwards or downwards rent review provision, the rent will decrease at the next review or when the lease ends, whichever occurs soonest. However, with an upwards-only review clause, the rent will stay the same or rise at each review and so if at the review date the market rent is more than the rent already being paid, the passing rent will not change. The tenant will pay more than the property is worth, possibly until the end of the lease if market levels beforehand never rise above the rent payable.

When valuing the freehold interest in an overrented property, the possible methods that can be used are:

- conventional term and reversion, with the term yield higher than the reversionary market rent all risks yield, to reflect the increased risk or insecurity of the term rent; or
- hardcore (which many practitioners consider the most appropriate method) with the higher-risk top slice being received in the term period rather than the reversion; or
- discounted cash flow, using projected rental growth (which will show when the market rent will equal or exceed the rent payable) and an equated yield.

In all cases correct selection of yield is critical and yet probably the most difficult aspect. Just how much riskier is the rental income above the market rent?

The leasehold capital value calculated for an overrented property would represent the 'reverse premium' that the leaseholder would at least need to pay to a prospective assignee in order to 'sell' or assign the lease. As the lease is a liability, causing a negative annual cash flow, nobody will be interested in taking an assignment without a capital payment incentive. Even then, it may be extremely difficult to arrange an assignment.

Example 1

A twenty-year FRI lease of a shop is to be assigned. The lease provides for upwards-only rent reviews every five years, and the lease has an unexpired term of eight years. The current rent payable is £100,000 per annum, but comparables indicate that the market rent is only £80,000 per annum. Nearby properties let at their market rent show a leasehold all risks yield of 9 per cent. A prospective assignee pays tax at 21 per cent. What reverse premium would need to be offered on assignment using a dual rate approach and a net 3 per cent sinking fund rate?

Rent receivable =	£	80,000 p.a.
Less Rent payable =	£	100,000 p.a.
	Net loss rent =	(£ 20,000) p.a.
× YP 8 yrs @ 9%+3% (tax 21%)		4.3039
	Capital value =	(£86,078)
Therefore SAY reverse premium payable =		**£86,100**

This is the sum the assignor should consider paying to the assignee as an inducement to take the lease. It ignores the value of fixtures and fittings, goodwill and/or 'key money', which are assets that could potentially be set off against the liability that the unexpired lease represents.

A more accurate and informative analysis can be undertaken using discounted cash flow. This approach is considered in the second example.

Example 2

A tenant is trying to assign its leasehold interest in a workshop industrial property. It is held on a twenty-year term granted six years ago with an upwards-only rent review after ten years. The rent payable is £40,000 per annum but the current estimated market rent is only £30,000 per annum. The tenant accepts it will need to offer an assignee an incentive to take the property. What reverse premium should it expect to pay assuming an all risks leasehold yield at a market rent of 12 per cent, a 3 per cent net sinking fund rate and 30 per cent tax? Also, what will be the value of the freehold assuming a 10 per cent all risks yield at market rent level?

Valuation of leasehold interest

Rent receivable =	£	30,000 p.a.
Less rent payable =	£	40,000 p.a.
Net loss rent =	−£	10,000 p.a.
× YP 14yrs @ 12% + 3% (tax 30%)		4.9114

$$\text{Negative capital value (reverse premium)} = £49,114$$
$$\text{SAY} = £49,200$$

(ignoring any offset value attributable to fixtures & fittings, goodwill and/or 'key money').

Note: if rent review was on upwards or downwards pattern, YP would only be for four years, that is until the next rent review when rent payable and rent receivable would become the same.

It is informative to consider the projected rent approach to show when the market rent will equal or exceed the rent payable due to future growth. The time taken for the market rent to rise above the current rent is obtained by use of the Amount of £1 formula or table. The ratio of present rent to current market rent is 1.333:1 (£40,000/£30,000 = 1.333). Therefore, the Amount of £1 at the estimated annual rental growth rate must equal or exceed 1.333 before the current rent will at least equal £40,000. Assuming rental growth at 3 per cent per annum is expected, it will take ten years as shown in the table.

Year	Amount of £1 @ 3%	Estimated market rent
1	1.0300	£30,900
2	1.0609	£31,827
3	1.0927	£32,782
4	1.1255	£33,765
5	1.1593	£34,778
6	1.1941	£35,822
7	1.2299	£36,896
8	1.2668	£38,003
9	1.3048	£39,143
10	1.3439	£40,317

Year	Amount of £1 @ 3%	Estimated market rent
11	1.3842	£41,527
12	1.4258	£42,773
13	1.4685	£44,056
14	1.5126	£45,378

As the lease being valued only has one rent review, in four years' time, before the lease ends in fourteen years, the ten-year period does not affect the years purchase used in the valuation. A prospective assignee of the leasehold interest would, however, have an expectation that the 'loss rent' would be suffered for the next ten years, but the remaining four years of the lease would provide a possible profit rent. The estimated cash flows would be:

Year	Rent receivable (ERV)	Rent payable (profit or loss rent)	Net cash flow
1	£30,000	£40,000	–£10,000
2	£30,900	£40,000	–£9,100
3	£31,827	£40,000	–£8,173
4	£32,782	£40,000	–£7,218
5	£33,765	£40,000	–£6,235
6	£34,778	£40,000	–£5,222
7	£35,822	£40,000	–£4,178
8	£36,896	£40,000	–£3,104
9	£38,003	£40,000	–£1,997
10	£39,143	£40,000	–£857
11	£40,317	£40,000	£317
12	£41,527	£40,000	£1,527
13	£42,773	£40,000	£2,773
14	£44,056	£40,000	£4,056
		Total = – £47,411	

In practice, the assignee could invest the reverse premium in a sinking fund and use the invested capital and accumulated interest each year to pay the projected shortfall between the rent receivable and rent payable. In this way, no loss will be incurred through taking on the lease of the overrented property. To find the minimum reverse premium needed to be receivable to achieve a break-even position, the net cash flow above needs to be discounted at the net annual sinking fund rate. Using a net rate of 2.5 per cent this discounted cash flow will be as follows:

Year	Net cash flow	PV @ 2.5%	DCF
1	−£10,000	0.975609756	−£9,756
2	−£9,100	0.951814396	−£8,662
3	−£8,173	0.928599411	−£7,589
4	−£7,218	0.905950645	−£6,539
5	−£6,235	0.883854288	−£5,511
6	−£5,222	0.862296866	−£4,503
7	−£4,178	0.841265235	−£3,515
8	−£3,104	0.820746571	−£2,547
9	−£1,997	0.800728362	−£1,599
10	−£857	0.781198402	−£669
11	£317	0.762144782	£242
12	£1,527	0.743555885	£1,135
13	£2,773	0.725420376	£2,011
14	£4,056	0.707727196	£2,871
			GPV −£44,631

Discounting the projected loss rent or profit rent figures at a lower sinking fund yield rate shows the expected overall loss in present value terms to be £44,631. This represents the minimum reverse premium that would be sought by the assignee. Effectively if this sum was invested in a sinking fund in year 0 and compound interest accumulates at 2.5 per cent net per annum then the losses incurred from paying more rent than the market level will be made good by withdrawing money from the fund. Therefore, the suggested reverse premium found from a conventional investment method calculation is shown to be an appropriate amount to cover all the potential losses expected over the remaining term of the lease.

When valuing the freehold interest in this same property, a number of approaches could be adopted, as suggested above.

Valuation of freehold interest using term and reversion
Term

Net income =	£	40,000 p.a.	
× YP for 14 yrs @ 13%		6.302	
		= £	252,100

Reversion to market rent

Net income =	£	30,000 p.a.	
× YP perp def 14 yrs @ 10%		2.633	
		= £	78,999
	Market value =	£	331,099
	SAY =	£	**331,000**

Note: term yield increased by 3 per cent to reflect additional risk associated with tenant paying a rent 33 per cent higher than current market level. There is a clear subjective element to this choice of yield as to whether it adequately reflects the additional risk involved of the tenant remaining in business and being able to continue maintaining rental payments substantially above current market rent level. Should the tenant default, particularly through becoming insolvent, the landlord will be left with an empty property that may be difficult to relet, even at the market rent, due to depressed market conditions, may require repair and will need continued maintenance and insurance even if empty.

By calculating the internal rate of return, the nominal equivalent yield at the above valuation is found to be 11.42 per cent, which is 1.42 per cent higher than the all risks yield (and is a true equivalent yield of 12.28 per cent quarterly in advance):

Term

Net income =	£	40,000 p.a.
× YP for 14 yrs @ 11.42%		6.830
		= £ 273,188

Reversion to market rent

Net income =	£	30,000 p.a.
× YP perp def 14 yrs @ 11.42%		1.927
		= £ 57,806
	Market value =	£ 330,994
	SAY =	£ 331,000

Were the property not overrented and was let at its current market rent, the market value would only by £300,000 (£30,000 p.a. × YP perp @ 10%). The calculated value is thus 10.33 per cent higher than this. Whether a potential investor purchaser considers this still too high a price given the additional risk or a reasonable figure is an individual assessment.

Valuation of freehold using layer approach

There is an argument for valuing the hardcore at the all risks yield since this is the market rent and the implied growth of this yield will apply to this part of the rental income. However, 1 per cent has been added to reflect the greater uncertainty of income associated with this letting and a further 4 per cent added to compensate for the even higher risk represented by the top slice 'overage'. Again, the choice of top slice yield is a matter of subjective judgement, based on returns on other forms of investment considered equally risky.

Hardcore

Net income	£	30,000 p.a.
× YP perp @11%		9.091
		= £ 272,727

Top slice

Net income	£	10,000 p.a.
× YP 14 yrs @ 15%		5.724
		= £ 57,245
	Market value =	£ 329,972
	SAY =	£ 330,000

This produces a similar valuation outcome to the term and reversion approach. A discounted cash flow analysis will also be informative.

Valuation of freehold interest using DCF:
- expected average annual rental growth rate from now in perpetuity based on market evidence = 3 per cent
- equated yield (annually in arrear) = 13 per cent
- true equated yield (quarterly in advance) = 14.13 per cent

The annually in arrear equated yield could be justified as being a risk free rate of 5 per cent plus 2 per cent property market risk plus 6 per cent lease covenant additional risk from the building being overrented or simply by the addition of the all risks yield and the estimated annual growth rate (10 per cent + 3 per cent). The resultant valuation taken over perpetuity rather than a holding period will be:

Years	Income	PV factor	DCF
1 to 14	£40,000	6.302	£252,100
15 to 19	£45,378	0.635	£28,837
20 to 24	£52,605	0.345	£18,144
25 to perp	£60,984	0.532	£32,444
		GPV =	£331,525
		SAY =	£331,000

Conclusion: all three approaches to valuing the freehold produce similar estimations of current market value and thus the figure can be adopted with a reasonable level of confidence.

Progress check questions

- Lease premiums can include payments for what other elements in addition to that paid for the rent being below the market level?
- What is the approach usually adopted to calculate a lease capital value or premium?
- What four valuations are undertaken for a lease surrender and renewal proposal?
- For what reasons may a freeholder or owner of a long leasehold consider a sale and leaseback transaction?
- When and why may marriage value arise and how is it calculated?
- In what two ways may a divorce value occur?
- How might a restriction in the lease user clause or presence of tenant's improvements affect the rental valuation of a property?
- Who devised the constant rent and the rental equivalent on an annual tenancy formula and when may it be appropriate to use either of them?

- What problems exist with allowing for and using an annual sinking fund in leasehold valuations?
- Why is it difficult to analyse accurately leasehold market transactions for comparison purposes?
- What arguments can be put forward for valuing leasehold interests using a single rate no tax years purchase approach, as used for freeholds?
- What is the 'double sinking fund error' in leasehold term and reversion valuations, how can it be solved and what effect is it likely to have on value?
- When and why are 'reverse premiums' paid?
- What is an 'equivalent' or 'net effective' rent?
- Why may inducements be offered as part of a lease deal and how can their effect on value be assessed?
- What is an overrented property, how does it become one and how can it be valued?

Chapter summary

There are many occasions when a valuation is required in connection with a landlord and tenant matter. These applied valuations will not only need to consider the appropriate valuation methodology to use, but also take into account many legal, management and business finance issues.

A premium or other capital payment may be required when a lease is being assigned, a lease is granted at below the market rent, the lease terms are changed or as part of a surrender and renewal arrangement. Rental valuations are needed on the grant or renewal of a lease or at rent review. Adjustments to rental values may be required where market evidence and the subject property are on dissimilar rent review periods. Capital valuations are undertaken when a freehold property is sold and leased back or where the merging or division of property interests or physical assets may release latent value.

Premiums can be difficult to assess as they may incorporate payment for other elements in addition to a potential profit rent. Analysing leasehold comparable evidence can be problematic as can the use of a traditional dual rate tax adjusted years purchase approach to valuing leasehold interests. This raises the question of whether it is simpler and more practical to treat the valuation of leaseholds in a similar way to that of freeholds, but adopting higher all risks yields to reflect their inherent disadvantages.

There are many types of lease inducement that can be included in an agreement between landlord and tenant. Their presence affects the 'equivalent' or 'net effective rent', which needs to be assessed for comparable purposes. There are a number of differing views and approaches over how this is to be done. There are times when other difficult valuation issues need to be addressed associated with property let at more than its current market rent.

Further reading

Askham, P. (1991) 'Mainly for students: constant rent', *Estates Gazette*, 9139 (5 Oct.): 145–6; repr. in P. Askham and L. Blake (eds), *The Best of Mainly for Students* (London: Estates Gazette, 1993), 354–63.

Banfield, A. (2005) *Stapleton's Real Estate Management Practice*, 4th edn, London: Estates Gazette.

Bannister, E. (2007) *Commercial Leases 2008: A Surveyor's Guide*, Coventry: RICS Books.

Baum, A. (1993) 'Professional briefing: dealing with merger, marriage value', *CSW The Property Week* (15 Apr.): 48.

—— (2002) *Commercial Real Estate Investment*, London: Estates Gazette.

—— and Crosby, N. (2007) *Property Investment Appraisal*, 3rd edn, London: Wiley Blackwell.

—— Nunnington, N. and Mackmin, D. (2006) *The Income Approach to Property Valuation*, 5th edn, London: Estates Gazette.

—— Sams, G., Stevens, D., Ellis, J. and Hampson, C. (2007) *Statutory Valuations*, 4th edn, London: Estates Gazette.

Beaumont, B. (2004) *Arbitration and Rent Review*, 3rd edn, London: Estates Gazette.

Bernstein, R. and Reynolds, K. (1995) *Essentials of Rent Review*, London: Sweet & Maxwell.

Butler, D. (1995) *Applied Valuation*, 2nd edn, London: Macmillan.

—— and Richmond. D. (1990) *Advanced Valuation*, London: Macmillan.

Davidson, A. and Darlow, C. (1993) *Boston Gilmore Rental Equivalent Tables and Lease Incentives*, London: Estates Gazette.

Estates Gazette (1994) 'Mainly for students: valuing over-rented property', 9421 (28 May): 120–2; repr. in P. Askham and L. Blake (eds), *The Best of Mainly for Students* (London: Estates Gazette, 1999), vol. 2, pp. 295–302.

—— (1995) 'Mainly for students: rental adjustments for rating use', 9517 (29 Apr.): 121–2; repr. in P. Askham and L. Blake (eds), *The Best of Mainly for Students* (London: Estates Gazette, 1999), vol. 2, pp. 316–21.

—— (1995) 'Mainly for students: rent-free periods and valuation' (13 May): 143–4; repr. in P. Askham and L. Blake (eds), *The Best of Mainly for Students* (London: Estates Gazette, 1999), vol. 2, pp. 310–15.

—— (1996) 'Mainly for students: valuation and compensation', 9637 (14 Sep.): 135–8; repr. in P.Askham and L. Blake (eds), *The Best of Mainly for Students* (London: Estates Gazette, 1999), vol. 2, pp. 345–50.

—— (1999) 'Mainly for students: the price of surrender', 9905 (6 Feb.): 152–3; repr. in L. Blake and A. Imber (eds), *The Best of Mainly for Students* (London: Estates Gazette, 2004), vol. 3, pp. 444–9.

Hayward, R. (ed.) (2008) *Valuation: Principles into Practice*, 6th edn, London: Estates Gazette.

Imber, A. (2001) 'Mainly for students: tenants, reviews, renewals and rent', *Estates Gazette*, 0142 (20 Oct.): 128–31; repr. in L. Blake and A. Imber (eds), *The Best of Mainly for Students* (London: Estates Gazette, 2004), vol. 3, pp. 213–22.

—— (2002) 'Mainly for students: seeing the truth beyond the horizon', *Estates Gazette*, 0248 (30 Nov.): 128–30; repr. in L. Blake and A. Imber (eds), *The Best of Mainly for Students* (London: Estates Gazette, 2004), vol. 3, pp. 473–9.

Johnson, T., Davies, K.I. and Shapiro, E. (2000) *Modem Methods of Valuation of Land, Houses and Buildings*, 9th edn, London: Estates Gazette.

Kingsmill, R. (2005) 'The journey from owner to occupier', *Estates Gazette*, 0531 (6 Aug.): 74–6.

Royal Institution of Chartered Surveyors (1997) *Commercial Investment Property: Valuation Methods*, Coventry: RICS Books.

—— (2007) *RICS Valuation Standards: 'The Red Book'*, 6th edn, Coventry: RICS Books.

Valuation Office Agency (2006) *Rating Manual*, vol. 5, *Rating Valuation Practice, All Classes of Hereditament*. Online: <http://www.voa.gov.uk/instructions/chapters/rating_manual/vol5/frame.htm> (accessed 25 June 2007).

Bibliography

Al-Khatib, L. (2006) 'Changing valuations', *RICS Residential Property Journal*, 9 (Sep.): 18.

Argus Software (2008a) *Development and Construction*. Online: <http://www.argussoftware.com/uk/solutions/development.aspx> (accessed 6 Nov. 2008).

—— (2008b) *Valuation and Appraisal*. Online: <http://www.argussoftware.com/uk/solutions/valuation.aspx > (accessed 6 Nov. 2008).

Armatys, J. (2008) 'Mainly for students: how to avoid zero points', *Estates Gazette*, 0827 (12 Jul.): 112–13.

Askham, P. (1989a) 'Mainly for students: report writing', *Estates Gazette*, 8908 (4 Mar.): 87–8; repr. in P. Askham and L. Blake (eds), *The Best of Mainly for Students* (London: Estates Gazette, 1993), 154–60.

—— (1989b) 'Mainly for students: insurance valuations', *Estates Gazette*, 8921 (27 May): 87–8; repr. in P. Askham and L. Blake (eds), *The Best of Mainly for Students* (London: Estates Gazette, 1993), 302–6.

—— (1989c) 'Mainly for students: depreciation in commercial buildings', *Estates Gazette*, 8933 (19 Aug.), 52–3; repr. in P. Askham and L. Blake (eds), *The Best of Mainly for Students* (London: Estates Gazette, 1993), 337–45.

—— (1990a) 'Mainly for students: conventional and contemporary methods of investment valuation', *Estates Gazette*, 9011 (17 Mar.): 84–6; repr. in P. Askham and L. Blake (eds), *The Best of Mainly for Students* (London: Estates Gazette, 1993), 346–53.

—— (1990b) 'Mainly for students: conventional and contemporary methods of investment valuation (2)', *Estates Gazette*, 9023 (9 Jun.): 76–8.

—— (1990c) 'Mainly for students: turnover rents', *Estates Gazette*, 9035 (1 Sep.): 77; repr. in P. Askham and L. Blake (eds), *The Best of Mainly for Students* (London: Estates Gazette, 1993), 94–100.

—— (1991a) 'Mainly for students: service charges', *Estates Gazette*, 9127 (13 Jul.): 131–2; repr. in P. Askham and L. Blake (eds), *The Best of Mainly for Students* (London: Estates Gazette, 1993), 101–7.

—— (1991b) 'Mainly for students: constant rent', *Estates Gazette*, 9139 (5 Oct.): 145–6; repr. in P. Askham and L. Blake (eds), *The Best of Mainly for Students* (London: Estates Gazette, 1993), 354–63.

—— (1992a) 'Mainly for students: valuation tables', *Estates Gazette*, 9205 (8 Feb.): 174–6; repr. in P. Askham and L. Blake (eds), *The Best of Mainly for Students* (London: Estates Gazette, 1993), 364–72.

—— (1992b) 'Mainly for students: public house valuation', *Estates Gazette*, 9217 (2 May): 99–100; repr. in P. Askham and L. Blake (eds), *The Best of Mainly for Students* (London: Estates Gazette, 1993), 326–32.

—— (1993a) 'Mainly for students: government stock', *Estates Gazette*, 9315 (17 Apr.): 139–40; repr. in P. Askham and L. Blake (eds), *The Best of Mainly for Students* (London: Estates Gazette, 1999), vol. 2, pp. 103–8.

—— (1993b) 'Mainly for students: yields and rates of interest', *Estates Gazette* (4 Sep.): 117–18; repr. in P. Askham and L. Blake (eds), *The Best of Mainly for Students* (London: Estates Gazette, 1999), vol. 2, pp. 109–16.

—— (1993c) 'Mainly for students: the valuation of agricultural property', *Estates Gazette* (2 and 30 Oct.): 121–2 and 107–8; repr. in P. Askham and L. Blake (eds), *The Best of Mainly for Students* (London: Estates Gazette, 1999), vol. 2, pp. 281–94.

—— (1994) 'Mainly for students: valuing over-rented property', *Estates Gazette*, 9421 (28 May): 120–2; repr. in P. Askham and L. Blake (eds) (1999) *The Best of Mainly for Students* (London: Estates Gazette, 1999), vol. 2, pp. 295–302.

—— (1995a) 'Mainly for students: rental adjustments for rating use', *Estates Gazette*, 9517 (29 Apr.): 121–2.

—— (1995b) 'Mainly for students: rent-free periods and valuation', *Estates Gazette*, 9519 (13 May): 143–4; repr. in P. Askham and L. Blake (eds), *The Best of Mainly for Students* (London: Estates Gazette, 1999), vol. 2, pp. 310–15.

—— (1996) 'Mainly for students: valuation and compensation', *Estates Gazette* (14 Sep.): 135–8; repr. in P. Askham and L. Blake (eds), *The Best of Mainly for Students* (London: Estates Gazette, 1999), vol. 2, pp. 343–50.

—— (1997) 'Mainly for students: examine your technique', *Estates Gazette*, 9720 (17 May): 139–42; repr. in P. Askham and L. Blake (eds), *The Best of Mainly for Students* (London: Estates Gazette, 1999), vol. 2, pp. 351–7.

—— (2003) *Valuation: Special Properties and Purposes*, London: Estates Gazette.

—— (2008) 'Mainly for students: it doesn't have to be a drama', *Estates Gazette*, 0829 (26 Jul.): 86–7.

—— and Blake, L. (eds) (1993) *The Best of Mainly for Students,* London: Estates Gazette.

—— and —— (1999) *The Best of Mainly for Students*, vol. 2, London: Estates Gazette.

Balchin, P., Kieve, J.L. and Bull, G.H. (1995) *Urban Land Economics and Public Policy,* 5th edn, Basingstoke: Palgrave Macmillan.

Ball, M., Lizieri, C. and MacGregor, B.D. (1998) *The Economics of Commercial Property Markets,* London: Routledge.

Banfield, A. (2005) *Stapleton's Real Estate Management Practice,* 4th edn, London: Estates Gazette.

Bannister, E. (2007) *Commercial Leases 2008: A Surveyor's Guide,* Coventry: RICS Books.

Baum, A. (1991) *Property Investment Depreciation and Obsolescence*, London: Routledge.

—— (1993) 'Professional briefing: dealing with merger, marriage value', *CSW: The Property Week* (15 Apr.): 48.

—— (2002) *Commercial Real Estate Investment,* London: Estates Gazette.

—— and Crosby, N. (2007) *Property Investment Appraisal*, 3rd edn, London: Wiley Blackwell.

—— Nunnington, N. and Mackmin, D. (2006) *The Income Approach to Property Valuation*, 5th edn, London: Estates Gazette.

—— Sams, G., Stevens, D., Ellis, J. and Hampson, C. (2007) *Statutory Valuations,* 4th edn, London: Estates Gazette.

BCIS Wessex (2008) *Comprehensive Building Price Book, Major and Minor Works 2009*, 26th edn, London: BCIS Wessex.

Beaumont, B. (2004) *Arbitration and Rent Review,* 3rd edn, London: Estates Gazette.

Bennett, E. and Rigby, S. (1993) 'A closer look at the turnover lease', *CSW: The Property Week* (15 Jul.): 57.

Berney, H.M. (1981a) 'Goodwill', *Estates Gazette*, 258 (20 Jun.): 1147–50.

—— (1981b) 'Goodwill II', *Estates Gazette*, 258 (27 Jun.): 1255–8.

Bernstein, R. and Reynolds, K. (1995) *Essentials of Rent Review*, London: Sweet & Maxwell.

Blake, L. and Imber, A. (eds) (2004) *The Best of Mainly for Students*, vol. 3, London: Estates Gazette.

Bond, P. and Brown, P. (2006) *Rating Valuation: Principles and Practice*, 2nd edn, London: Estates Gazette.

Bornand, D. (1988) 'Mainly for students: internal rate of return and external rates', *Estates Gazette*, 843 (29 Oct.): 88; repr. in P. Askham and L. Blake (eds), *The Best of Mainly for Students* (London: Estates Gazette, 1993), 111–15.

Bowcock, P. (1978) *Property Valuation Tables*, London: Macmillan.

—— and Bayfield, N. (2000) *Excel for Surveyors*, London: Estates Gazette.

—— and —— (2004) *Advanced Excel for Surveyors*, London: Estates Gazette.

—— and Rose, J.J. (1979) *Valuing with the Pocket Calculator*, Oxford: Technical Press.

Boyd, H. and Walker, K. (2003) 'Mainly for students: to build or not to build?', *Estates Gazette*, 0344 (3 Nov.): 164–5.

—— and —— (2004) 'Mainly for students: to develop or not to develop?', *Estates Gazette*, 0406 (7 Feb.): 140–1.

Brett, M. (1998) *Property and Money*, 2nd edn, London: Estates Gazette.

—— (2002) *Valuation Standards for the Global Market*, London: RICS.

—— (2004) *Property Under IFRS: A Guide to the Effects of the New International Financial Reporting Standards*, London: RICS.

Britton, W., Conellan, O.P. and Crofts, M.K. (1991) *The Cost Approach to Valuation*, London: RICS and Kingston Polytechnic.

Brown, G.R. and Matysiak, G. (1999) *Real Estate Investment: A Capital Market Approach*, London: Financial Times/Prentice Hall.

Brown, P. (1992) 'Turnover rents', *IRRV Valuation Journal* (Jul.): 164–5.

Building Cost Information Service (2006) *Life Expectancy of Building Components*, London: BCIS.

—— (2008a) *Guide to House Rebuilding Costs 2008*, London: BCIS/RICS.

—— (2008b) *Independent Cost Information for the Built Environment*, London: BCIS/RICS. Online: <http://www.bcis.co.uk/> (accessed 2 Nov. 2008).

Butler, D. (1995) *Applied Valuation*, 2nd edn, London: Macmillan.

—— and Richmond. D. (1990) *Advanced Valuation*, London: Macmillan.

Cadman, D., Austin-Crowe. L., Topping, R. and Avis, M.R. (1995) *Property Development*, 4th edn, London: Spon Press.

Carr, W.D. (1994) 'What do machinery and equipment valuers do exactly?', *CSM* (Jun.): 42–3.

Cavanagh, E., Attabiyeh, M. and Batham, M. (2005) 'Out of town retail and factory outlets', *Estates Gazette*, 0534 (27 Aug.): 79–84.

Chapman, D.H. (1973) *Technical Terms for Property People*, London: Estates Gazette.

Chartered Surveyor Weekly (1985) 'Research ranks retail centres' (25 Apr.): 263–5.

Cherry, A. (2009) *A Valuer's Guide to the RICS Red Book 2009*, Coventry: RICS Books.

Chidell, R.A. (2007) 'A capital idea', *RICS Business*, 2: 24–5.

Christaller, W. (1933) *Die Zentralen Orte in Süddeutschland*, Jena, Germany: Gustav Fischer.

Cockcroft, L. and Shirley, A. (2005) 'Rural and agricultural focus', *Estates Gazette*, 0522 (4 Jun.): 89–95.

Colborne, A. (1989) 'The profits method', *Estates Gazette*, 8936 (9 Sep.): 24–6.

College of Estate Management (1995) Development Appraisal 3 (GP/P&D/QS), unpublished course study notes, Diploma in Surveying, Reading: College of Estate Management.

—— (2004) Valuation and Investment Level 2, unpublished course study notes, BSc External Degree, Reading: College of Estate Management.

—— (2006) Applied Valuation: Level 3, unpublished course study notes, BSc External Degree, Reading: College of Estate Management.

Connellan, O. and James, H. (1996) *Estimated Realisation Price (ERP) by Neural Networks*, University of Glamorgan, University of Portsmouth and RICS. Online: . <http://www.rics.org/Property/Residentialproperty/Residentialpropertyvaluationandappraisal/estimated_realisation_price_19960101.htm> (accessed 29 Sep. 2007).

Cooke, H. and Wiggins, K. (eds) (1995) Property management, unpublished unit study notes for postal courses, Diploma in Surveying: Measurement and Valuation, Reading: College of Estate Management.

Coulson, J., Carr, C.T., Hutchinson, L. and Eagle, D. (eds) (1975) *The Oxford Illustrated Dictionary*, 2nd edn, Oxford: Oxford University Press.

Creamer, M. (1999) 'Valuations: make them quarterly in advance', *CSM* (Nov./Dec.): 26.

Crosby, N. (1991) 'Valuing commercial investment property', *CSM* (Oct.): 9–12.

—— (2006) 'Obsolescence to depreciation: the impact on appraisal models', unpublished paper presented at RICS International Valuation Conference, London, 15 Nov.

—— and French, N. (1996) 'The effect of metrication on arithmetic zoning of shops', *CSM* (Jan.): 46–7.

—— and Matysiak, G. (2002) 'Seek carefully and ye shall find', *Estates Gazette*, 0234 (24 Aug.): 76–7.

Cunningham, D. and Kivlehan, N.P. (2005) 'Automotive property focus', *Estates Gazette*, 0527 (9 Jul.): 65–71.

Curwin, J. and Slater, R. (1996) *Quantitative Methods for Business Decisions*, 4th edn, London: International Thomson Business Press.

Cushman Wakefield Healey & Baker (2008) *Marketbeat UK June 2008*, London: Cushman & Wakefield.

Darlow, C. (ed.) (1983) *Valuation and Investment Appraisal*, London: Estates Gazette.

Davidson, A. and Darlow, C. (1993) *Boston Gilmore Rental Equivalent Tables and Lease Incentives*, London: Estates Gazette.

Davidson, A.W. (ed.) (2002) *Parry's Valuation and Conversion Tables*, 12th edn, London: Estates Gazette.

Davis Langdon (ed.) (2008) *Spon's Architects' and Builders Price Book 2009*, 134th edn, London: Spon Press.

Denyer-Green, B. (2005) *Compulsory Purchase and Compensation*, 8th edn, London: Estates Gazette.

Dixon, A. (2008) *Retail Figures Predict Monthly Rents will Become 'the Norm'*, Online: <http://www.egi.co.uk/Articles/Article.aspx?liArticleID=690322 (accessed 22 Oct. 2008).

Domb, P. (2000) 'The rebirth of the sale/leaseback', *Real Estate Forum* (Sep.): 53.

Downie, M.L. and Dobson, G. (2008) 'Added value or alternative?', *RICS Residential Property Journal* (Mar./Apr.): 6–9.

Dubben, N. and Sayce, S. (1991) *Property Portfolio Management: An Introduction*, London: Routledge.

Egan, D. (1995) 'Mainly for students: property cycles explained', *Estates Gazette*, 9547 (25 Nov.): 147–8; repr. in P. Askham and L. Blake (eds), *The Best of Mainly for Students* (London: Estates Gazette, 1999), vol. 2, pp. 117–23, London: Estates Gazette.

Enever, N. and Isaac, D. (2002) *The Valuation of Property Investments*, 6th edn, London: Estates Gazette.

Estates Gazette (1976a) 'Mainly for students: Discounted cash flow techniques and model building – Part I', 238 (26 Jun.): 984–5.

—— (1976b) 'Mainly for students: discounted cash flow techniques and model building – Part II', 239 (10 Jul.): 140–1.

—— (1984) 'Mainly for students: analysis of shop rents', 269 (28 Jan.): 333–6.

—— (1985) 'Mainly for students: investment terms', 273 (26 Jan.): 311–12.

—— (1987) 'Mainly for students: DCF appraisals', 284 (31 Oct.): 639–40.

—— (1992a) 'Mainly for students: investing in leasehold property', 9225 (27 Jun.): 129.

—— (1992b) 'Mainly for students: why study accounts?', 9237 (19 Sep.): 122–4.

—— (1992c) 'Mainly for students: why study accounts?', 9241 (17 Oct.): 105–7.

—— (1994) 'Mainly for students: valuing over-rented property', 9421 (28 May): 120–2; repr. in P. Askham and L. Blake (eds), *The Best of Mainly for Students* (London: Estates Gazette, 1999), vol. 2, pp. 295–302.

—— (1995a) 'Mainly for students: spreadsheets and valuation', 9503 (21 Jan.): 116–19.

—— (1995b) 'Mainly for students: rental adjustments for rating use', 9517 (29 Apr.): 121–2; repr. in P. Askham and L. Blake (eds), *The Best of Mainly for Students* (London: Estates Gazette, 1999), vol. 2, pp. 316–21.

—— (1995c) 'Mainly for students: rent-free periods and valuation' (13 May): 143–4; repr. in P. Askham and L. Blake (eds), *The Best of Mainly for Students* (London: Estates Gazette, 1999), vol. 2, pp. 310–15.

—— (1995d) 'Mainly for students: agricultural lettings', 9525 (24 Jun.): 153–4.

—— (1996a) 'Mainly for students: economic role of the valuer', 9633 (17 Aug.): 82–3.

—— (1996b) 'Mainly for students: valuation and compensation', 9637 (14 Sep.): 135–8; repr. in P. Askham and L. Blake (eds), *The Best of Mainly for Students* (London: Estates Gazette, 1999), vol. 2, pp. 345–50.

—— (1997) 'Mainly for students: inns and outs', 9724 (14 Jun.): 117–20.

—— (1998a) 'Mainly for students: ups and downs of turnover rents', 9812 (21 Mar.): 127–9.

—— (1998b) 'Mainly for students: surveying the field', 9840 (3 Oct.): 176–8.

—— (1999a) 'Mainly for students: the price of surrender', 9905 (6 Feb.): 152–3; repr. in L. Blake and A. Imber (eds), *The Best of Mainly for Students* (London: Estates Gazette, 2004), vol. 3, pp. 444–9.

—— (1999b) 'Mainly for students: a price on the potential', 9925 (26 Jun.): 178–9; repr. in L. Blake and A. Imber (eds), *The Best of Mainly for Students* (London: Estates Gazette, 2004), vol. 3, pp. 450–5.

—— (2000a) 'Mainly for students: how to reflect price paid', 0019 (13 May): 136–8.

—— (2000b) 'Mainly for students: times of change', 0027 (8 Jul.): 134–5, and 0028 (15 Jul.): 137.

—— (2000c) 'Mainly for students: search for the elusive figure', 0043 (28 Oct.): 172–3.

—— (2004) 'Mainly for students: the value of standards', 0430 (24 Jul.): 104–5.

—— (2005) 'Mainly for students: pricing adjustments', 0523 (11 Jun.): 136–7.

Evans, A. (2004a) *Economics and Land Use Planning*, Oxford: Blackwells.

—— (2004b) *Economics, Real Estate and the Supply of Land*, Oxford: Blackwells.

Fawcett, S. (2003) *Designing Flexible Cashflows*, London: Estates Gazette.

Fraser, W.D. (1993) *Principles of Property Investment and Pricing*, 2nd edn, Basingstoke: Palgrave Macmillan.

—— (2004) *Cash-flow Appraisal for Property Investment*, Basingstoke: Palgrave Macmillan.

French N. (1998) 'Word play', *Estates Gazette*, 9803 (17 Jan.): 130–2.

—— (2004) 'The big picture prevails', *Estates Gazette* (18 Dec.): 82–3.

—— (2005a) 'The little Red Book', *Estates Gazette*, 0541 (15 Oct.): 200–1.

—— (2005b) 'Compare notes', *Estates Gazette*, 0543 (29 Oct.): 184–5.

Gaskell, C. (1998) 'Words on worth', *Estates Gazette*, 9820 (16 May): 125–8; repr. in P. Askham and L. Blake (eds), *The Best of Mainly for Students* (London: Estates Gazette, 1999), vol. 2, pp. 369–75.

George, H. (1879) *Progress and Poverty*, Garden City, NY: Doubleday Page & Co. Online: <http://www.econlib.org/library/YPDBooks/George/grgPP1.html> .

Gilbertson, B., Preston, D. and Howarth, A. (2006) *A Vision for Valuation* (RICS Leading Edge Series), London: RICS. Online: <http://www.rics.org/NR/rdonlyres/BBEBD43B-11CA-4A2E-BDAC-B37BB2505CA1/0/vision_for_valuation.pdf>.

Gill, S. (2008) 'Take the measure of the figures', *Estates Gazette*, 0821 (31 May): 133.

Grenfall, W. (2003) 'Perfect pricing', *Estates Gazette*, 0348 (29 Nov.): 122–3.

Harford, T. (2007) *The Undercover Economist*, London: Abacus.

Harper, D. (2007) *Valuation of Hotels for Investors*, London: Estates Gazette.

—— (2008) 'Mainly for students: an opinion of balance', *Estates Gazette*, 0811: 88–9.

Harris, D. (2004) 'Funds in pole position', *Estates Gazette* (10 Jul.): 112–13.

Harris, R. (2005) *Property and the Office Economy*, London: Estates Gazette.

Harvey, J. (2000) *Urban Land Economics: The Economics of Real Property,* 5th edn, Basingstoke: Palgrave Macmillan.

Hattersley, M. (1981) 'How to value existing and new hotels', *Chartered Surveyor* (Jul.): 784–5.

Havard, T. (2002) *Investment Property Valuation Today,* London: Estates Gazette.

—— (2007) 'Valuation software and valuation methods: converging futures?', unpublished paper presented at RICS International Valuation Conference, London, 6 Nov.

Hayward, R. (ed.) (2008) *Valuation: Principles into Practice,* 6th edn, London: Estates Gazette.

Hillman, P. and Dalby. R. (2000) 'Stranger than fiction', *Estates Gazette,* 0039 (30 Sep.): 136.

HM Revenue and Customs (2008) *VAT.* Online: < http://www.hmrc.gov.uk/vat/index.htm> (accessed 6 Nov. 2008).

HM Treasury (2006) *Stern Review on the Economics of Climate Change,* London: HMSO. Online: <http://www.hm-treasury.gov.uk/independent_reviews/stern_review_economics_climate_change/stern_review_report.cfm>.

Hoesli, M. and MacGregor, B. (2000) *Property Investment: Principles and Practice of Portfolio Management,* London: Longman.

Hutchison, N. and Nanthakumaran, N. (1996) *The Cutting Edge: Estimation of the Investment Worth of Commercial Property,* London: University of Aberdeen and RICS Research.

Ifediora, B.U. (2005) *Valuation Mathematics for Valuers and Other Financial and Investment Analysts,* Enugu: Immaculate Publications.

Imber, A. (1997) 'Mainly for students: value of theory and practice', *Estates Gazette,* 9744 (1 Nov.): 195–8; repr. in P. Askham and L. Blake (eds), *The Best of Mainly for Students* (London: Estates Gazette, 1999), vol. 2, pp. 358–68.

—— (1998a) 'Mainly for students: the loan arrangements', *Estates Gazette,* 9804 (24 Jan.): 144–6; repr. in P. Askham and L. Blake (eds), *The Best of Mainly for Students* (London: Estates Gazette, 1999), vol. 2, pp. 124–9.

—— (1998b) 'Mainly for students: called to accounts', *Estates Gazette,* 9836 (5 Sep.): 142–4; repr. in L. Blake and A. Imber (eds), *The Best of Mainly for Students* (London: Estates Gazette, 2004), vol. 3, pp. 30–8.

—— (1999a) 'Mainly for students: the price of surrender', *Estates Gazette* (6 Feb.): 152–3; repr. in L. Blake and A. Imber (eds), *The Best of Mainly for Students* (London: Estates Gazette, 2004), vol. 3, pp. 444–9.

—— (1999b) 'Mainly for students: valuable lessons in the art of calculation', *Estates Gazette* (18 Sep.): 164–7; repr. in L. Blake and A. Imber (eds), *The Best of Mainly for Students* (London: Estates Gazette, 2004), vol. 3, pp. 435–43.

—— (2000) 'Mainly for students: time to move forward', *Estates Gazette,* 0023 (10 Jun.): 136–8; repr. in L. Blake and A. Imber (eds), *The Best of Mainly for Students* (London: Estates Gazette, 2004), vol. 3, pp. 465–72.

—— (2001) 'Mainly for students: tenants, reviews, renewals and rent', *Estates Gazette,* 0142 (20 Oct.): 128–31; repr. in L. Blake and A. Imber (eds), *The Best of Mainly for Students* (London: Estates Gazette, 2004), vol. 3, pp. 213–22.

—— (2002) 'Mainly for students: seeing the truth beyond the horizon', *Estates Gazette,* 0248 (30 Nov.): 128–30; repr. in L. Blake and A. Imber (eds), *The Best of Mainly for Students* (London: Estates Gazette, 2004), vol. 3, pp. 473–9.

International Valuation Standards Committee (2007a) *Determination of Fair Value of Intangible Assets for IFRS Reporting Purposes,* London: IVSC. Online: <http://www.ivsc.org/pubs/comment/intangibleassets.pdf>.

—— (2007b) *International Valuation Standards,* 8th edn, London and Chicago: International Valuation Standards Committee.

Investment Property Forum and Imber, A. (2001) 'Mainly for students: making more of money', *Estates Gazette*, 0114 (7 Apr.): 138–41; repr. in L. Blake and A. Imber (eds), *The Best of Mainly for Students* (London: Estates Gazette, 2004), vol. 3, pp. 187–95.

Isaac, D. (1996) *Property Development: Appraisal and Finance*, Basingstoke: Palgrave Macmillan.

—— (1997) *Property Investment*, Basingstoke: Palgrave Macmillan.

—— (2001) *Property Valuation Principles*, Basingstoke: Palgrave Macmillan.

—— and Steley, T. (1999) *Property Valuation Techniques*, 2nd edn, Basingstoke: Palgrave Macmillan.

James, D. and Imber, A. (2002) 'Mainly for students: how much will I make?', *Estates Gazette*, 0232 (10 Aug.): 80; repr. in L. Blake and A. Imber (eds), *The Best of Mainly for Students* (London: Estates Gazette, 2004), vol. 3, pp. 196–202.

Johnson, Tim. (2008) 'Walking the line', *RICS Residential Property Journal* (Mar./Apr.): 22–4.

Johnson, Tony, Davies, K.I. and Shapiro, E. (2000) *Modern Methods of Valuation of Land, Houses and Buildings*, 9th edn, London: Estates Gazette.

Johnstone, V., Kane, B. and Koranteng, A. (2004) *Guidelines for the Use of Automated Valuation Models for U.K. RMBS Transactions*, New York: Standard & Poor's/McGraw-Hill. Online: <http://www.rics.org/NR/rdonlyres/8FCDD20C-7FAC-4549–86FB-3930CD0CBC05/0/StandardandPoorsReportonAVMs.pdf>.

Jowsey, E. and Harvey, J. (2003) *Urban Land Economics*, 6th edn, Basingstoke: Palgrave Macmillan.

KEL Computing Limited (2008) *KEL*. Online. <http://www.kel.co.uk/index.php> (accessed 6 Nov. 2008).

—— (2008) *KEL Delta*. Online: <http://www.kel.co.uk/pdfs/KEL%20Delta.pdf> (accessed 21 Oct. 2008).

Kingsmill, R. (2005) 'The journey from owner to occupier', *Estates Gazette*, 0531 (6 Aug.): 74–6.

Kivelhan, N.P., Elghamry, N. and Attabiyeh, M. (2005) 'Out-of-town retail', *Estates Gazette*, 0516 (23 Apr.): 75–87.

Lam, E. T-K. (1996) *Modern Regression Models and Neural Networks for Residential Property Valuation*, London: RICS. Online: <http://www.rics.org/Property/Residentialproperty/Residentialpropertyvaluationandappraisal/modern_regression_models_19960101.htm> (accessed 29 Sep. 2007).

Law, J. (ed.) (2006) *Oxford Dictionary of Business and Management*, 4th edn, Oxford: Oxford University Press.

Lean, W. and Goodall, B. (1966) *Aspects of Land Economics*, London: Estates Gazette.

Levy, D. and Anderson, J. (2006) 'Practice notes: exactly what bit is usable?', *Estates Gazette*, 0624 (17 Jun.): 174.

Louargand, M. (2007) 'The world is our oyster', *RICS Business* (Oct.): 24–5.

Lumby, S. (1994) *Investment Appraisal and Financing Decisions: A First Course in Financial Management*, 5th edn, London: Chapman & Hall.

Lynch, T. and Clark, K. (eds) (2006) *Real Estate Transparency Index*, Chicago: Jones Lang LaSalle.

Mackmin, D. (1981) 'How effective are tables?', *Estates Gazette*, 259 (4 Jul.): 23–6.

—— (2007) *The Valuation and Sale of Residential Property*, 3rd edn, London: Estates Gazette.

Marshall, H. and Williamson, H. (1997) *Law and Valuation of Leisure Property*, 2nd edn, London: Estates Gazette.

Marshall, P. (1976) 'Equated yield analysis: a valuation method of the future?' *Estates Gazette*, 239 (14 Aug.): 493–7.

—— (1988) *Donaldsons' Investment Tables*, 3rd edn, London: Donaldsons.

Marx, K. (1887) *Das Kapital*, Moscow: Progress Publishers. Online: <http://www.marxists.org/archive/marx/works/1867–c1/index.htm>.

Matysiak, G., Rodney, B., Crosby, N., French, N. and Newell, G. (1996) 'The variable value of property investment valuation reports' *CSM* (Apr.): 38–9.

May, R. (2006) *The Profits Method of Valuation: Does it Value the Property or the Business?* Online: <http://www.rics.org/NR/rdonlyres/E41B0818–AAE1–4BD2–9962–CB7FAEC488F6/0/RICSarticleJan06.pdf> (accessed 16 Jan. 2008).

—— (2008) 'A different approach', *RICS Commercial Property Journal* (Feb./Mar.): 16–17.

Mendell, S. (1994) 'Hotel valuations: a modern approach', *Estates Gazette*, 9406 (12 Feb.): 126–7.

Mill, J.S. (1909) *Principles of Political Economy*, 7th edn, London: Longmans, Green & Co. Online: <http://www.econlib.org/library/Mill/mlP1.html>.

Millington, A.F. (2000) *An Introduction to Property Valuation*, 5th edn, London: Estates Gazette.

Morley, A. (2008) 'It all adds up in the end', *Estates Gazette*, 0806 (9 Feb.): 162–3.

Morley, S. (1995) 'The future for rental growth', *Estates Gazette*, 9503 (21 Jan.): 110–12.

Mott, G. (1997) *Investment Appraisal*, 3rd edn, London: Pitman.

Murdoch, J. (2004) 'The value of valuation', *Estates Gazette*, 0431 (31 Jul.): 91.

Murdoch, S. (2006) 'Legal notes: caught in the net', *Estates Gazette*, 0609 (4 Mar.): 181.

Murphy, G. (1999) 'Valuation of farm buildings', *CSM* (Jul./Aug.): 26–7.

Office of the Deputy Prime Minister (2006) *The Building Regulations: F Ventilation. Ventilation of Habitable Rooms through Another Room or a Conservatory*. Online: <http://www.thenbs.com/BuildingRegs/knowledgeCentre/ShowContents.asp?section=F&topic=b_0602_APPDOC_F_00140> (accessed 23 Sep. 2007).

Park, R.E., Burgess, E. and McKenzie, R. (1925) *The City*, Chicago, IL: University of Chicago Press.

Parsons, G. (ed.) (2004) *The Glossary of Property Terms*, 2nd edn, London: Jones Lang LaSalle and Estates Gazette.

Parnham, P. and Rispin, C. (2000) *Residential Property Appraisal*, London: Spon Press.

Peto, R. (2007a) 'New RICS Red Book', email (22 Aug.).

—— (2007b) 'RICS Valuation Faculty review 2007', unpublished paper presented at RICS International Valuation Conference, London, 6 Nov.

Plimmer, F.A. (1998) *Rating Law and Valuation*, London: Longman.

Ratcliffe, J. (1974) *An Introduction to Town and Country Planning*, London: Hutchinson.

Rayner, M. (1989) *Asset Valuation*, London: Macmillan.

Ricardo, D. (1821) *On the Principles of Political Economy and Taxation*, 3rd edn, London: John Murray. Online: <http://www.econlib.org/library/Ricardo/ricP1.html>.

Rich, J. (1999) 'How to calculate the true equivalent yield', *Estates Gazette*, 9949 (11 Dec.): 84–5.

Richmond, D. (1994) *Introduction to Valuation*, 3rd edn, London: Macmillan.

Rose, J. (1976) *Property Valuation Tables*, Oxford: Freeland Press.

—— (1994) 'A valuation revolution? Arithmetic v. algebra', *Estates Gazette*, 9445 (12 Nov.): 141–2.

Royal Institution of Chartered Surveyors (1994) *The Mallinson Report: Commercial Property Valuations*, London: RICS.

—— (1997a) *Commercial Investment Property: Valuation Methods*, Coventry: RICS Books. .

—— (1997b) *RICS Appraisal and Valuation Standards: 'The Red Book'*, 4th edn, Coventry: RICS Books.

—— (1998) *Agricultural Tenancies Act 1995: Supplementary Guidance Note*, London: RICS. Online: <http://www.isurv.com/site/scripts/download_info.aspx?downloadID=128> (accessed 21 Sep. 2007).

—— (1999) *Guide to Carrying out Reinstatement Cost Assessments: RICS Guidance Note*, Coventry: RICS.

—— (2002a) *Code of Measuring Practice: A Guide for Surveyors and Valuers*, 5th edn, Coventry: RICS Books.

—— (2002b) *Property Valuation: The Carsberg Report*, London: RICS.

—— (2003a) *The Capital and Rental Valuation of Petrol Filling Stations in England, Wales and Scotland*, Valuation Information Paper No. 3, Coventry: RICS Books.

—— (2003b) *The Capital and Rental Valuation of Restaurants, Bars, Public Houses and Nightclubs in England, Wales and Scotland*, Valuation Information Paper No. 2, Coventry: RICS Books.

—— (2003c) *RICS Appraisal and Valuation Standards: 'The Red Book'*, 5th edn, Coventry: RICS Books.

—— (2003d) *Rural Property Valuation*, Valuation Information Paper No. 5, Coventry: RICS Books.

—— (2003e) *Specification for Residential Mortgage Valuation*, Coventry: RICS Books.

—— (2003f) *Valuation of Owner-occupied Property for Financial Statements*, Valuation Information Paper No. 1, Coventry: RICS Books.

—— (2003g) *The Valuation of Surgery Premises used for Medical or Health Services*, Valuation Information Paper No.4, Coventry: RICS Books.

—— (2004) *The Capital and Rental Valuation of Hotels in the UK*, Valuation Information Paper No. 6, Coventry: RICS Books.

—— (2005a) *Building Surveys and Inspections of Commercial and Industrial Property: Section 2.2.3*, 3rd edn, Coventry: RICS Books.

—— (2005b) *Leasehold Reform in England and Wales*, Valuation Information Paper No. 7, Coventry: RICS Books.

—— (2005c) *The RICS Homebuyer Survey and Valuation Service 2005 Practice Notes*, Coventry: RICS Books.

—— (2006a) *The Analysis of Commercial Lease Transactions*, Valuation Information Paper No. 8, Coventry: RICS Books.

—— (2006b) *The Capital and Rental Valuation of Restaurants, Bars, Public Houses and Nightclubs in England, Wales and Scotland*, Valuation Information Paper No. 2, 2nd edn, Coventry: RICS Books.

—— (2006c) *Carsberg Report on Property Valuations*. Online: <http://www.rics.org/Practiceareas/Property/Valuation/Carsberg+Report+on+Property+Valuations.htm> (accessed 5 Dec. 2008).

—— (2006d) *Land and Buildings Apportionments for Lease Classification under International Financial Reporting Standards*, Valuation Information Paper No. 9, Coventry: RICS Books.

—— (2007a) *About RICS*, London: RICS. Online: <http://www.rics.org/AboutRICS/about_rics.htm> (accessed 29 Sep. 2007).

—— (2007b) *Code of Measuring Practice: A Guide for Property Professionals*, 6th edn, Coventry: RICS Books.

—— (2007c) *History of RICS*, London: RICS. Online: <http://www.rics.org/AboutRICS/history_of_rics.htm> (accessed 29 Sep. 2007).

—— (2007d) *RICS Valuation Standards: 'The Red Book'*, 6th edn, Coventry: RICS Books.

—— (2007e) *The Depreciated Replacement Cost Method of Valuation for Financial Reporting*, Valuation Information Paper No. 10, Coventry: RICS Books.

—— (2007f) *The Valuation and Appraisal of Private Care Home Properties in England, Wales and Scotland*, Valuation Information Paper no.11, Coventry: RICS Books.

—— (2007g) *The Valuation of Development Land*, Valuation Information Paper No. 12, Coventry: RICS Books.

—— (2007h) *Welcome to isurv*. Online: <http://www.hostref1.com/NXT/GATEWAY.DLL/Environment/News?f=templates&fn=default.htm> (accessed 22 Sep. 2007).

—— (2008a) *Confirmation of Rental Evidence: Commercial and Industrial Premises*, London: RICS. Online: <http://www.isurv.com/site/scripts/download_info.aspx?fileID=375> (accessed 17 Oct. 2008).

—— (2008b) *Confirmation of Rental Evidence: Offices*, London: RICS. Online: <http://www.isurv.com/site/scripts/download_info.aspx?fileID=377> (accessed 17 Oct. 2008).

—— (2008c) *Confirmation of Rental Evidence: Shops*, London: RICS. Online: <http://www.isurv.com/site/scripts/download_info.aspx?fileID=379> (accessed 17 Oct. 2008).

——(2008d) *RICS Red Book*. Online: <http://www.rics.org/Knowledgezone/RICSRedBook/spotlight.htm> (accessed 5 Dec. 2008).

—— (2009a) *About Faculties*. Online: <http://www.rics.org/Networks/Faculties/Faculties.htm> (accessed 1 Jan. 2009).

—— (2009b) *Consultation Draft Valuation Information Paper: Reflecting Sustainability in Commercial Property Valuations*, London:RICS. Online:<http://www.rics.org/NR/rdonlyres/07B0710A-7BB4-4F3A-85C5-029571035E64/0/ConsultationDraftVIPReflectingSustainabilityinCommercial PropertyValuations.pdf> (accessed 4 Jan. 2009).

—— and Investment Property Forum (1997) *Calculation of Worth: An Information Paper*, London: RICS. Online: <http://www.rics.org/NR/rdonlyres/B59119AF-A281-4E8A-A079-1FD98394CD09/0/ Calculation_of_Worth_1997.pdf> (accessed 5 Mar. 2009).

—— and Investment Property Databank (2006) *Valuation Sale Price Variance Report 2006*, London: RICS and IPD.

—— and —— (2007) *Valuation Sale Price Report 2007*, London: RICS. Online: <http://www.rics.org/ Practiceareas/Property/Valuation/ipdvaluation2007_c_151107.html>.

—— and —— (2008) *Valuation And Sale Price Report 2008*, London: RICS. Online: <http://www.rics.org/ NR/rdonlyres/E2B714B6-C1EB-4299-B3FB-112A9AB461D9/0/430_Valuation_And_Sale_Pricefinal. pdf> (accessed 30 Dec. 2008).

Royal Institution of Chartered Surveyors Valuation Faculty (2002) *Response to Carsberg Report*, London: RICS.

Salway, F. (1986) *Depreciation of Commercial Property*, Reading: Centre for Advanced Land Use Studies / College of Estate Management.

Santo, P. and Hall, B. (2008) 'Question time', *RICS Residential Property Journal* (Jun./Jul.):9.

Saunders, O. (2001) 'Earth, wind and water', *Estates Gazette*, 0106 (10 Feb.): 152–4.

Sayce, S., Smith, J., Cooper, M.R.R. and Venmore-Rowland, P. (2006). *Real Estate Appraisal: From Value to Worth*, Oxford: Blackwell.

Scarrett, D. and Smith, M. (2007) *Property Valuation: The Five Methods*, 2nd edn, London: Routledge.

Selftrade Investments (2008) *Market Data: Gilts and Bonds*. Online: <http://www.selftrade.co.uk/market-data/gilts-bonds/gilts.php> (accessed 27 Oct. 2008).

Smith, A. (1776) *An Inquiry into the Nature and Causes of the Wealth of Nations*, London: Adam Smith Institute. Online: <http://www.adamsmith.org/smith/won-index.htm>.

Stevens, J. and Steventon, K. (2001) 'Valuation', in A. Baum (ed.) *Freeman's Guide to Property Finance*, London: Freeman, 17–46.

Thorne, C. (2007) 'Valuation standards and the new Red Book', unpublished paper presented at RICS International Valuation Conference, London, 6 Nov.

Thünen, J.H. von. (1966) *Isolated State: An English Edition of Der isolierte Staat*, Oxford and New York: Pergamon Press.

Trott, A. (ed.) (1986) *Property Valuation Methods*, London: Royal Institution of Chartered Surveyors and Polytechnic of the South Bank.

United Kingdom Debt Management Office (2007) *Gilt Market*. Online: <http://www.dmo.gov.uk/index. aspx?page=About/About_Gilts> (accessed 19 Nov. 2007).

Valuation Office Agency (2006a) *Rating Manual*, vol. 4, section 5, practice note 1, *Rental Adjustment*. Online: <http://www.voa.gov.uk/instructions/chapters/rating_manual/vol4/frame.htm> (accessed 25 Jun. 2007).

—— (2006b) *Rating Manual*, vol. 4, section 6, *The Receipts and Expenditure Method*. Online: <http:// www.voa.gov.uk/instructions/chapters/rating_manual/vol4/frame.htm> (accessed 25 Jun. 2007).

—— (2006c) *Rating Manual*, vol. 4, section 7, *The Contractor's Basis of Valuation*. Online: <http://www. voa.gov.uk/instructions/chapters/rating_manual/vol4/frame.htm> (accessed 25 Jun. 2007).

—— (2006d) *Rating Manual*, vol. 5, *Rating Valuation Practice, All Classes of Hereditament*. Online: <http:// www.voa.gov.uk/instructions/chapters/rating_manual/vol5/frame.htm> (accessed 25 Jun. 2007).

—— (2007) *Revaluation 2005 Valuation Scales*. Online: <http://www.voa.gov.uk/business_rates/ ValuationScales/> (accessed 30 Oct. 2007).

—— (2008a) *Non-domestic Rating Lists*. Online: <http://www.voa.gov.uk/business_rates/index.htm> (accessed 6 Nov. 2008).

—— (2008b) *Property Market Report July 2008*. Online: <http://www.voa.gov.uk/publications/property_market_report/pmr-Jul-08/index.htm> (accessed 21 Oct. 2008).

Walter, G. (1992) 'Turnover rents: prospects and implications', *CSM* (Jul.): 5–6.

Weatherhead, M. (1997) *Real Estate in Corporate Strategy*, Basingstoke: Macmillan Press.

Westbrook, R.W. (1983) *The Valuation of Licensed Premises,* London: Estates Gazette.

Whipple, R.T.M. (1995) *Property Valuation and Analysis,* Sydney: Law Book Co of Australasia.

Williams, R.G. (2008) *Agricultural Valuations: A Practical Guide,* 4th edn, London: Estates Gazette.

Wyatt. P. (2007) *Property Valuation in an Economic Context,* Oxford: Blackwells.

Index